Organic
Photochemistry

Organic Photochemistry

Volume 11

Edited by

ALBERT PADWA

DEPARTMENT OF CHEMISTRY
EMORY UNIVERSITY
ATLANTA, GEORGIA

CRC Press
Taylor & Francis Group
Boca Raton London New York

CRC Press is an imprint of the
Taylor & Francis Group, an **informa** business

First published 1991 by Marcel Dekker, Inc.

Published 2019 by CRC Press
Taylor & Francis Group
6000 Broken Sound Parkway NW, Suite 300
Boca Raton, FL 33487-2742

© 1991 by Taylor & Francis Group, LLC
CRC Press is an imprint of Taylor & Francis Group, an Informa business

First issued in paperback 2019

No claim to original U.S. Government works

ISBN-13: 978-0-367-45054-0 (pbk)
ISBN-13: 978-0-8247-8561-1 (hbk)

Visit the Taylor & Francis Web site at
http://www.taylorandfrancis.com

and the CRC Press Web site at
http://www.crcpress.com

LIBRARY OF CONGRESS CARD NUMBER 66-11283

Preface

Organic photochemistry is the science arising from the application of photochemical methods to organic chemistry and organic chemical methods to photochemistry. It is an interdisciplinary frontier.

Intense activity in organic photochemistry in the last decade has produced so vast an accumulation of factual knowledge that chemists in general have viewed it with awe. Even those chemists engaged in the study of organic photochemistry will find the rate of development of the field perplexing to a high degree. This series originated to fill the need for a critical summary of this vigorously expanding field with the purpose of drawing together seemingly unrelated facts, summarizing progress, and clarifying problems.

Volume 11 continues to fulfill the original, essential role of this unique self-series by providing a convenient review of the structural aspects of organic photochemistry. As with earlier volumes, this new book offers the research findings of distinguished authorities. It stresses timely aspects of organic photochemistry—previously scattered throughout the large body of literature—for which necessary critical review has been lacking.

This volume of the series emphasizes the mechanistic details of the di-π-methane rearrangement, the synthetic aspects of the oxa-di-π-methane reaction, the photochemistry of carbenium ions and related species, photoinduced hydrogen atom abstraction by carbonyl compounds, and matrix photochemistry of nitrenes, carbenes, and excited triplet states. Complete with numerous illustrations and bibliographic citations of the literature, this book explores these important processes to the advantage of organic chemists, as an aid to research and as a source for supplementary knowledge on particular topics.

Albert Padwa

Contents

Contents

Contributors

Ronald F. Childs   Department of Chemistry, McMaster University, Hamilton, Ontario, Canada

Martin Demuth   Max Planck Institut für Strahlenchemie, Mülheim an der Ruhr, Germany

Karl Haider   Department of Chemistry, Yale University, New Haven, Connecticut

Elisa Leyva   Department of Chemistry, Ohio State University, Columbus, Ohio

Bong-Ser Park   Department of Chemistry, Michigan State University, East Lansing, Michigan

Matthew S. Platz   Department of Chemistry, Ohio State University, Columbus, Ohio

Gary B. Shaw   Department of Chemistry, University of Alberta, Edmonton, Alberta, Canada

Peter Wagner   Department of Chemistry, Michigan State University, East Lansing, Michigan

Howard E. Zimmerman   Department of Chemistry, University of Wisconsin, Madison, Wisconsin

Contents of Other Volumes

The Trapping of Biradicals and Related Photochemical Intermediates   *R. Marshall Wilson, Department of Chemistry, University of Cincinnati, Cincinnati, Ohio*

VOLUME 8

Photochemical Transformations on the Primitive Earth and Other Planets *James P. Ferris, Department of Chemistry, Rensselaer Polytechnic Institute, Troy, New York*

Photochemistry with Short UV Light   *Mark G. Steinmetz, Department of Chemistry, Marquette University, Milwaukee, Wisconsin*

Matrix Isolation Photochemistry   *Robert S. Sheridan, Department of Chemistry, University of Wisconsin-Madison, Madison, Wisconsin*

The Influence of the Molecular Crystalline Environment on Organic Photorearrangements   *John R. Scheffer, Miguel Garcia-Garibay, and Omkaram Nalamasu, Department of Chemistry, University of British Columbia, Vancouver, Canada*

VOLUME 9

The Photochemistry of Substances Containing the C = N Moiety With Emphasis on Electron Transfer Processes   *Patrick S. Mariano, Department of Chemistry, University of Maryland, College Park, Maryland*

Photocyclizations and Intramolecular Cycloadditions of Conjugated Olefins   *Wim H. Laarhoven, Department of Organic Chemistry, Catholic University, Nijmegen, The Netherlands*

Photolytic Deprotection and Activation of Functional Groups   *V. N. Rajasekharan Pillai, Department of Chemistry, University of Calicut, Kerala, India*

VOLUME 10

Applications of Intramolecular 2 + 2-Photocycloadditions in Organic Synthesis   *Dan Becker and Nizar Haddad, Department of Chemistry, Technion-Israel Institute of Technology, Haifa, Israel*

Photocyclizations and Intramolecular Photocycloadditions of Conjugated Arylolefins and Related Compounds   *Wim H. Laarhoven, Department of Organic Chemistry, Catholic University of Nijmegen, The Netherlands*

1

The Di-π-Methane Rearrangement

HOWARD E. ZIMMERMAN

University of Wisconsin
Madison, Wisconsin

A. Introduction

The Di-π-Methane Rearrangement is one of those particularly general
organic photochemical reactions with fascinating mechanistic complexity in
conjunction with remarkable consistency and beauty.

The requirement for this photochemical reaction is a reactant with two π
groups bonded to a single carbon atom. The most common variations lead a
reactant divinylmethane to a vinylcyclopropane or an arylvinylmethane reactant
to an arylcyclopropane. More often than not, the photochemical products are
not readily prepared by alternative approaches. Two typical examples of
the di-π-methane rearrangment are given in equations [1] and [2]. The
first reaction[1] can be superficially pictured mnemonically as a 1,2-shift of
carbon 2 of the diphenylvinyl group in (1) from C-3 to C4 followed by
three-membered ring closure between C-3 and C-5. However, the actual reaction
mechanism is not so simple, as will be discussed. The second example (i.e. in
equation [2]) is the rearrangement of barrelene (3) to semibullvalene (4).[2]
Here the rearrangement seems even more complex.

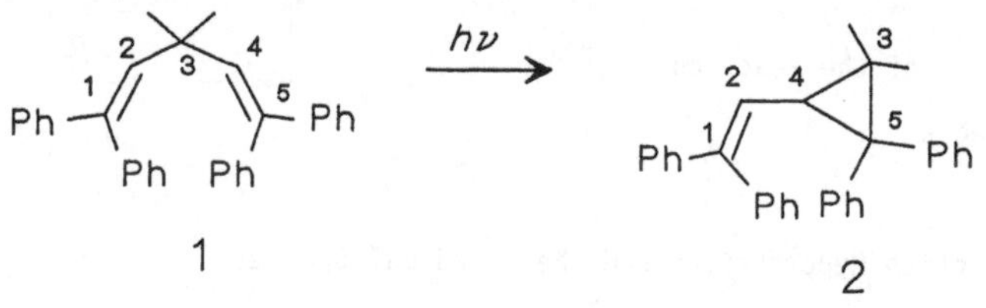

[1]

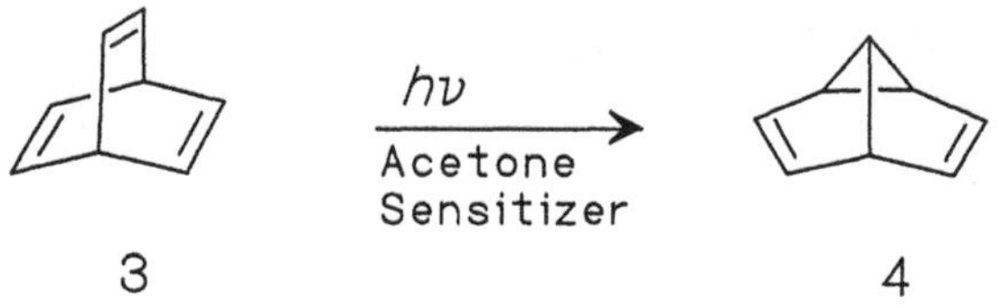

1   2

[2]

3   4

B. General Formulation of the Reaction

Two general equations ([3] and [4]) suffice to depict the course of the
di-π-methane rearrangement. Thus, the excited state of divinylmethane (5)

bridges to form a new sigma bond and afford a species we may write as 1,4-diradical (6) which proceeds onwards to 1,3-diradical (7). This then closes to form the three-membered ring product (8). In parallel fashion the excited

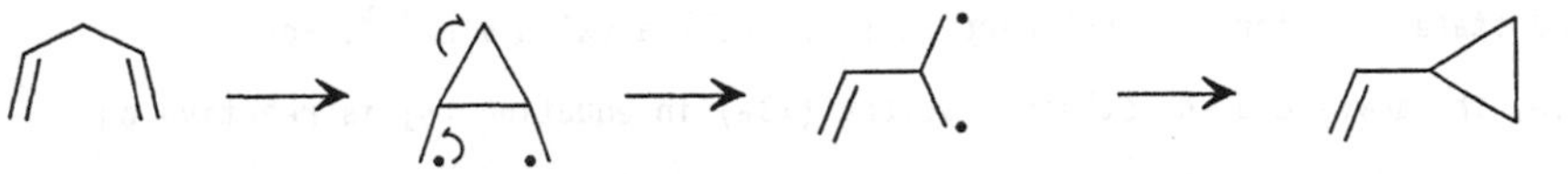

5   6   7   8   [3]

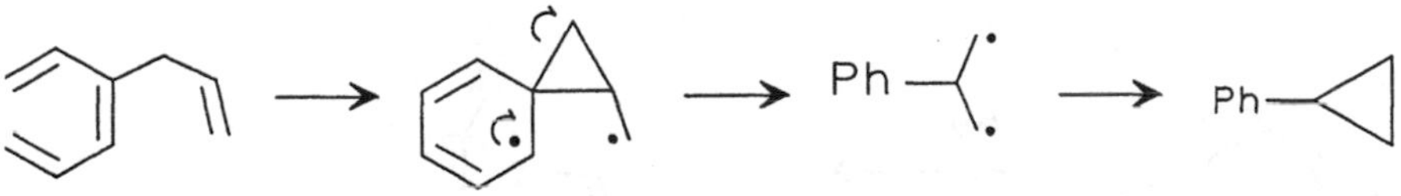

9   10   11   12   [4]

state of aryl-vinyl reactant (9) bridges to afford 1,4-diradical (10) which proceeds onward to 1,3-diradical (11). This cyclizes to give rise to the aryl cyclopropane product (12). Species such as (6), (7), (10) and (11) are the useful valence-bond structures which have proven so powerful in mechanistic organic chemistry in permitting organic chemists to intuitively rationalize, understand and even predict the course of reactions. As has been repeatedly noted by the author[2c,3], which electronic state is implied by these structures will vary with the reaction being considered. Thus, as noted below, for some di-π-methane rearrangements only the singlet excited state is reactive, while for others the triplet is the reacting species. Furthermore, whether the diradical species drawn are energy minima or just points on the reaction hypersurface is not implied. However, these structures nicely convey the nature and geometry of the molecule as it proceeds onwards towards product. In addition, each of the conversions indicated by arrows has firm precedent in ground state organic chemistry, if only in the microscopic reverse in the 5->6 and 9->10 conversions.

C. Multiplicity Effects

One of the striking facets of the di-π-methane rearrangement is its dependence on reaction multiplicity. Originally, it was felt that acyclic di-π-methane systems were capable of rearrangement only via the singlet excited state, and for the most part this is still a valid view[4,5]. For example, the acyclic di-π-methane reactant (13a) in equation [5] is reactive on direct irradiation, via S_1.[6,7] However, as T_1 (i.e. in sensitized runs) it does not give the di-π-methane rearrangement but instead the reactant merely undergoes cis-trans isomerization to afford (13b). It is clear in this

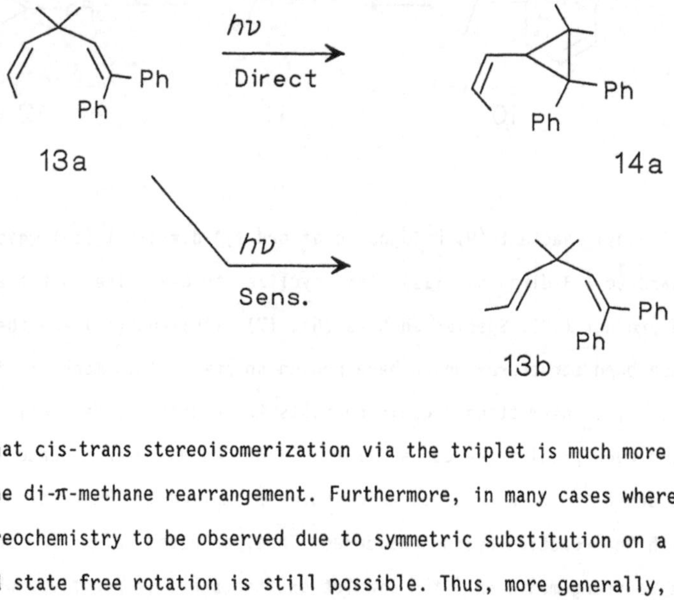

case that cis-trans stereoisomerization via the triplet is much more rapid than the di-π-methane rearrangement. Furthermore, in many cases where there is no stereochemistry to be observed due to symmetric substitution on a π-bond, excited state free rotation is still possible. Thus, more generally, di-π-methane reactants which have a potential "free-rotor"[4,7] tend to be unreactive as triplets.

The mechanism of free-rotor energy dissipation is readily understood when one considers the potential energy surfaces of alkene ground states and triplets.[8,9] Generally, the ground state (i.e. S_0) has an energy maximum near a 90 degree twist while the triplet (T_1) has an energy minimum, and crossing of the singlet and triplet curves occurs. The net result is a very rapid conversion of the twisted triplet to a vibrationally excited S_0. Of

course, the more rigid cyclic di-π-methane systems are less capable of

energy dissipation by this mechanism.

Nevertheless, one is dealing with competitive rates, namely the rate

of internal conversion to ground state versus the rate of triplet di-π-

methane rearrangement. With a sufficiently facile triplet rearrangement rate,

the di-π-methane rearrangement will take place. Also, in sufficiently

sterically congested acyclic di-π-methane reactants, free rotation may be

inhibited to the extent that the rearrangement again takes place. For example,

the diisopropyl diene (15) does undergo the rearrangement as a triplet as a

consequence of the cluttered nature of the molecule which leads to inhibition

of free-rotoring; note equation [7].[10]

Another situation in which acyclic di-π-methane systems have proven

reactive as triplets is where the central, "methane" carbon is substituted with

a group which is strongly odd-electron stabilizing, such as cyano[11],

phenyl[12] or carbomethoxy.[13] Three such examples from our studies are

15   16

[7]

illustrated in equation [8].[11-13] In these publications it has been noted

that triplet quantum efficiencies are quite high (ϕ= 0.42 for X = Ph, ϕ =

0.53 for X = CN, and ϕ = 0.32 for X = COOMe). This unusual reactivity for

acyclic di-π-methane systems was ascribed in part to enhanced three-membered

ring opening of the cyclopropyldicarbinyl diradical (e.g. (20) in Scheme 1) as

a consequence of stabilization of the 1,3-diradical species formed (e.g. (21)).

The mechanism for this rearrangement is shown in Scheme 1 with the 3,3-

dicyano diene (17b) being used as an example. While the evidence is that

the initial bridging step of 1,4-dienes such as 1,1,5,5-tetraphenyl-3,3-

Scheme 1. Mechanism of Rearrangement of 1,1,5,5,-Tetraphenyl-3,3-dicyano-
1,4-pentadiene.

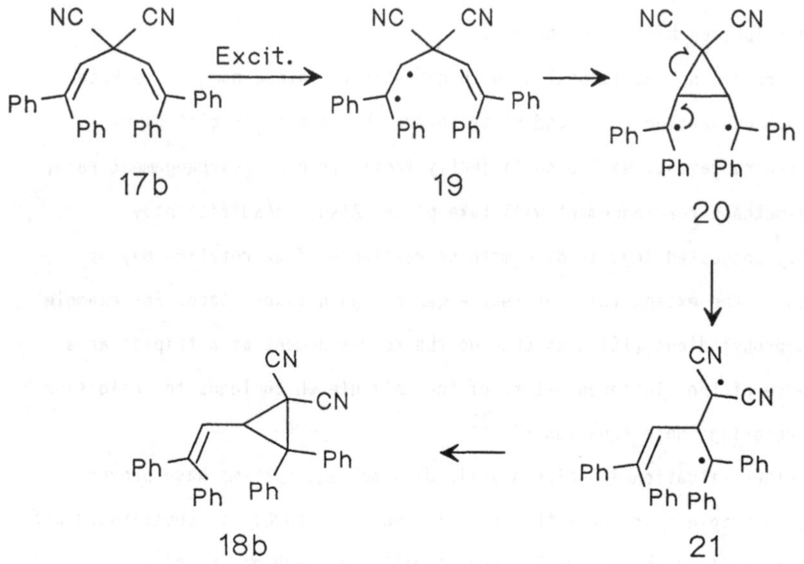

17b   19   20

18b   21

dimethyl-1,4-pentadiene (1) has a major impact on the rate of the S_1
rearrangement (vide infra), the presence of strongly odd-electron stabilizing

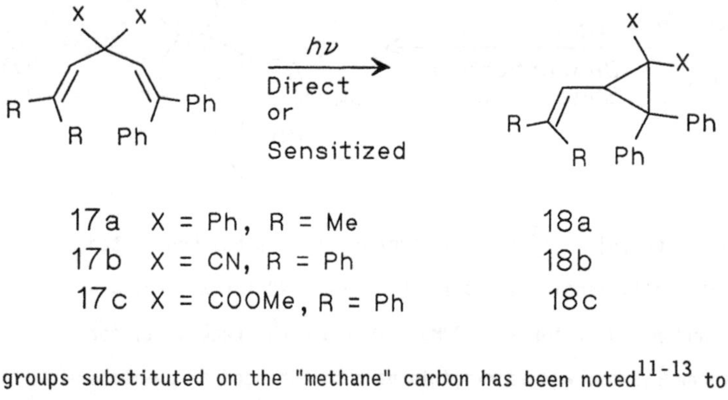

[8]

17a X = Ph, R = Me   18a
17b X = CN, R = Ph   18b
17c X = COOMe, R = Ph   18c

groups substituted on the "methane" carbon has been noted[11-13] to provide a
major enhancement of the reaction efficiency for the triplet. The singlet
reaction does not show this dramatic effect (note again below). Further
evidence supporting enhancement of triplet rearrangement by such stabilizing
groups substituted on the "methane carbon" has been reported by Paquette in the
di-π-methane rearrangement of dibenzobarrelenes;[14] barrelene examples are
discussed below.

One interesting observation is that central substitution by electron withdrawing and delocalizing groups such as cyano and carbomethoxy actually inhibit the rates of singlet excited di-π-methane rearrangement in contrast to the triplet situation described above[13]. This suggests that the energy versus reaction coordinate contours for the singlet processes differ from the triplet counterparts. A further point has been noted[13], namely that there is a linear correlation between the ground state Hammett sigma constants for groups substituted on the methane carbon and the rate of S_1 rearrangement. S_1 of the 1,1,5,5-tetraphenyl diene (1) with central methyl groups (note equation [1]) reacts more rapidly than the counterpart with central carbomethoxy groups and the latter, in turn, reacts more rapidly than the S_1 analog with central cyano substituents. From this we can deduce that the rate limiting stage of the acyclic singlet di-π-methane rearrangement occurs early before three-membered ring opening has begun to occur appreciably and before odd-electron stabilization in the 1,3-diradical (e.g. (7)) is important.

A dramatic example of multiplicity control of the course of the di-π-methane rearrangement is outlined in Scheme 2. In this example,[13] the tetra-

Scheme 2. Multiplicity Dependence of the Di-π-methane Rearrangement of 1,1,3,3,-Tetraphenyl-5,5-dicyano-1,4-pentadiene.

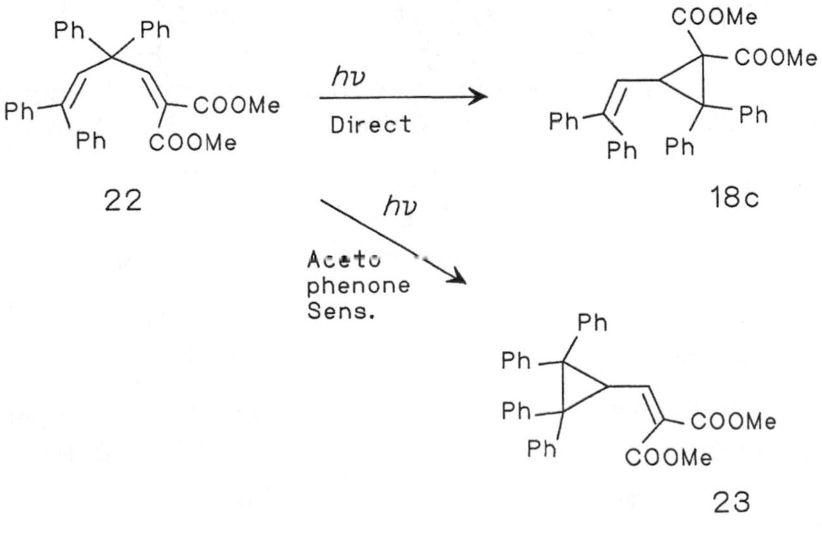

phenyl diester (22) gives a totally different photoproduct on direct irradiation
than in sensitized photolyses. The carbomethoxy groups appear as cyclopropane
ring substituents (note (18c)) in the direct irradiation while they are found on
the product vinyl group (note (23)) in sensitized photolyses. When one considers
the di-π-methane rearrangement mechanism, as in Scheme 3, he recognizes that
the direct irradiation and sensitized photoproducts originate from related 1,4-
diradical species, namely the singlet and triplet, (25s) and (25t). Thus,
depending on the relative spin of the odd electrons in diradicals (25), the
three-membered ring opens on one side or the other (arrows s for the singlet
and arrows t for the triplet). This fascinating result is an example of a
more general phenomenon.

We can approach our discussion of this diradical behavior[11,13,15,16]
either from a quantum mechanical vantage-point or by beginning with the
experimental observation. The latter dictates that the rate-controlling stage
of the reaction of singlet diradical (25s) has a lower energy barrier for

Scheme 3. Multiplicity Dependence of the Modes of the Cyclopropyldicarbinyl
Three-membered Ring Opening.

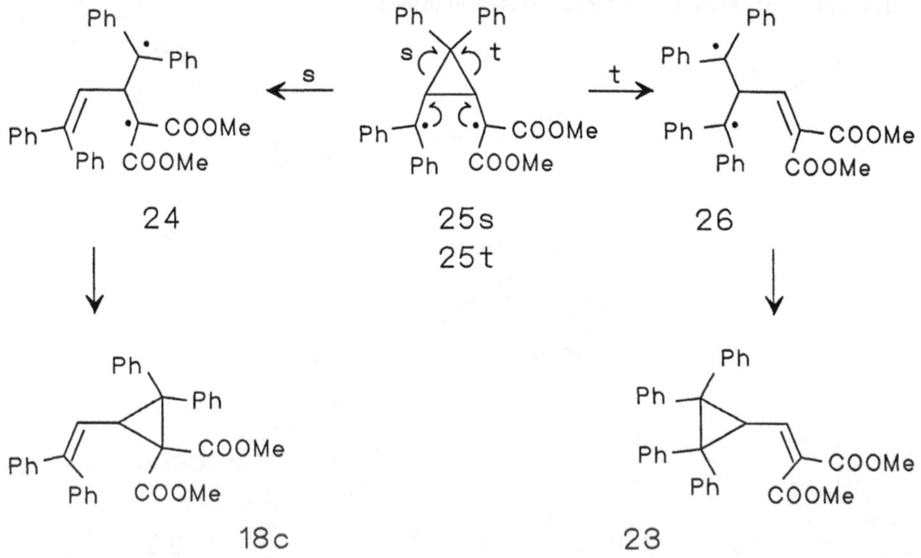

24   25s   26
              25t

18c   23

three-membered ring opening with the arrows labeled s̲ rather than those
labeled t̲. Conversely, the energy barrier for three-ring opening of triplet
diradical (25t) must be lower for the process using arrows t̲; here refer to
Scheme 3.

We shall term the singlet transition state a "small K process" and the
transition state utilized by the triplet a "large K process" where we have yet
to define "K". Also, from our knowledge of the regioselectivity of the singlet
and the triplet processes we can place the energy level for each reaction course
actually taken lower than that of the course not utilized; this is a truism.
These observations are stated schematically and qualitatively in Figure 1 which
clarifies that a singlet would select a "small K" variety of reaction, since
this has the lower energy available to a singlet, while a triplet reacting
would select a "large K" transition state because this is lower in energy for a
triplet. In setting up this diagram it is known that for corresponding
electronic configurations of the diradicals the triplet will be of lower
energy. It is seen that the energy separation for each transition state is
labeled "2K". This is because for a set of corresponding triplet and singlet

Figure 1. Singlet and Triplet Energy Levels Governing Reaction Type.

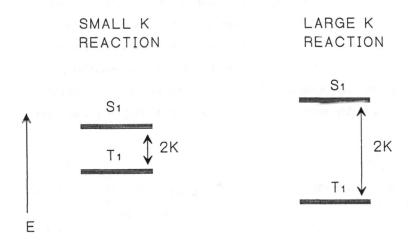

diradicals, represented as sums and differences of Slater determinants, the energy difference is twice the exchange integral, which is commonly termed K.

The only matter we have not discussed is if one can decide in advance which are "small K" reactions and which are of the "large K" variety. It has been noted[11,13,15,16] that diradicals tend to have large exchange integrals. However, one exception is where one odd electron center is substituted with electronegative groups. Another useful generalization is that pericyclic processes tend to have small values of K and are often selected as the preferred reaction of a reacting singlet. A fourth generalization is that where the excitation process can be described as reasonably approximated by excitation from HOMO to LUMO and these have electron densities weighted in different parts of the molecule, then K will be be small.

D. Regioselectivity

The di-π-methane rearrangement is particularly susceptible to control of its regioselectivity by substituents. The subject is most conveniently approached by consideration first of the singlet. Most simply, it may be stated that electron donating groups tend to appear on the residual vinyl group while electron withdrawing groups tend to appear on the product three-membered ring. Three relevant examples are found in equations [9], [10] and [11]. In the case of the bis-dimethylaminophenyl di-π-reactant (27) only the product with the electron donating dimethylaminophenyl groups substituted on the residual double bonds of product (28) is formed;[17] note equation [9]. Conversely, as depicted in equation [11], the electron withdrawing p-cyanophenyl groups of reactant (32) appear only on the three-membered ring of photoproduct

p-Me₂N-Ph ... p-Me₂N-Ph ... Ph ... Ph $\xrightarrow[\text{Direct}]{h\nu}$ p-Me₂N-Ph ... p-Me₂N-Ph ... Ph ... Ph [9]

27   28

(33).[18] With anisyl (i.e. An) groups present as in equation [10], two

photoproducts, (30) and (31) result with the isomer having the anisyl

substituents located on the double bond predominating 3.3:1.[17] Considering

that anisyl is less effective as an electron donor than dimethylaminophenyl,

this is quite consistent. Further examples are known where a methoxy electron

donor group,[19] or a cyano[19] or carbomethoxy[13] electron withdrawing group

is substituted directly on one double bond.[19] These examples follow the same

pattern of regioselectivity.

The rationale given for this substituent control relied on truncated

SCF-CI calculations which revealed the carbinyl centers of the cyclopropyl

dicarbinyl diradical (i.e. (6)) to be electron rich and and thus destabilized

by electron donating groups such as dimethylamino and methoxy.[20-22] Thus,

that ring opening which dissipates this electron density at the carbinyl

carbon bearing the electron donating groups is favored. This qualitative

argument was confirmed[21] with SCF-CI calculations of cyclopropyldicarbinyl

diradicals in which either bond a or bond b (note Figure ?) had been stretched.

Interestingly, in agreement with this qualitative argument and with experiment,

bond a stretching provided a lower energy species when the para substituent X

was cyano while bond b stretching gave the lower energy when para substituent

X was methoxy. The reasoning, again, is that excess electron density at a

carbinyl center is destabilized by donating groups and stabilized by electron

withdrawing substituents.

Figure 2. SCF-CI Prediction of Ease of Bond Stretching

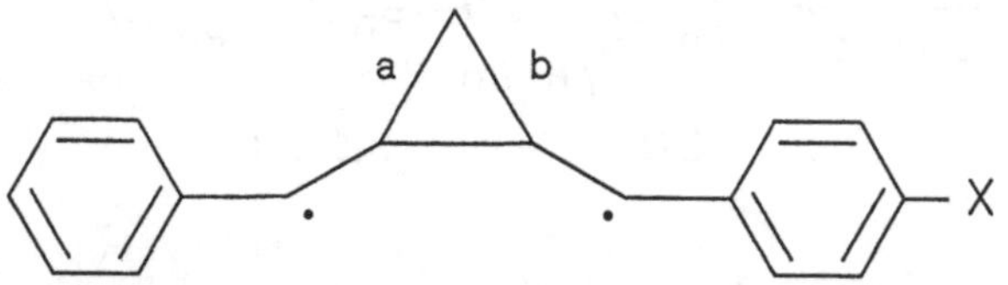

FOR X = CN BOND a STRETCHING FAVORED
FOR X = MEO BOND b STRETCHING FAVORED

This is in agreement with the observation that the rate of the singlet di-π-methane rearrangement is inhibited with para electron donating groups such as dimethylamino and accelerated with para substitution by electron withdrawing groups such as cyano.[20]

Somewhat different control is found in instances such as in equations [12] and [13]. In each case it is the π-bond with no terminal electron delocalizing substituents which survives. This behavior is understood on the basis of

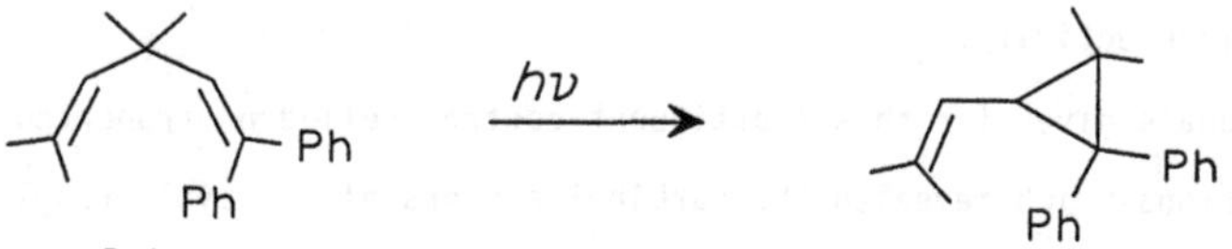

[12]

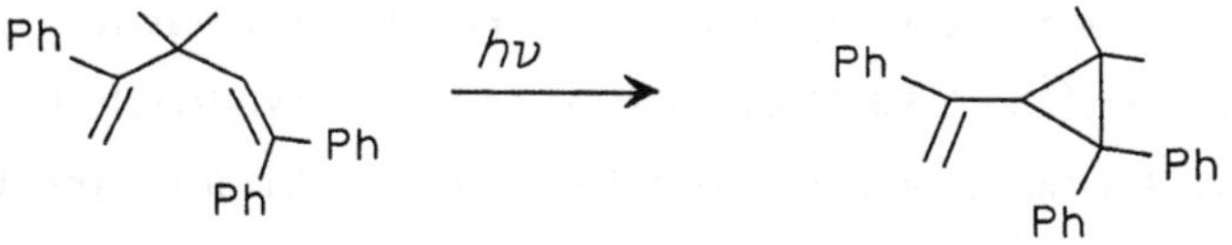

[13]

the usual di-π-methane reaction mechanism as outlined in Scheme 4. It is seen seen that at the cyclopropyldicarbinyl stage of the rearrangement (i.e. species (40)), there are two, alternative "unzipping" options available, a and b. However, process a utilizes the odd electron at the benzhydryl center and

Scheme 4. Delocalization Control of Regioselectivity of Cyclopropyldicarbinyl
Ring Opening.

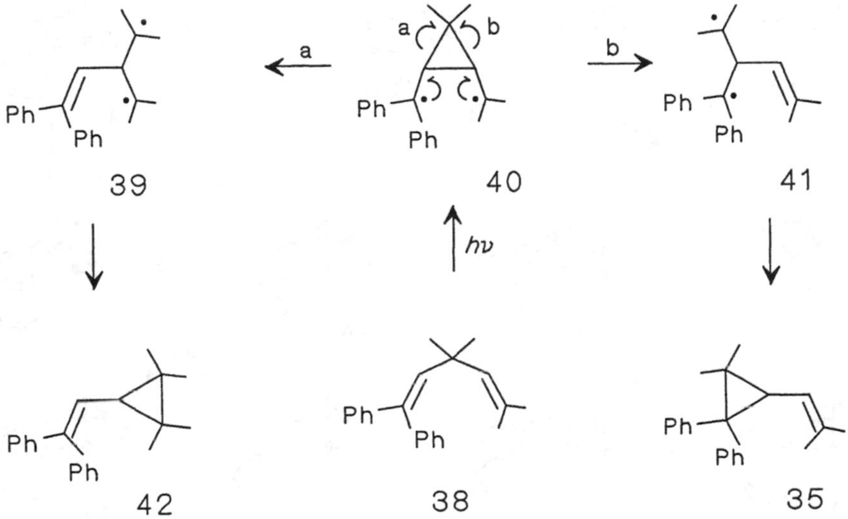

this opening of the three-membered ring to form 1,3-diradical (39) results in
loss of benzhydryl delocalization. In contrast, process b, which uses the
less stabilized odd-electron, goes on to form 1,3-diradical (41) which still
possesses this delocalization energy. The experimentally observed regioselective formation of vinylcyclopropane (35) is therefore not unexpected.

When the same reasoning is applied to the di-π-methane rearrangement of
diene (36), we predict the observed regioselective formation of vinylcyclopropane (37). Note equation [13].[23]

Interesting regioselectivity is found in bicyclic di-π-methane
rearrangements as well. One example[24] is found in the triplet rearrangement
of 1,2-naphthobarrelene (43). This selectivity was studied by labeling of the
two bridgehead atoms of the original naphthobarrelene (43) with hydrogen atoms
and the remaining sp^2-hybridized carbons with deuterium. Note Scheme 5 where
the hydrogen labeled carbons are labeled with solid circles. The excited state
bridges selectively between the vinyl group and the α-position of the naphthyl

Scheme 5. Regioselectivity of the Di-π-Methane Rearrangement of α-Naphtho-barrelene.

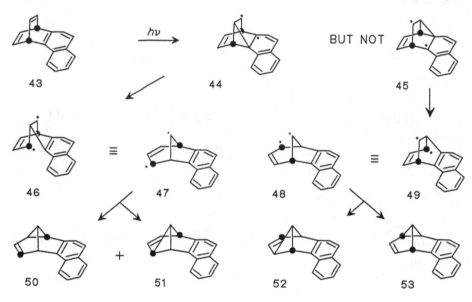

moiety, thus affording species (44). No bridging to the β-naphthyl carbon to afford diradical (45) is encountered. The preference for bridging to the α-position, of course, has ground-state parallel. This results from the greater delocalization present in the diradical (44) resulting from α-bridging compared with diradical (45) which arises from β-bridging. Additionally, there is a preference for aryl-vinyl bridging over the alternative possibility of vinyl-vinyl bridging. This contrasts with the case of 2,3-naphthobarrelene where only bridging to the β-carbon is a possibility and vinyl-vinyl bridging results.[24] Also, in the cases of benzobarrelene[25a] and 2,3-anthraceno-barrelene,[25b] only vinyl-vinyl bridging occurs in preference to benzo-vinyl bridging. Finally, in benzonaphthobarrelene there is approximately equal competition between benzo-vinyl and β-naphtho-vinyl bridging.[26]

Thus we can conclude that α-naphtho-vinyl bridging of the triplet excited state is preferred over vinyl-vinyl bridging, which in turn is preferred over β-naphtho-vinyl and benzo-vinyl bridging. This regioselectivity has been

explained[26] as arising from the energy of the various partially bridged
triplet species in which two π-moieties have approached one another.

   In a particularly elegant study, Bender and Coworkers[27] described the
regioselectivity of the di-π-methane rearrangement of 2-cyanobenzobarrelene
(54). Four pathways are a priori possible, two beginning with vinyl-vinyl
bridging and two with benzo-vinyl bonding. There are two potential mechanisms

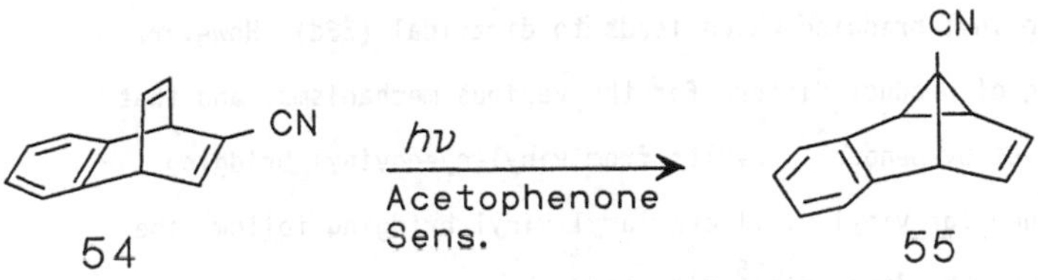

[14]

which would lead to the observed cyanobenzosemibullvalene. By using labeling
with deuterium and hydrogen the mechanism was shown to be that depicted in
Scheme 6, where the solid circled carbons were labeled with deuterium. Thus
cyanobenzobarrelene (54d) affords photoproduct (55d).

   There are two aspects to this regioselectivity. The first is the
preference for that π-π bridging to afford a diradical in which the cyano

Scheme 6. The Effect of Cyano Substitution on Regioselectivity.

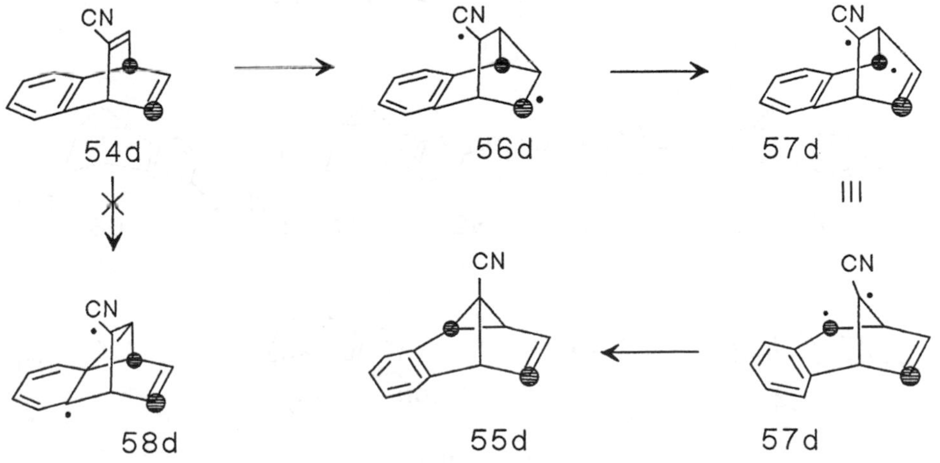

group is at an odd-electron center, thus affording extra delocalization.
A further factor is that the carbinyl carbons of the cyclopropyldicarbinyl
species have been noted[20-22] to be electron rich. Thus the electron
withdrawing cyano group affords stabilization of charge as well. If one writes
a mechanism with π-π bridging in which the cyano substituent does not find
itself at an odd-electron center, totally different products result. Second,
there is a choice between vinyl-cyanovinyl bridging which affords diradical
(56d) and benzo-cyanovinyl bridging which leads to diradical (58d). However,
the hydrogen labeling of product differs for the various mechanisms, and that
experimentally observed by Bender[27] results from vinyl-cyanovinyl bridging.
The observed preference for vinyl-vinyl over aryl-vinyl bridging follows the
general pattern observed by Zimmerman[25] discussed above.

　　We can now turn to the matter of substitution in aromatic rings of
a cyclic aryl-vinyl di-π-methane reactant. Meta-substitution is instructive.
Three examples are given in equations [15][28], [16][29] and [17][29]. It is
seen that π-π bridging occurs para to the electron withdrawing nitro and
cyano groups and meta to an electron donor group such as methoxy. Notably, this

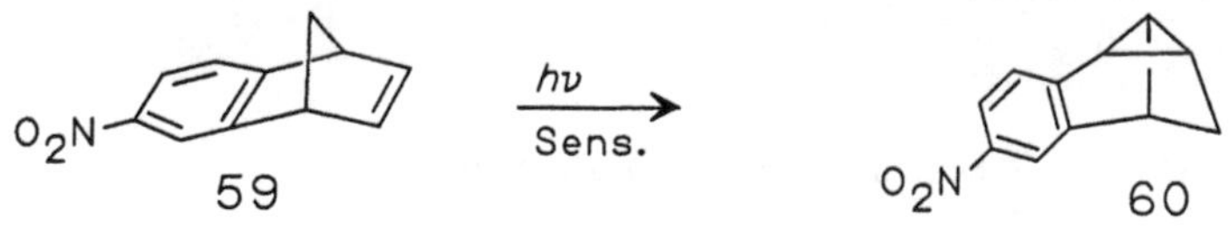

[15]

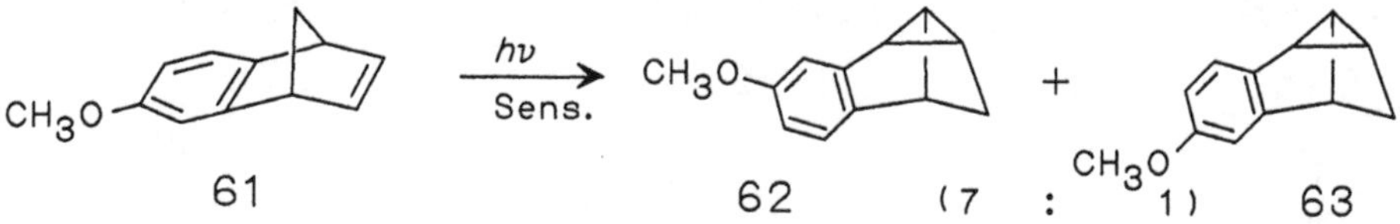

[16]

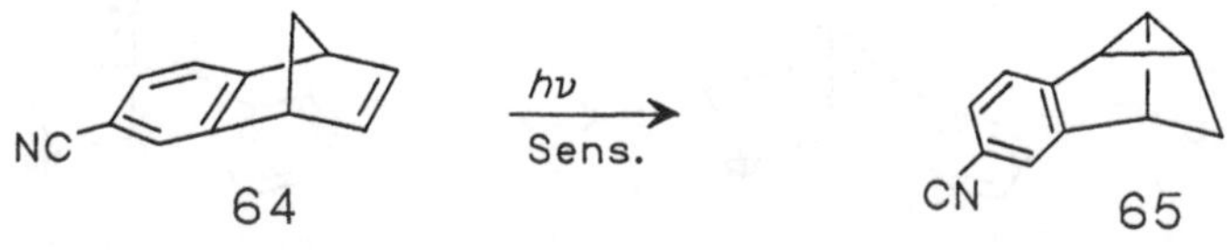

[17]

is the same type of effect seen earlier in connection with the regioselectivity
encountered in the rearrangements of di-anisyl, bis-dimethylaminophenyl and
di-p-cyanophenyl acyclic di-π-methane systems (note equations [9], [10] and
[11] and accompanying discussion). Thus, that bridging is favored which
positions a cyano group so as to stabilize the excess negative charge of
the carbinyl carbon of the cyclopropyldicarbinyl diradical. Conversely, a
p-methoxy group will avoid such a position. This is seen in Figure 3, where the
carbons in the six-membered ring which are conjugated with the electron rich
carbinyl position are reluctant to bear a methoxy group while stabilized by a
cyano substituent. In any event, the behavior in the bicyclic systems is
consistent with that observed in the earlier studies of the Wisconsin group as
cited above. Additionally, this behavior has been noted to be in agreement with
molecular orbital considerations of the perturbation type by Houk and
Paquette.[30] For both the methoxy and the cyano examples, it was noted[30]
that MO considerations predict bridging to be in the experimentally observed
direction based on the π-π bridging step.

Turning to the effect of central (i.e. the "methane substituted") carbon
substitution, we are reminded of the control by central phenyl substitution on
the facility of di-π-methane triplet rearrangements; for example, note Schemes

Figure 3. Dicyclopropyldicarbinyl Diradical Species and Substituent Effects

66   PREFERRED   OVER   67

68   PREFERRED   OVER   69

* High density sites

2 and 3. In this study by Zimmerman and Factor[13] the reaction efficiencies
were markedly enhanced for the triplet processes ($\phi_{singlet}$ = 0.39 and
$\phi_{triplet}$ = 0.92). Similarly, in this study central carbomethoxyl
substitution was found to enhance the quantum yields. Also, in a still earlier
study by Zimmerman, Armesto and Johnson[11] both phenyl and cyano substitution
were found to enhance triplet reactivity. Note equations [18] and [19].

In contrast, acyclic systems lacking these central substituents reveal
little or no reactivity since the rate of triplet rearrangment is sufficiently
slow compared with free-rotor energy dissipation.

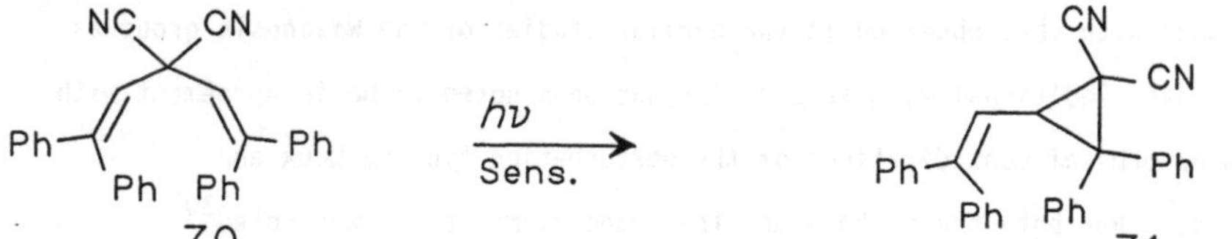

$$\text{70} \xrightarrow[\text{Sens.}]{h\nu} \text{71} \qquad [18]$$

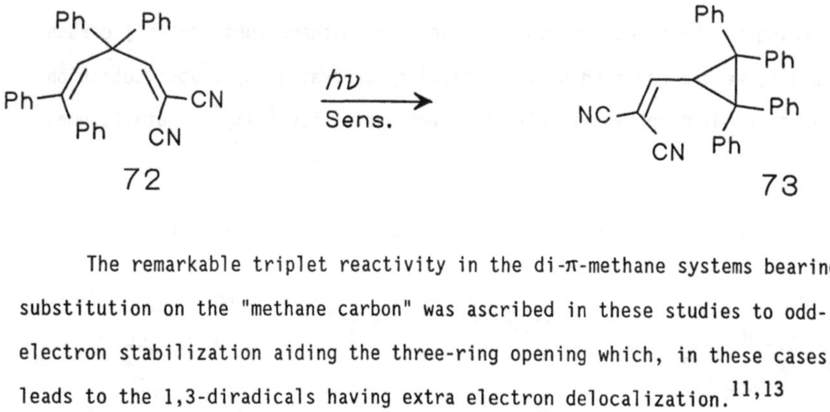

$$\text{72} \xrightarrow[\text{Sens.}]{h\nu} \text{73} \qquad [19]$$

The remarkable triplet reactivity in the di-π-methane systems bearing
substitution on the "methane carbon" was ascribed in these studies to odd-
electron stabilization aiding the three-ring opening which, in these cases,
leads to the 1,3-diradicals having extra electron delocalization.[11,13]

However, the reactivity enhancement seen for the triplets is in sharp
contrast to the situation observed for singlet (i.e. S_1) reactivity. Here[13]
the reaction rates for S_1 rearrangement decreased in the order dimethyl >
dicarbomethoxy > dicyano, where these groups were placed on the central
methane carbon. Additionally, when these cyano and carbomethoxy groups were
placed on an end of the di-π-methane system, the rate of singlet rearrange-

ment, instead, was increased. More generally, the presence of electron stabilizing groups on the termini of di-π-methane systems was found[21] to enhance the excited singlet rearrangement rates.

These results suggest that for the singlet rearrangement, the rate-limiting stage of the di-π-methane rearrangement is π-π bridging (i.e. the first step in equations [3] and [4]) while for the triplet counterpart, three-ring opening of the cyclopropyldicarbinyl diradical (i.e. (6) in equation [3] and (10) in equation [4]) plays a role. Thus, it is clear that both the bridging phase and also the three-ring opening control the overall reactivity in the triplet reaction.

One interesting effect is seen in regioselectivity arising when there is a bridgehead deuterium atom. Scheme 7 provides one example,[14,31] that of benzonorbornadiene. Photoproducts (79) and (80) are formed in a 1.3:1 ratio, thus showing a preference for having the deuterium in a methine position rather than on the three-membered ring. The same preference is encountered in a similar experiment involving dibenzobarrelene with a single bridgehead

Scheme 7. Regioselectivity of Bridgehead Labeled Di-π-Methane Rearrangement.[14,31]

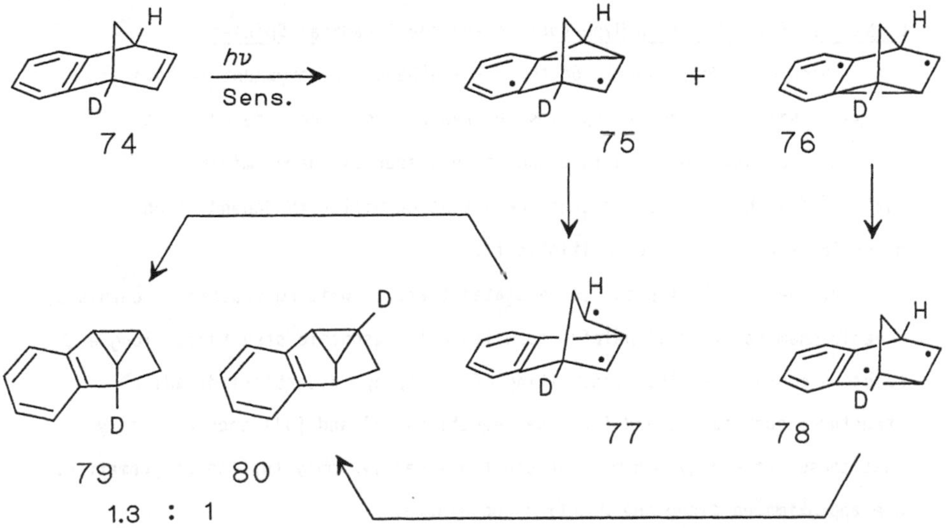

deuterium atom[14,31] and also in work by Hemetsberger[32] in which
mono bridgehead-deuterated triptycene was photolyzed.

Referring to Scheme 7, we note that in the formation of diradical species
(75) and (76) there is the choice of whether the bridgehead carbon bearing
deuterium will remain sp^3 hybridized (i.e. in forming (75)) or will become
close to sp^2 hybridized by virtue of being on a cyclopropane ring if (76) is
formed. Somewhat further along the reaction hypersurface we encounter 1,3-
diradicals (77) and (78). Here the hybridization of the orbital bonding to
deuterium remains sp^3 hybridized if we reach (77) and remains close to sp^2
hybridized if we form diradical (78). Thus the transformation (74^*) --> (75)
--> (77) maintains the sp^3 hybridization of the orbital bonding deuterium while
the transformation (74^*) --> (76) --> (78) results in conversion of sp^3 to sp^2
hybridization as soon as (76) is reached and maintains this as (78) is attained.

The regioselectivity follows a modestly often encountered secondary
deuterium isotope effect in which deuterium prefers to bond to orbitals with
minimum s-character.[33-35] Our interpretation and the latter statement differs
from that of the literature[14,31] which stated the reverse dependence on
s-character.

E. Nature of the Reaction Hypersurface and the Diradical Species

Throughout our research on the di-π-methane rearrangement we have
stressed that the diradical species we draw are to be considered points
on an excited state hypersurface and not necessarily energy minima or
maxima.[4,5] We have suggested that the situation will vary depending on
molecular structure and on multiplicity.

At the outset it needs to be stated that, in writing reaction mechanisms,
organic chemists find it helpful to write all reasonable structures along a
reaction coordinate. The mere writing of cyclopropyldicarbinyl diradical
structures such as (6) and (10) (see equations [3] and [4]) does not imply
that these are energy minima. The equations merely imply that these geometries
are approximated along the reaction coordinate.

This discussion is relevant to the suggestion[14,31] that there are two mechanisms possible for the di-π-methane rearrangement, a step-wise pathway and a "direct" mechanism.[14,31] The "direct mechanism" is said to avoid intervention of the cyclopropyldicarbinyl diradical species such as (6) and (10). To the extent that the discussion is whether or not such diradicals are transition states or are intermediates, one has a valid question. To the extent that one is questioning the occurrence of the cyclopropyldicarbinyl diradical species on the hypersurface of the reaction, the point is not valid.

Additionally there is the question of whether or not <u>both</u> π-systems play a role. This is equivalent to questioning whether one is dealing with a $\sigma + \pi$ (i.e. a "sigma rearrangement") or the "di-π-methane" process which does require <u>two</u> π-systems. This question was addressed much earlier by Zimmerman and Little[36] where it was shown that with one double bond removed from di-π-methane reactant (1), different reactivity was observed (note equation [20]). The rearrangement proceeded with diminished efficiency and excited state rate, thus signifying that any undetected di-π-methane process proceeded with still less reactivity. In comparing the dihydro reactant (81) with the ordinary di-π-methane reactant (1), we concluded that if the di-π-methane rearrangement proceeds via a $\sigma + \pi$ process, the dihydro reactant (81) should rearrange more readily than the ordinary di-π-reactant (1). In the dihydro reactant (81) it is sigma bond 2-3 which is involved and will be weakened in proceeding to the transition state. Bond 2-3 is sp^3-sp^3 hybridized. In contrast, in the di-π-methane system (1), it is a stronger, sp^2-sp^3 bond (i.e. bond 2-3 again) which will be weakened. Hence, if a $\sigma + \pi$ rearrangement is possible, the dihydro reactant (81) should be more reactive than the

      81   82   83

di-π-methane system (1). Since it is not, the second π-bond is clearly
needed for rearrangement.

A consideration is whether with two π-systems present, a σ + π mechanism
is a meaningful alternative to having involvement of both π-systems. Figure 4
depicts the change in orbital positioning of relevant orbitals during the
conversion of the di-π-reactant to the first diradical (e.g. (6) or (10)),
termed "diradical 1" and thence to the second diradical (e.g. (7) or (11))
which we term "diradical 2". Only the most relevant basis orbitals are drawn.
Thus p-orbitals _a_ and _b_ might comprise the π-system of a vinyl group or the
ipso-ortho p-orbitals of an aromatic ring. Orbitals _g_ and _h_ provide a vinyl
group. Orbitals _c_ and _d_ are used to form the initial sigma bond of reactant
while _e_ and _f_ are included for consistency but are not used directly in
the reaction. The vinyl group, consisting of the π-system _a_-_b_ and the
sp^2 hybridized orbital _c_ begins to migrate from the "methane carbon" and
orbital _d_ towards orbital _g_ of the second π-bond eventually reaching

Figure 4. Geometric Repositioning of Orbitals of the Di-π-methane System
in Proceeding From Reactant to Diradical 2.

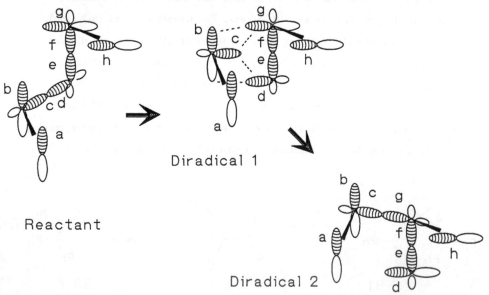

the geometry of "diradical 2". When half-migrated we have the geometry of "diradical 1". It is seen that at this point, the overlap is not just between the sp^2 hybrid $\underline{c}$ and the orbitals $\underline{d}$ and $\underline{g}$. Additionally, due to proximity, there is overlap of the π-system orbital $\underline{b}$ with $\underline{d}$ and $\underline{g}$. While this limited set of orbitals oversimplifies the picture, it does point out that one has interaction of the two π-systems at half reaction. Thus, in forming "diradical 2" the molecule necessarily proceeds through the geometry and electronic structure of "diradical 1".

This is very much like the 1,2-rearrangement of a phenyl group to a cationic, odd-electron or carbanion center on the adjacent carbon in the sense that at half-reaction, one has a bridged species. One cannot intellectually inhibit overlap. Now, whether at half reaction one has an energy minimum or a maximum, will vary with the system under consideration.

Perhaps the most convincing evidence for strong interaction of the π-system of the migrating vinyl or aryl group with the second π-bond, thus giving rise to the cyclopropyldicarbinyl diradical (i.e. "Diradical 1"), is the role played by substituents on the migrating vinyl or aryl group. Thus, as noted above in the singlet rearrangement of the acyclic systems, reactivity is controlled by substitution on both π bonds; note equations [9], [10] and [11] for example. In general, para substitution on a migrating aryl group or directly on a migrating vinyl group does affect the reaction rate.[17,19-21,37,38] In a $\sigma + \pi$ rearrangement one would not anticipate an appreciable effect by substitution on the migrating vinyl or aryl group.

In the case of our rearrangement of 1,2-naphthobarrelene (note Scheme 5) there is a complete preference for migration of the α-naphthyl carbon rather than the β-naphthyl carbon.[24] If this were just a sigma migration, the regioselectiv would be difficult to rationalize. Similarly, in subsequent studies by Hahn and Johnson[28] and also by Paquette and Coworkers[29] (note equations [15], [16] and [17]) regioselectivity is controlled by remote substitution on the aryl group migrating. Again, this is understood on the basis of π-π interaction and bridging to afford a cyclopropyldicarbinyl diradical (i.e. a "Type 1"

diradical) species. In brief, clearly there is π-π bonding in the initial
phases of the di-π-methane rearrangement.

Another item deals with the tendency of "Type 1" diradicals to
proceed in the forward direction as opposed to reverting to reactant. Triplet
"diradical 1" in the cases of barrelene, benzobarrelene and 2,3-naphthobarrelene
was generated as outlined in equations [21], [22], and [23]. In each case, the
diradical proceeded in the forward direction to afford the corresponding
semibullvalene product. In this study by Zimmerman and Coworkers[39] only minor
reversion to the barrelenes was encountered when the triplet diradicals (85),
(89) and (93) were generated. In contrast to this behavior, it was observed
that thermolysis of the same azo compounds - (84), (88) and (92) - led only to
the corresponding barrelenes. Still different, direct irradiation of the same
azo compounds led to both barrelenes and semibullvalenes. Note Table I. This
result was interpreted to signify that the "Type 1" T_1 diradicals proceeded
efficiently to semibullvalenes along the di-π-methane hypersurface but S_0
(and perhaps S_1) "Type 1" diradicals undergo a 1,4-(2,3) scission[11,40]

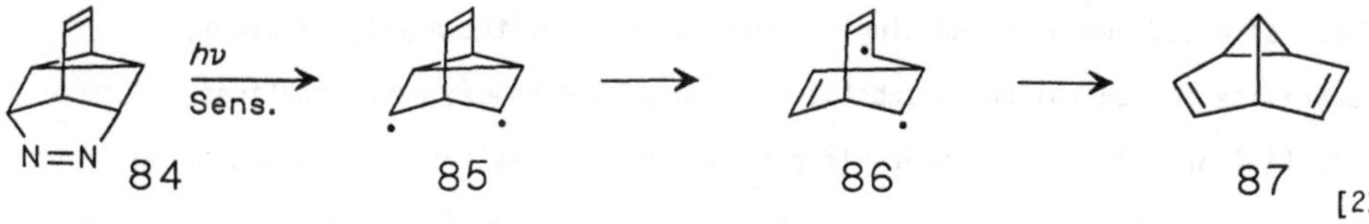

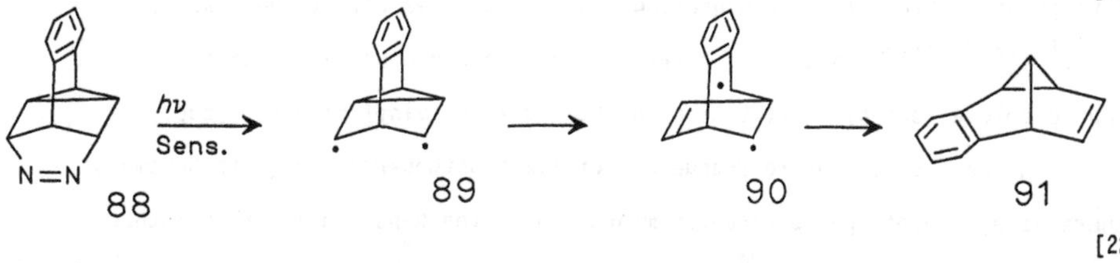

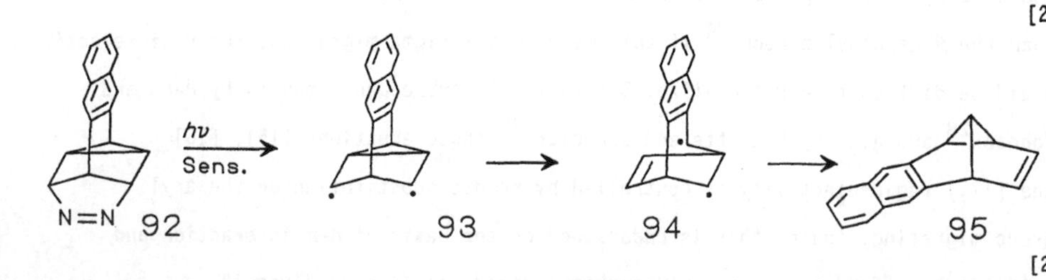

Table I. Photochemistry of Azo Barrelenes.

Azo Reactant	Conditions	Result	
		Barrelene	Semibullvalene
Barrelene Azo	Thermolysis (S$_0$)	100%	0%
Compound (84)	Direct Irradiation (S$_1$)	24%	73%
	Sensitized Irradiation (T$_1$)	0%	100%
Benzobarrelene	Thermolysis (S$_0$)	100%	0%
Azo Compound (88)	Direct Irradiation (S$_1$)	$\phi = .70$	$\phi = .21$
	Sensitized Irradiation (T$_1$)	$\phi = .08$	$\phi = .58$
2,3-Naphthobarrelene	Thermolysis (S$_0$)	100%	0%
Azo Compound (92)	Direct Irradiation (S$_1$)	$\phi = .78$	$\phi = .21$
	Sensitized Irradiation (T$_1$)	$\phi = .1$	$\phi = .63$

quite similar to a Grob fragmentation. The proclivity of S$_0$ diradicals to undergo 1,4-(2,3)-fragmentation with loss of the central bond of a variety of 1,4-valence deficient diradicals and zwitterions has been advanced in a variety of publications[5]. The species of interest as relevant to the di-π-methane rearrangement, however, is T$_1$ for these bicyclic systems. In conclusion, the observation for the T$_1$ diradical species is a strong tendency to follow a forward pathway from T$_1$ of the various barrelenes to T$_1$ of the "diradical 1" to T$_1$ of "diradical 2" and thence with intersystem crossing to S$_0$ of the semibullvalene.

Studies by Adam[41] in still different bicyclic systems arrived at the same conclusion of lack of reversibility. In this connection it is relevant to note that diazenyl diradicals have been proposed to intervene in nitrogen loss of azo compounds[41]. However, such nitrogen-containing diradicals seem likely to lead onward to the free carbon diradicals in the case of triplets. For example, recently Adam and Wirz[42] have utilized azo compounds to produce 1,3-diradicals successfully, both on direct and sensitized irradiations.

Often in literature publications dealing with this question, evidence obtained for S$_0$ (i.e. thermal) and S$_1$ (i.e. often direct irradiation) chemistry is erroneously applied to discussions of T$_1$ mechanisms.

One study by Zimmerman and Coworkers is very relevant to the present

matter of di-π-methane diradical species[2b,c]. Thus, in our original study
wherein the mechanism of the di-π-methane rearrangement was established, we
utilized barrelene which had been labeled at the bridgehead positions with
hydrogen and with deuterium elsewhere. This is outlined in Scheme 8 where the
bridghead labels are depicted with filled circles. It was observed that the
semibullvalene product (4) had its label in a 1:1 distribution between that for
(4a) and for (4b). In the opening of the cyclopropyldicarbinyl 1,4-diradical
species (i.e. "diradical 1 (96)") it can be seen that the resulting 1,3-diradical
(i.e. "diradical 2") <u>initially</u> will have the conformation (97). However, the
observation of equal amounts of (4a) and (4b) means that "diradical 2" has time
to unfold to conformation (98) with allylic delocalization and a plane of
symmetry (see (99)). This conformational equilibration is needed to account for
the equal bonding at the two allylic odd-electron centers. The net result is
that we identified our "diradical 2" as having sufficient lifetime for the
conformational change and thus "diradical 2" was taken to be a true
intermediate of modest lifetime.

Hence we can conclude that for the triplet di-π-methane rearrangements
of these bicyclic systems, diradicals of "Type 2" are true intermediates. This

Scheme 8. The Triplet Rearrangement of Bridgehead Labeled Barrelene.

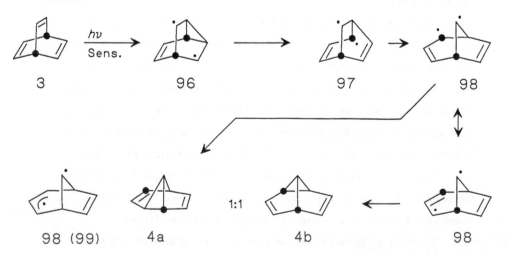

contrasts with the suggestion by Zimmerman and Coworkers that the singlet (i.e.
S_1) rearrangement of acyclic systems most likely is concerted[4] in view of
the reaction stereospecificity, but is in agreement with our view that the
di-π-methane rearrangements differ in mechanistic subtleties depending on
precise structure and multiplicity.

We turn now to the matter of "Type 1" diradicals. If the singlet
rearrangements are, indeed, concerted as we have suggested, then the cyclopropyldicarbinyl diradical structures will not correspond to energy minima.
These structures are important, nevertheless, since all the evidence is that
the bridging process controls the excited state (i.e. S_1) reaction rate.

The case of the triplet rearrangements is more problematic. Demuth,
Schaffner and Coworkers[43] have presented spectroscopic evidence for the
utilization of two biradicals in an example of the di-π-methane rearrangement,
one taken to be the cyclopropyldicarbinyl diradical. Further triplet cyclopropyldicarbinyl diradicals of finite lifetime have been detected
spectroscopically by Caldwell[44] and also Sciano[45] and their Coworkers,
although these diradicals were not involved in di-π-methane chemistry.

On the theoretical side, Borden, Davidson and Coworkers[46] have reported
MCSCF ab initio computations on the parent 1,4-pentadiene di-π-methane
system in which it was concluded that the cyclopropyldicarbinyl diradical is a
true intermediate. In work at Wisconsin[47a] MNDO-CI computations with geometry
optimization of the reaction hypersurface of the triplet rearrangement of
barrelene to semibullvalene suggested a circa 10 kcal/mole barrier leading
forward from "diradical 1". Ab initio computation using MCSCF with STO-3G and
geometry optimization gave a similar barrier. However, UHF with geometry
optimization and also MCSCF - 3-21G using MNDO optimized geometry indicated a
smaller energy barrier approximating 5 kcal/mole.[47b]

F. Reversibility: Examples of the Reverse Di-π-Methane Rearrangement

In our discussion above we have commented that a number of bicyclic examples
of the di-π-methane rearrangement have shown a tendency to proceed in the

forward direction along the triplet hypersurface. However, the di-π-methane
rearrangement is not invariably irreversible, and a number of examples of the
"reverse di-π-methane rearrangement" have been described.[11,13,40,48] Three
are outlined in equations [24], [25] and [26]. In each of the first two
examples[11,13] an isomeric vinylcyclopropane (i.e. (18b) or (18c)) is also

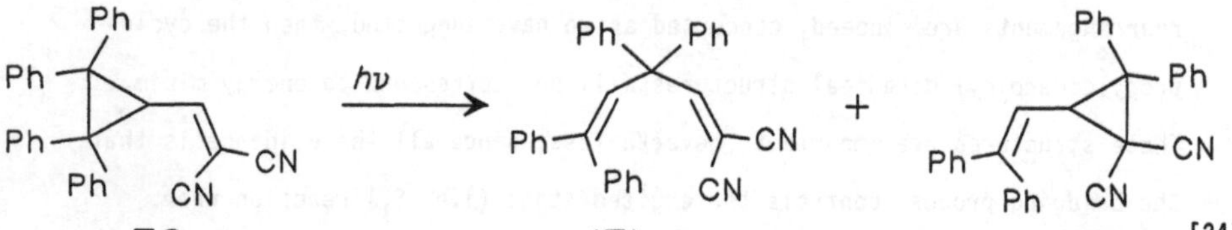

$$\varphi = 0.34 \qquad \varphi = 0.12 \qquad [24]$$

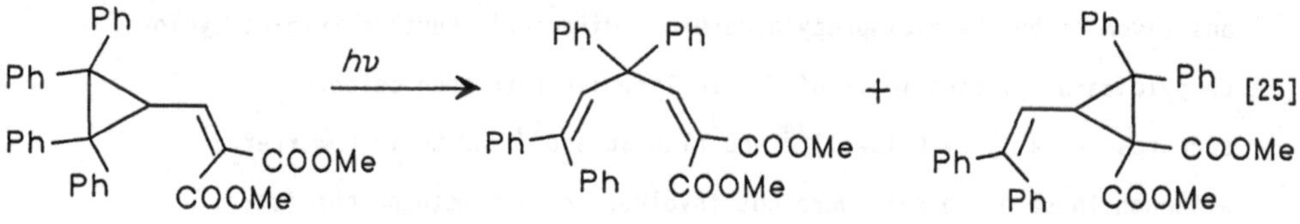

$$\varphi = 0.069 \qquad \varphi = 0.10 \qquad [25]$$

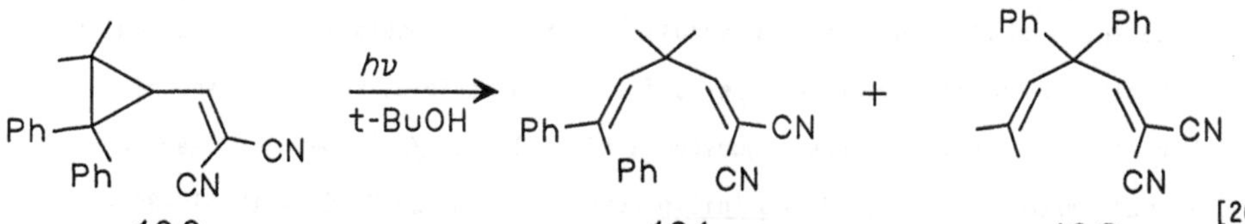

$$\varphi = 0.0076 \qquad \varphi = 0.0079 \qquad [26]$$

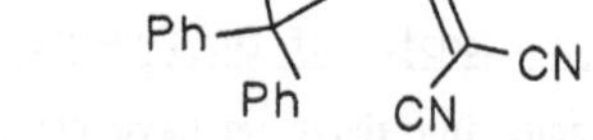

$$103 \qquad \varphi = 0.017$$

formed in the reaction in addition to the di-π-methane diene (i.e. (17b) and

(22), respectively). In the third example[48] (note equation [26]) two di-π-

methane dienes ((101) and (102)) were isolated but no isomeric vinylcyclo-

propane was observed. Additionally a Griffin[49] disproportionation product

(103) was formed, this arising from three-membered ring opening and hydrogen

transfer in the resulting diradical.

The mechanism of the "reverse di-π-methane rearrangment" may be written

as in Scheme 9 where X represents the various terminal substituents in

equations [24], [25] and [26]. The reaction may be considered to be part of

the "Bicycle Rearrangement"[13,40,50] in which bicycling terminates once the

cyclopropyldicarbinyl diradical (105) is reached. Such 1,4-diradicals as S_0

species have been noted to undergo central bond scission (vida supra). Hence

the amount of reverse di-π-methane rearrangement -- in this case to give diene

(107), relative to complete bicycling affording (106) -- depends on the

relative rate of internal conversion of S_1 cyclopropyldicarbinyl diradical

(105) to its ground state S_0 configuration.

Scheme 9. Mechanism of the Reverse Di-π-Methane Rearrangement and the
Bicycling Rearrangement.

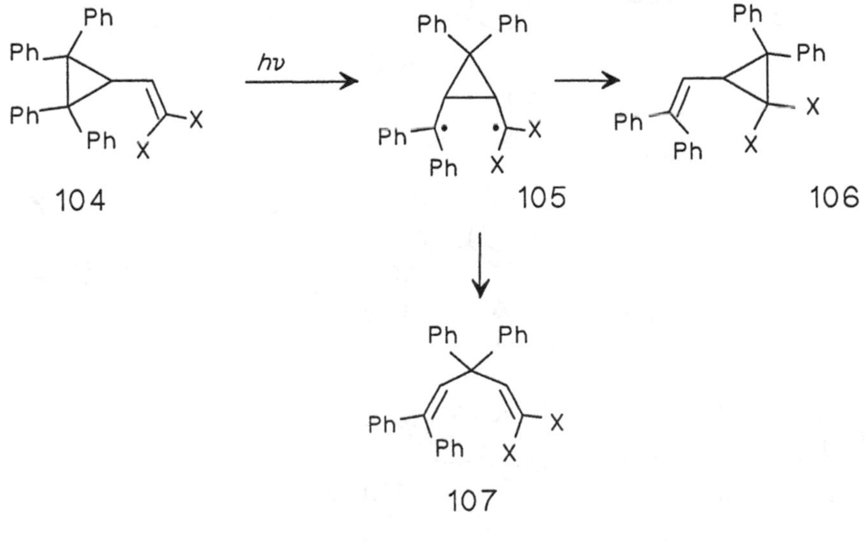

While these examples have electronegative groups on the termini of the di-π-methane system, this is not a requisite as seen in the study[40] outlined in Scheme 10 which depicts both a "reverse di-π-methane rearrangement" and its di-π-methane counterpart, and provides a rationale for an apparent paradox. Thus cyclopropyldicarbinyl diradical (109) is generated from two different starting materials by direct irradiation. Yet, depending on the direction of its formation, different products arise. This emphatically demonstrates the important role played by the <u>electronic state</u> of the cyclopropyl dicarbinyl diradical in determining its reactivity. Throughout di-π-methane photochemistry and here, one sees that the S_0 configuration of cyclopropyldicarbinyl diradicals undergoes the 1,4-(2,3)-fragmentation (a Grob-like scission of bond 2,3) while the diradical with a S_1 configuration undergoes other processes such as bicycling.[14,40,50] In this particular example (note Scheme 10), the S_1 processes which lead to S_1 species are depicted by heavy arrows. Those

Scheme 10. An Example of the "Reverse Di-π-Methane Rearrangement" and Its Di-π-Methane Counterpart. Dependence of the Cyclopropyldicarbinyl Diradical Reactivity on its Electronic Configuration.

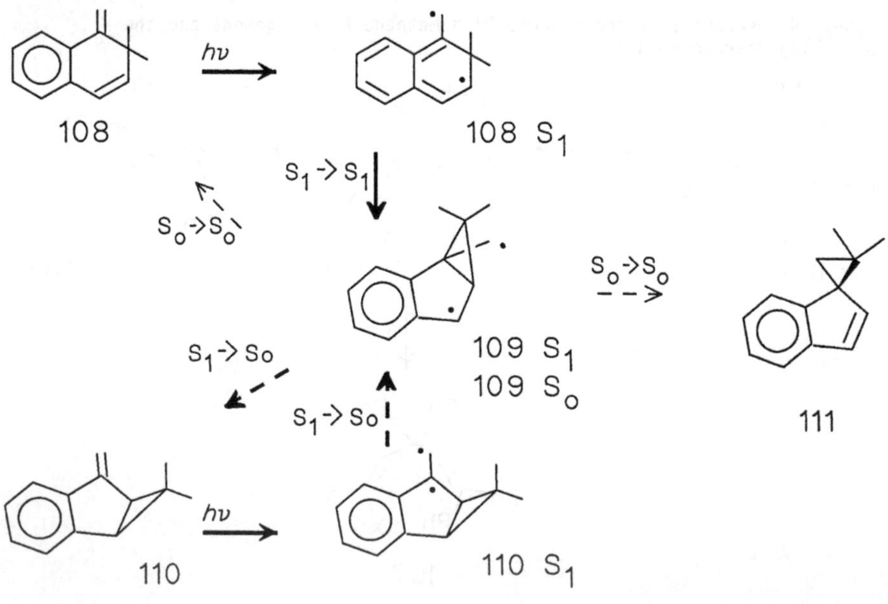

transformations which begin with ground state species (i.e. S_0) and afford

ground state species are shown by dotted arrows. Finally, those processes which

originate with S_1 species and lead to S_0 products are depicted with a

broken, heavy arrow.

Scheme 10 shows that it is the ground state diradical (109 S_0) which

gives rise to the "reverse di-π-methane Rearrangement" (i.e. product (108))

as well as the spiro product (111). In contrast it is the excited singlet

(i.e. (109 S_1)) which leads in the di-π-methane direction and affords (110).

These conclusions were reached from computational study of the hypersurfaces

involved. The source of the different probabilities in getting ground state

versus excited cyclopropyldicarbinyl diradical was found to derive from an

S_0-S_1 avoided crossing which was close to (110) on the reaction

hypersurface.[40]

In summary regarding di-π-methane reversibility, it may be stated that

the reaction directionality is determined both by structural features and

also by electronic factors including multiplicity.

Further Variations of the Di-π-Methane Rearrangement.

The scope of this treatment of the di-π-methane rearrangement is

intended to limit itself to the parent reaction. Nevertheless, it would be

incomplete not to mention that the reaction has two particularly important

variations. One is the "oxa- di-π-methane" rearrangement and the other is the

"aza-di-π-methane" process. Equations [27][51] and [28][52a] typify these

reactions. The mechanisms are quite similar to that of the di-π-methane

rearrangement itself. In the case of the oxa-di-π-methane rearrangement, we

Ph

Ph

112

$\xrightarrow{h\nu}$

Ph

Ph

113

[27]

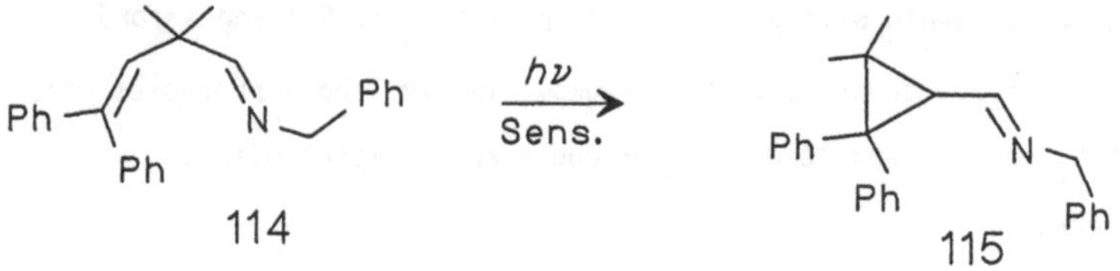

[28]

note that this triplet process was first suggested by Swenton[53] while the mechanism is due to Givens[54] and the "oxa-di-π-methane" designation derives from Dauben.[55] The reaction has been studied extensively and reviewed.[4,56] The "aza-di-π-methane rearrangement was uncovered by Armesto and Horspool and has been delineated in a series of extensive and intriguing publications.[52]

G. Conclusion

This review has attempted to bring coverage of the di-π-methane rearrangement up-to-date and also to deal with questions which have arisen since the last review. It is clear that the di-π-methane rearrangement is one of the most general and intensively studied of photochemical reactions. It is certain that it is of considerable synthetic utility and additionally is mechanistically intriguing.

H. Acknowledgement

This research would not have been possible with the support of the National Science Foundation and this is gratefully acknowledged. Most appreciated is the philosophy of supporting basic, untargeted research. Without this approach, science will soon lack the broad base needed even for targeted aims.

References

1. H. E. Zimmerman and P. S. Mariano, J. Am. Chem. Soc., 91, 1718-1727 (1969).

2.(a) H. E. Zimmerman and G. L. Grunewald, J. Am. Chem. Soc., 88, 183-184 (1966); (b) H. E. Zimmerman, R. W. Binkley, R. S. Givens and M. A. Sherwin, J. Am. Chem. Soc., 89, 3932-3933 (1967); (c) H. E. Zimmerman, R. W.

Binkley, R. S. Givens, G. L. Grunewald and M. A. Sherwin, J. Am. Chem. Soc., 91, 3316-3323 (1969).

3. H. E. Zimmerman and A. P. Kamath, J. Am. Chem. Soc., 110, 900-911 (1988).

4. S. S. Hixson, P. S. Mariano and H. E. Zimmerman, Chem. Revs, 73, 531-551 (1973).

5. H. E. Zimmerman in "Rearrangements in Ground and Excited States," Vol. 3, Edited by P. DeMayo, Academic Press, New York, 1980.

6. H. E. Zimmerman and A. C. Pratt, J. Am. Chem. Soc., 92, 1409-1411 (1970).

7. H. E. Zimmerman and A. C. Pratt, J. Am. Chem. Soc., 92, 6267-6271 (1970).

8. P. Borrell and H. H. Greenwood, Proc. Roy. Soc. (London), A 298, 453-466 (1967).

9. J. Saltiel, J. Am. Chem. Soc., 90, 6394-6400 (1968).

10. H. E. Zimmerman and D. N. Schissel, J. Org. Chem., 51, 196-207 (1986).

11. H. E. Zimmerman, D. Armesto, M. G. Amezua, T. P. Gannett and R. P. Johnson, J. Am. Chem. Soc., 101, 6367-6383 (1979).

12. H. E. Zimmerman, R. J. Boettcher and W. Braig, J. Am. Chem. Soc., 95, 2155-2163 (1973).

13. H. E. Zimmerman and R. E. Factor, Tetrahedron, 37, Supplement 1, 125-141 (1981).

14. L. A. Paquette and E. Bay, J. Org. Chem., 47, 4597-4599 (1982).

15. H. E. Zimmerman, Accounts Chem. Res., 10, 312-317 (1982).

16. H. E. Zimmerman, J. H. Penn and M. R. Johnson, Proc. Natl. Acad. Sci. USA, 78, 2021-2025 (1981).

17. H. E. Zimmerman and W. T. Gruenbaum, J. Org. Chem., 43, 1997-2005 (1978).

18. H. E.Zimmerman and B. R. Cotter, J. Am. Chem. Soc., 96, 7445-7453 (1974).

19. H. E. Zimmerman and R. T. Klun, Tetrahedron, 43, 1775-1803 (1978).

20. H. E. Zimmerman, W. T. Gruenbaum, R. T. Klun, M. G. Steinmetz and T. R. Welter, J.C.S. Chemical Communications, 228-230 (1978).

21. H. E. Zimmerman and T. R. Welter, J. Am. Chem. Soc., 100, 4131-4145 (1978).

22. H. E. Zimmerman and R. L. Swafford, J. Org. Chem., 48, 3069-3083 (1984).

23. H. E. Zimmerman and A. A. Baum, J. Am. Chem. Soc., 93, 3646-3653 (1971).

24. (a) H. E. Zimmerman and C. O. Bender, J. Am. Chem. Soc., 91, 7516-7518 (1969); (b) H. E. Zimmerman and C. O. Bender, J. Am. Chem. Soc., 92, 4366-4376 (1970).

25. (a) H. E. Zimmerman, R. S. Givens and R. Pagni, J. Am. Chem. Soc., 90, 6096-6108 (1968); (b) H. E. Zimmerman, D. R. Amick and H. Hemetsberger, J. Am. Chem. Soc., 95, 4606-4610 (1973).

26. H. E. Zimmerman and M-L Viriot-Villaume, J. Am. Chem. Soc., 95, 1274-1280 (1973).

27. C. O. Bender, D. W. Brooks, W.Cheng, D. Dolman, S. F. O'Shea and S. S. Shugarman, Can. J. Chem., 56, 3027-3037 (1979).

28. R. C. Hahn and R. P. Johnson, J. Am. Chem. Soc., 99, 1508-1513 (1978).

29.(a) L. A. Paquette, D. M. Cottrell, R. A. Snow, K. B. Gifkins and J. Clardy, J. Am. Chem. Soc., 97 3275-3276 (1975); (b) L. A. Paquette, D. M. Cottrell, R. A. Snow, J. Am. Chem. Soc., 99, 3723-3733 (1977).

30. (a) C. Santiago and K. N. Houk, J. Am. Chem. Soc., 98, 3380-3381 (1976); (b) C. C. Santiago, K. N. Houk, R. A. Snow and L. A. Paquette, J. Am. Chem. Soc., 98, 7443-7445 (1976).

31. L. A. Paquette and E. Bay, J. Am. Chem. Soc., 106, 6693-6701 (1984).

32. H. Hemetsberger and F.-U. Neustern, Tetrahedron, 38, 1175-1182 (1982).

33.(a) Note an early example by Streitwieser: A. Streitwieser, Jr., R. H. Jagow, R. C. Fahey and S. Suzuki, J. Am. Chem. Soc., 80, 2326-2331 (1958), where out-of-plane bending was felt to be a controlling factor. (b) A simplified discussion relating s-character to the secondary deuterium isotope effect is found in "Mechanism and Theory in Organic Chemistry", T. H. Lowry and K. S. Richardson, 3rd. Ed., Harper and Row, New York (1987), pp. 238-240. (c) A very recent paper considers more molecular vibrations but, again, leads to the same directionality of the effect: D-H. Lu, D. Maurice and D. G. Truhlar, J. Am. Chem. Soc., 112, 6206-6214 (1990). (d) See E. A. Halevi, "Secondary Deuterium Isotope Effects" in Prog. in Phys. Org. Chem., 1, 109 (1963) for an early discussion.

34. Experimentally, an example revealing deuterium preferring an sp^3 methine center rather than a vinyl sp^2 one is found in the equilibration of [4.2.0] bicyclooctatriene systems: L. A. Paquette, W. Kitching, W. E. Heyd and R. H. Meisinger, J. Am. Chem. Soc., 96, 7371-7372 (1974).

35. Molecular mechanics treatment of cyclopropane and 1,1-dideuterio-cyclopropane leads to a small (circa 1 kcal/mole) preference for the undeuterated compound. The 1,1-dideuteriocyclopropane shows a (circa 4 deg) smaller D-C-D angle than for H-C-H, indicating a preference for lower s-character for the external three-membered ring orbitals. Most of the energy difference is attributed to bending vibrations. These very approximate results are of qualitative value at most (unpublished work by H. E. Zimmerman).

36. H. E. Zimmerman and R. D. Little, J. Am. Chem. Soc., 96, 5143-5152 (1974).

37. H. E. Zimmerman, M. G. Steinmetz and C. L. Kreil, J. Am. Chem. Soc., 100, 4146-4162 (1978).

38. S. S. Hixson, J. Am. Chem. Soc., 94, 2507-2508 (1972).

39. (a) H. E. Zimmerman, R. J. Boettcher, N. E. Buehler and G. E. Keck, J. Am. Chem. Soc., 97, 5635-5637 (1975); (b) H. E. Zimmerman, R. J. Boettcher, N. E. Buehler, G. E. Keck and M. G. Steinmetz, J. Am. Chem. Soc., 98, 7680-7689 (1976).

40. H. E. Zimmerman and R. E. Factor, J. Am. Chem. Soc., 102, 3538-3548 (1980).

41. (a) W. Adam, M. Dörr, J. Kron and R. J. Rosenthal, J. Am. Chem. Soc., 109, 7074-7081 (1987); (b) W. Adam, O. De Lucchi and M. Dörr, J. Am. Chem. Soc., 111, 5209-5213 (1989).

42. W. Adam, G. Reinhard, H. Platsch and J. Wirz, J. Am. Chem. Soc., 112, 4570-4571 (1990).

43. (a) M. Demuth, D. Lemmer and K. Schaffner, J. Am. Chem. Soc., 102, 5407-5409 (1980); (b) M. Demuth, C. O. Bender, S. E. Braslavsky, H. Görner, U. Burger, W. Amrein and K. Schaffner, Helv. Chim. Acta, 62, 847-851 (1970).

44. R. A. Caldwell and S. C. Gupta, J. Am. Chem. Soc., 111, 740-742 (1989).

45. L. Johnston, J. C. Scaiano, J. W. Sheppard and J. P. Bays, Chem. Phys. Lett., 124, 493-498 (1986).

46. K. Quenemoen, W. T. Borden, E. R. Davidson and D. Feller, <u>J. Am. Chem.</u> <u>Soc.</u>, <u>107</u>, 5054-5059 (1985).

47. (a) H. E. Zimmerman, unpublished; (b) Clearly geometry optimization with better basis orbitals is needed, and this work is proceeding.

48. H. E. Zimmerman, F. L. Oaks and P. Campos, <u>J. Am. Chem. Soc.</u>, <u>111</u>, 1007-1018 (1989).

49. H. Kristinsson and G. W. Griffin, <u>Tetrahedron Lett</u>. 3259-3265 (1966).

50. (a) H. E. Zimmerman, <u>Chimia</u>, <u>36</u>, 423-428 (1982); (b) H. E. Zimmerman and T. P. Cutler, <u>J. Org. Chem.</u>, <u>43</u>, 3283-3303 (1978).

51. H. E. Zimmerman and J. M. Cassel, <u>J. Org. Chem.</u>, <u>54</u>, 3800-3816 (1989).

52. For leading references see: (a) D. Armesto, W. M. Horspool, M. Apoita, M. G. Gallego and A. Ramos, <u>J. Chem. Soc. Perkin Trans. I</u>, 2035-2038 (1989); (b) D. Armesto, M. G. Gallego and W. M. Horspool, <u>Tetrahedron Letts.</u> 2475-2478 (1990).

53. J. S. Swenton, <u>J. Chem. Ed.</u>, <u>46</u>, 217-226 (1969).

54. R. S. Givens and W. F. Oettle, <u>Chem. Commun.</u>, 1164-1165 (1969).

55. W. G. Dauben, M. S. Kellogg, J. I. Seeman and W. A. Spitzer, <u>J. Am.</u> <u>Chem. Soc.</u>, <u>92</u> 1786-1787 (1970).

56. (a) Note D. I. Schuster in "Rearrangements in Ground and Excited States," Vol. 3, Edited by P. DeMayo, Academic Press, New York, 1980; (b) K. N. Houk, <u>Chem. Rev.</u>, <u>76</u>, 1 (1976).

2

Synthetic Aspects of the Oxadi-π-Methane Rearrangement

MARTIN DEMUTH

Max-Planck Institut für Strahlenchemie
Mülheim an der Ruhr, Germany

I.   INTRODUCTION

Knowledge of the photochemistry of β,γ-unsaturated carbonyls, and of their rearrangement pathways in particular, has grown at a very rapid pace during the past 10-15 years following the last comprehensive accounts of this field of research.[1-4] The main focus in the area of β,γ-enone photochemistry has clearly concentrated on the oxadi-π-methane (ODPM) rearrangement, a reaction the merits of which have also been recognized by synthetic chemists. Hence, successful applications in synthesis, e.g., for the preparation of natural products,[5-11] have contributed to broader acceptance of the value of this reaction. Responsible for this development have been the numerous detailed mechanistic investigations which have led to the evolution of a convincingly consistent and predictive picture of the mechanistic events of this rearrangement.

For synthetic purposes the reaction can be readily controlled by the use of appropriate triplet sensitizers and selection of the wavelength. A generally excellent reproducibility is assured for ODPM reactions since the optimum reaction conditions are not sensitive to small changes of the experimental parameters, such as reaction vessel, light source, concentration, irradiation time, and temperature. In this context it should be pointed out that the ODPM reaction is typically run at ambient temperature bringing about rearrangements even of

substrates with complex functionality patterns; very
often the rearrangement can be achieved with unprotected
functional groups. Remarkably, this latter aspect is
certainly not restricted to the area of ODPM chemistry but
is a valuable bonus of a large number of phototransformations as compared with most ground state processes,
offering a solution to attaining a very prominent goal of
modern synthetic methodology.

The chemical yields obtained in ODPM reactions are
generally high, i.e. yields of 80-95 % of isolated and
homogeneous materials are achieved in many cases together
with quantum efficiencies of 0.5-1.0. In summary this
phototransformation has the essential characteristics of
a general synthetic method permitting easy and safe
handling and has the potential to be employed in large
scale applications, at least, on a multigram to kilogram
scale.

II. MECHANISTIC BACKGROUND

β,γ-Unsaturated ketones (1) rearrange quite generally
upon triplet sensitization to cyclopropyl ketones (2) via
a formal 1,2-acyl shift, the oxadi-π-methane (ODPM)
rearrangement.[1] Specific arrangements of chromophores
undergo this reaction also upon direct excitation.

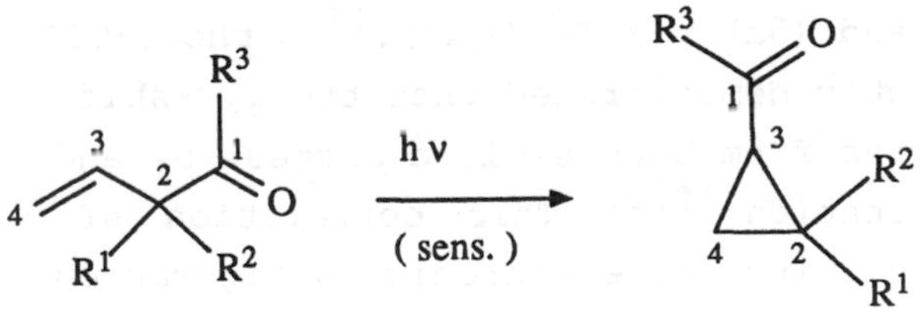

(1)   **(2)**

A. Excited State - Reactivity Correlation

The ODPM rearrangement evidently occurs from the lowest electronically excited β,γ-enone triplet state as shown for the transformations of the cyclopentenyl methyl ketones (3a-d) to mixtures of (4a-d) and (5a-d) (cf. Scheme 1).[12,13] The π,π^* configuration has been assigned to this state on the basis of CNDO-MO calculations,[12,14] phosphorescence studies at 77K[15,16] and mechanistic examinations.[16,17] The phosphorescence data together with sensitization experiments have revealed a T_1 energy range of 289-310 kJ/mol for (3a-c) and a much lower T_1 of 253 kJ/mol for (3d).[16]

Further investigations with (3a-c) have provided even more detailed information on the reactive excited states of β,γ-unsaturated ketones. The intersystem crossing from the singlet to the triplet state, albeit inefficient, populates the T_2 (n,π^*) state from which the 1,3-acyl shift, induced by Norrish Type I cleavage, occurs more readily than internal conversion to T_1.[2] T_2 has been estimated to lie within 8 kJ/mol of the acetone triplet (335 kJ/mol).[18] That the 1,3-acyl shift is coupled not only with singlet-state but also T_2-state reactivity has been unequivocally shown by two methods. Thermal decomposition of a dioxetane, a derivative of (3a), generates predominantly the T_2 state from which the 1,3-acyl shift product is formed concurrently with the ODPM products (4a) and (5a) via T_1 (π,π^*).[17] A photo-CIDNP study has independently demonstrated that the 1,3-shift with (3a-c) originates from both excited states, S_1 and T_2, on direct excitation.[19,20] This correlation of excited-state multiplicity and electronic configuration, at least with regard to the ODPM reaction, has recently been supported by more examples such as the rearrangements

Scheme 1

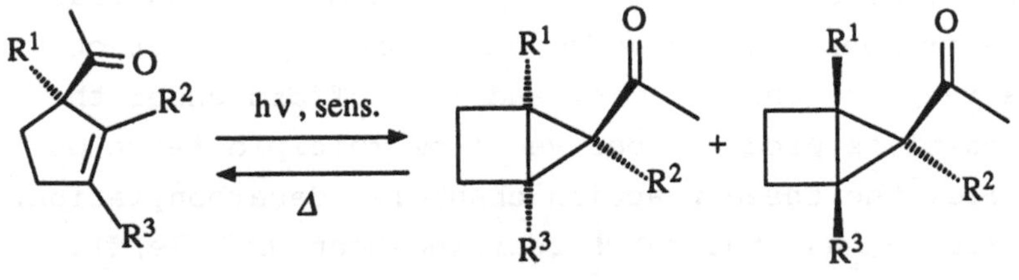

(3a-d)   (4a-d)   (5a-d)

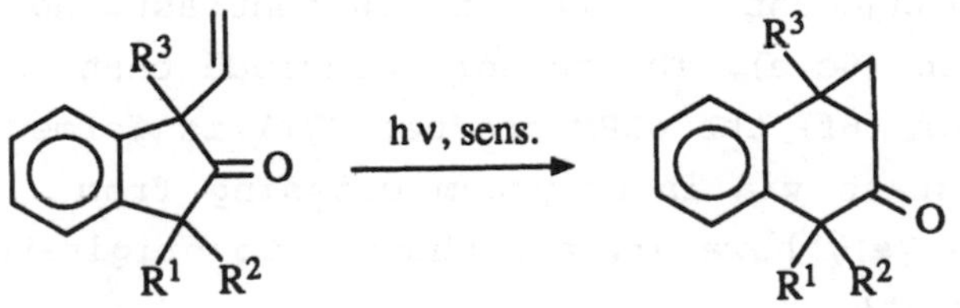

(6c,e)   (7c,e)

(a) $R^1=CH_3$, $R^2=R^3=H$ (c) $R^1=R^2=R^3=CH_3$

(b) $R^1=R^2=CH_3$, $R^3=H$ (d) $R^1=R^2=H$, $R^3=C_6H_5$

                                      (e) $R^1=R^2=H$, $R^3=CH_3$

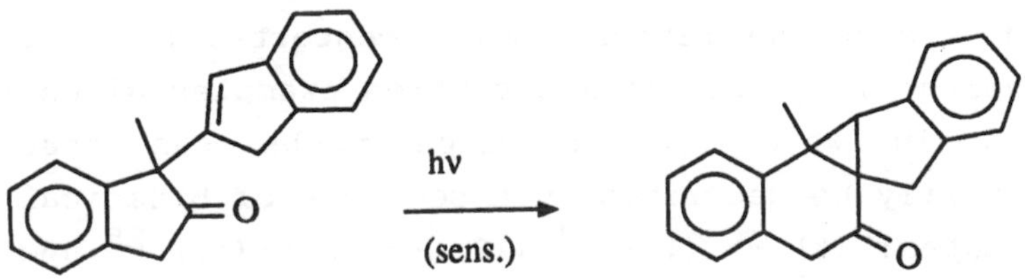

of (6c,e) to the corresponding cyclopropyl ketones
(7c,e).[21] Furthermore, this latter investigation has
uncovered a novel facet of ODPM photochemistry. Whereas
(6c) rearranges exclusively to (7c) upon direct
irradiation, the analogs (6e) and (6f) afford under the
same conditions products derived from multiple reaction
paths, i.e. the three reaction channels: decarbonylation,
1,3-migration and the ODPM rearrangement to (7e,f).
Interestingly, the formation of (7f), but not of (7c) and
(7e), is efficiently suppressed by triplet quenchers, a
finding which parallels the results of experiments
employing triplet sensitizers in which (6f) cleanly
undergoes the rearrangement to (7f) but, in contrast, no
reaction occurs with (6c,e). The authors conclude that on
direct excitation of (6f) the ODPM product (7f) is formed
from the $T_1(\pi,\pi^*)$ state via intersystem crossing from
$S_1(n,\pi^*)$. (7c) and (7e), however, are thought to originate
from the $S_2(\pi,\pi^*)$ state.

In summary, all the knowledge available up to the
present time of the energetic order, spin multiplicity,
and configuration of excited states involved in the
photochemistry of β,γ-enones is given in Scheme 2.

In sole conflict with this coherent picture of the
correlation of excited states and reactivity in the ODPM
context is the claim that a representative of the
1-methoxy-substituted bicyclo[2.2.2]octenones [cf. (66a)
in Scheme 15, Section III.D.2] may not necessarily
rearrange from the T_1 state but rather from T_2.[22,23]

With regard to the reaction mechanism evidence in
favor of a stepwise rather than a concerted mode of
rearrangement is gained from the three examples given in
Scheme 3 in which the reaction paths can most
satisfactorily be described by a sequence of biradical
intermediates: (8) -> (11),[24] (12) -> (14)+(15),[25] and

Scheme 2

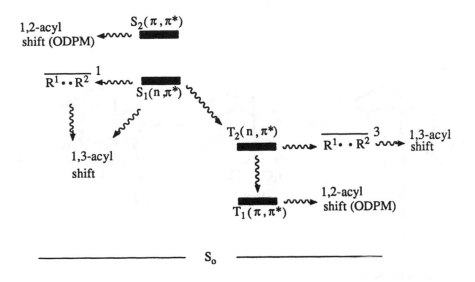

(16) -> (18)+(19).[26,27] In the first case, complete stereochemical scrambling of the isotopically labelled geminal methyls of the products (11a,b) ($R^1 = R^2 = CH_3$ and CD_3; 1:1 ratios) during the ODPM rearrangement requires the formulation of a stepwise reaction mode. The depicted sequences of biradicals via (9a,b) and (10a,b) are proposed as the likely reaction cascades fitting the experimental findings. Similarly, the ODPM products resulting from the rearrangements of (-)-(12) and (+)-(16) demand, in analogy to the previously postulated intermediacy of (10), transients such as (13) and (17), respectively, to account for the scrambling of the substituents prior to permanent bond formation, i.e. the assembly of the cyclopropane [-> (-)-(14) + (-)-(15) and (±)-(18) + (±)-(19)]. In summary, one may suggest that the

Scheme 3

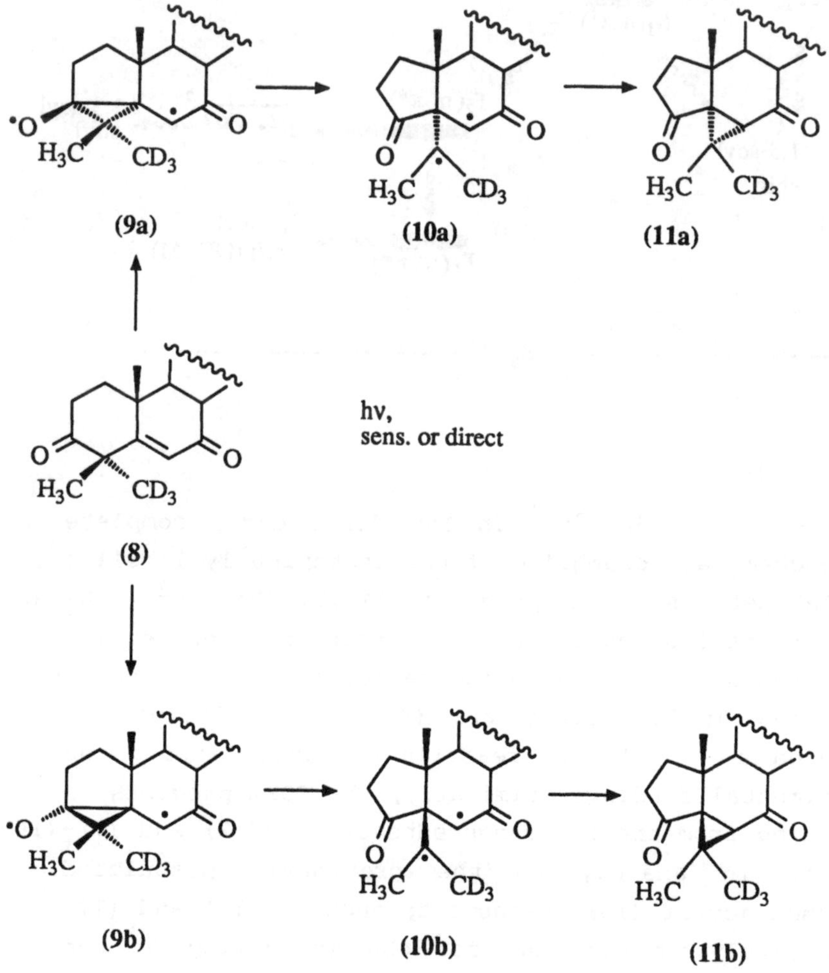

(9a)   (10a)   (11a)

hv,
sens. or direct

(8)

(9b)   (10b)   (11b)

Scheme 3 (contn'd)

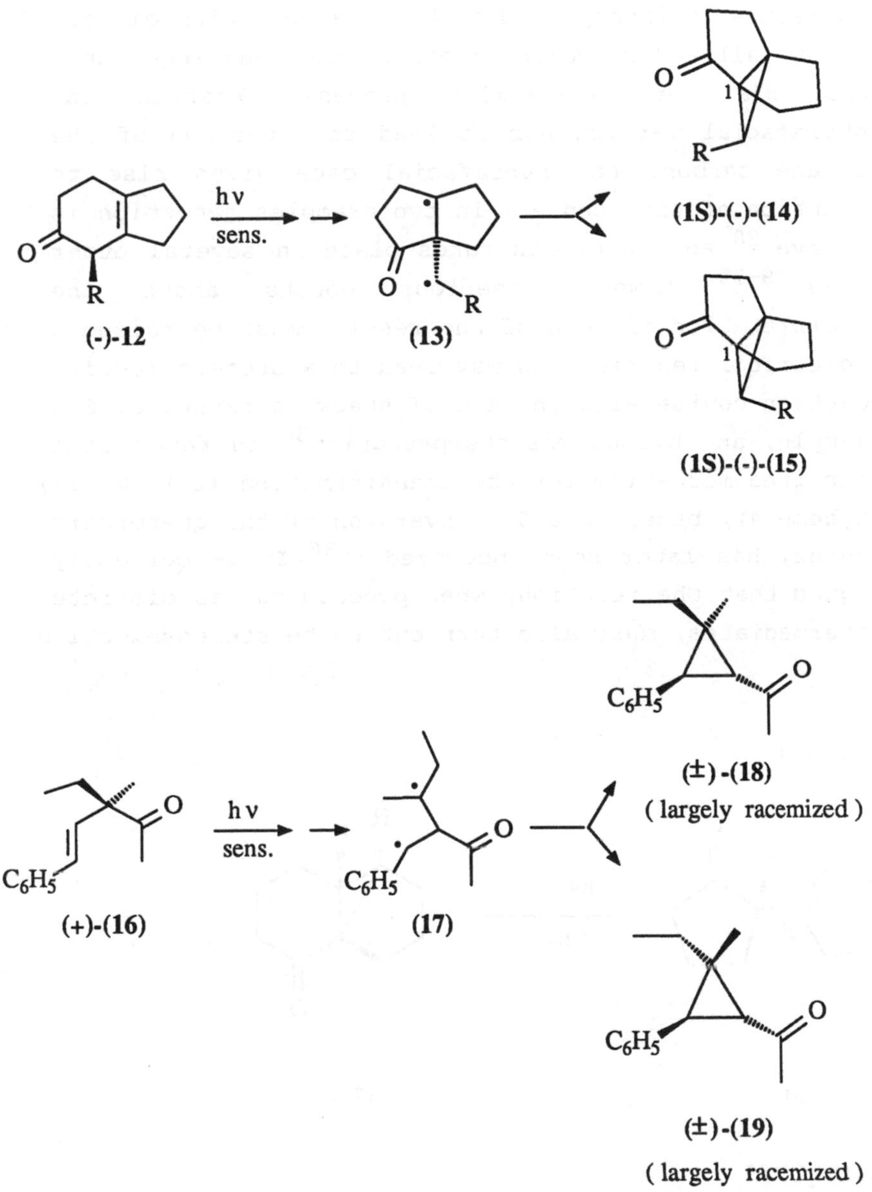

(-)-12   (13)

(1S)-(-)-(14)

(1S)-(-)-(15)

(+)-(16)   (17)

(±)-(18)
(largely racemized)

(±)-(19)
(largely racemized)

majority of ODPM reactions be classified as stepwise
processes.

The stereochemical course of other examples in the
literature is interpreted to be in accord with either of
the two allowed concerted modes of the rearrangement, a
[σ2a + π2a]   or   [σ2s + π2s]   process.   Whereas   the
antarafacial version has to lead to inversion of the
methane carbon, the suprafacial case gives rise to
retention of this center: in two examples retention is
observed[28] and inversion takes place in several other
cases.[29-35]   However,   serious   doubts   about   the
interpretation of some of the results must be raised since
geometrical restrictions may lead to a stereoselective
reaction course also in case of stepwise pathways. For
example, an obvious misinterpretation[35] in favor of a
concerted mechanism for the transformation (20) -> (21)
(Scheme 4), based on a 90% inversion of the quaternary
center, has later been uncovered.[1,36] It is correctly
argued that the reaction, when proceeding via discrete
intermediates, must also turn out to be stereoselective

Scheme 4

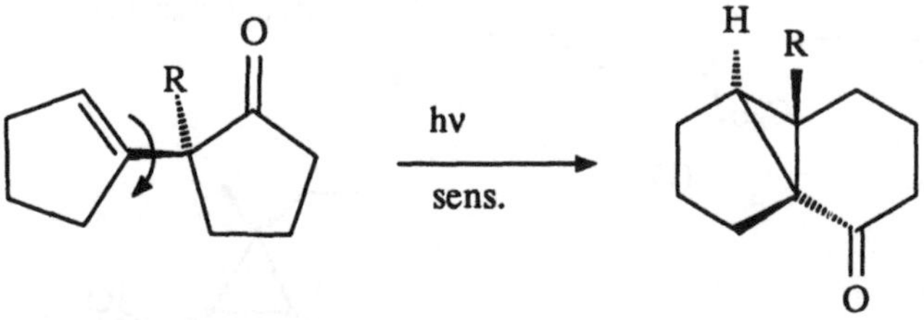

(20)   (21)

R=CH₂CO₂H

since inversion and retention of the methane carbon derive
from diastereoisomeric rotamers of (20).

A discussion of more detailed mechanistic aspects
of the 1,2-acyl migration, especially regarding the
potentially concerted variants, is not given in this
chapter since it would duplicate a comprehensive account
in reference 4.

B. Efficiency and Selectivity of the Rearrangement

Generally, the highest quantum and chemical yields
of ODPM rearrangements are obtained with substrates in
which the β,γ-enone chromophore is part of a
conformationally rigid molecular assembly which at the
same time guarantees adequate orbital overlap of the C=C
and C=O chromophore sites. Accordingly, bicyclic and
bridged β,γ-unsaturated ketones are the most "eager" ODPM
reactants while acyclic β,γ-unsaturated ketones rearrange
inefficiently, if at all, since other channels of energy
dissipation from the triplet state usually predominate.
Exceptions to this rule are substrates, in which the C=C
bond is part of a styrene or an α,β-enone moiety, and which
do undergo 1,2-bridging (cf. Section III.A). In this
context it should also be noted that β,γ-unsaturated
aldehydes, except for one case, are ODPM-unreactive
(cf.Section III.B). The principle of efficiency and
selectivity control of the 1,2-acyl shift can best be
demonstrated by the bicyclo[2.2.2]octenone
transformations given in Scheme 5.

As pointed out in Section II.A, the energy of the
lowest, and at the same time, ODPM-reactive triplet state
of structurally plain β,γ-unsaturated ketones is in the
range of 289-310 kJ/mol.[12-17] It means that sensitization

Scheme 5

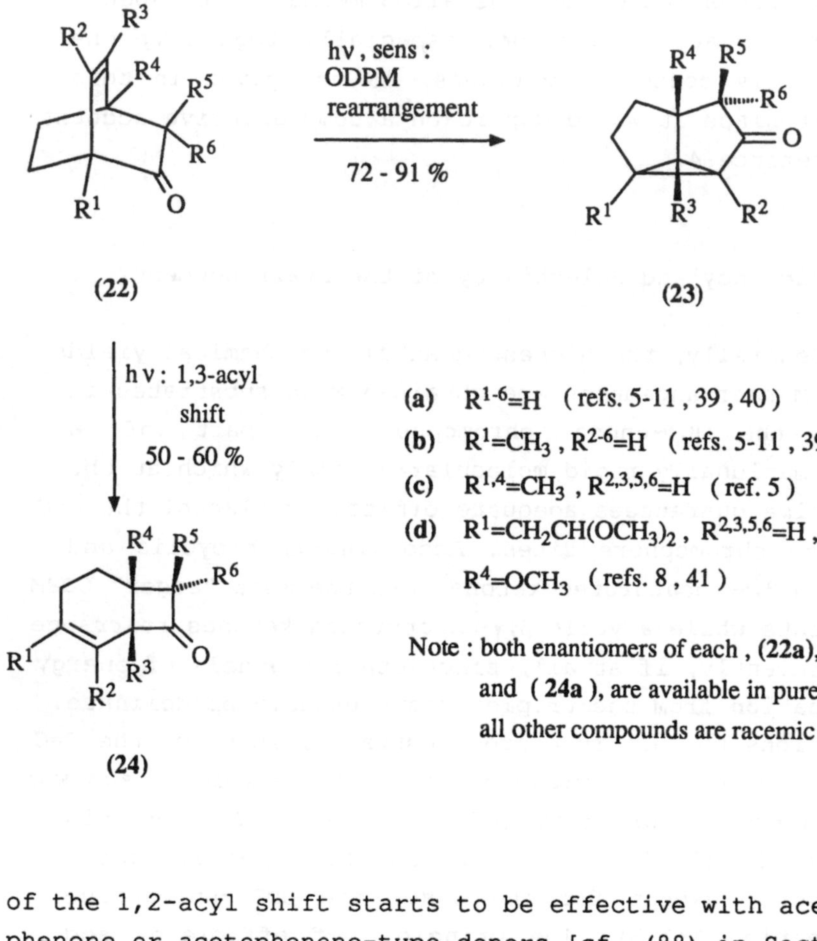

hv, sens :
ODPM
rearrangement

$\xrightarrow{\qquad\qquad}$

72 - 91 %

(22)   (23)

hv : 1,3-acyl
shift

50 - 60 %

(a) $R^{1-6}=H$ (refs. 5-11 , 39 , 40)

(b) $R^1=CH_3$, $R^{2-6}=H$ (refs. 5-11 , 39)

(c) $R^{1,4}=CH_3$, $R^{2,3,5,6}=H$ (ref. 5)

(d) $R^1=CH_2CH(OCH_3)_2$, $R^{2,3,5,6}=H$,

$R^4=OCH_3$ (refs. 8 , 41)

Note : both enantiomers of each , (22a), (23a)
and (24a), are available in pure form ;
all other compounds are racemic

(24)

of the 1,2-acyl shift starts to be effective with acetophenone or acetophenone-type donors [cf. (88) in Section
V, Scheme 20], known to possess a T_1 energy of 302-310
kJ/mol.[37,38] As expected on theoretical grounds, energy
transfer is most efficient with sensitizers transferring
their energy exothermically; e.g., a quantum yield close
to unity is determined for the conversion (22a) -> (23a)
in 1-2% acetone solution [$E(T_1)$ of acetone approx. 335
kJ/mol] when irradiating at $\lambda_{irr.}$ 300 nm; $\Phi=0.3$ is found

for the same transformation when sensitized by acetophenone [20% of acetophenone added to acetone, benzene or cyclohexane solutions of (22a); $\lambda_{irr.}$ > 340 nm].[39] The chemical yields are uniformly high, 86-91%, under either condition at 94-96% conversions of starting enone. The reaction run in pure acetone solution at $\lambda_{irr.}$ 300 nm is, in addition to its efficiency, very attractive from a practical viewpoint. Acetone serving both as solvent and sensitizer renders the work-up and purification preparatively easy: no separation procedure is required. It should be noted, however, that the substrate concentration under the latter conditions should not exceed 2-4% because another reaction path, the 1,3-shift, induced by Norrish Type I cleavage, becomes noticeably competitive affording the cyclobutanone (24a). This side reaction is a consequence of direct light absorption by (22a) competing increasingly with absorption by the acetone at the wavelength employed. For the same reason, even greater direct absorption of light renders the yield of (23a) much lower (34%), when the acetone solution of (22a) is irradiated with 254-nm light.[40]

The unwelcome cyclobutanone formation becomes more noticeable in the acetone-sensitized rerrangements of the homologs (22b-d) to (23b-d) in which approximately 10% of (24b-d) are already formed as by-products when 1-2% acetone solutions are irradiated. For these substrates, and generally for all 1-substituted bicyclooctenones of type (22), it is therefore advantageous to employ acetophenone sensitization at >340 nm, where the absorbance of the enone chromophore is negligible. Preparatively attractive high substrate concentrations of up to 10% of (22) can be employed under the latter conditions.

III. THE BASIC SET OF PHOTO-REACTIVE β,γ-ENONES

Certain spatial arrangements of the chromophoric sites and well defined substituent patterns allow efficient and stereoselective control of ODPM isomerizations. Based on the actual mechanistic knowledge and body of experimental data, predictions of the reactivity of new substrates can be made with reasonable assurance. Semicyclic, bicyclic and among these especially bridged β,γ-unsaturated ketones give rise to ODPM products in generally good to excellent yields. This does not only apply to the simplest case of ODPM chromophores such as schematically represented by (A) in Scheme 6, but also to chromophores that are conjugated additionally either at the ketonic (B) or at the olefinic site (C) as well as to intramolecularly competitive β,γ-enone arrangements (D). The full circles in the drawings represent the methane carbons without indication

Scheme 6

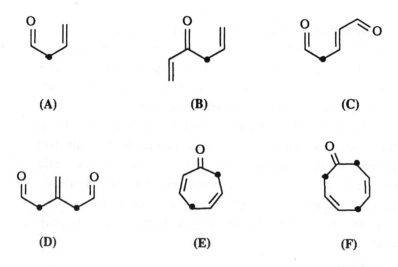

(A)   (B)   (C)

(D)   (E)   (F)

of substituents or ring arrangements: to guarantee high
ODPM reactivity of the substrates these centers should
either be fully substituted or - even more advantageously
- be bridgeheads. Examples referring to the cases (A)-(C)
and (D) are discussed in the Sections B.-D.2 and D.3,
respectively. These chromophores generally differ with
respect to their excitation properties; types (A) and (D)
typically need be sensitized whereas (B) and (C),
arrangements in which one chromophoric moiety is further
conjugated, are undergoing efficient singlet -> triplet
intersystem crossing and 1,2-acyl migration upon direct
irradiation.

Besides   exclusive   ODPM   arrangements   mixed
chromophores have been investigated in which additional
di-π-methane(DPM)-reactive sites are present as in the
examples (E) and (F). At this point, already, it should
be mentioned that these combinations of chromophores react
highly selectively via the DPM route rather than following
the ODPM option (cf. Section D.4).

With respect to the preparative usefulness of the
ODPM reaction it should be noted that the chemical yields
depend   strongly   on   the   skeletal   arrangement   and
substituents of the reactants; the yields increase, at
least as a general trend, in the order of sections A-D.

A.   Acyclic β,γ-Unsaturated Carbonyls

Acyclic β,γ-unsaturated aldehydes, in general, seem
not   to   undergo   the   ODPM   reaction   and   acyclic
β,γ-unsaturated ketones rearrange efficiently only if the
C=C bond is further conjugated, e.g. part of a styrene
moiety (see the examples below). In all the other cases
where the vinyl group is less favorably substituted other

channels of deactivation from the triplet state are more important than the ODPM path. The main process then becomes twisting around the carbon-carbon double bond resulting in *cis-trans* isomerization. This form of energy dissipation has been called the 'free rotor effect' for β,γ-unsaturated ketones adopting an earlier formulation made for unconstrained DPM arrangements.[42-44] This process is the main event in the case of irradiation of the isomers (25) and (26) (R = D) which interconvert with a quantum efficiency of 0.12; the potential ODPM product (27) is not formed at all.[45] Successful competition of the 1,2-acyl migration with the free rotor effect is suggested to depend on the degree of orbital mixing in the T_1 state which in turn is reflected by enhancement of the n -> π^* absorption.[1]

As mentioned above, if the vinyl group is part of a styrene moiety, ODPM reactivity, usually in competition with *cis-trans* isomerization, is again encountered. One example in this context has already been given earlier, the transformation of (16) (cf. Scheme 3) which proceeds with a moderate yield of 31% only to a mixture of (18) and (19) when chrysene is employed as the sensitizer (E_T 239 kJ/mol); 46% of starting material is recovered from this reaction as a *cis-trans* isomeric mixture.[26,27] Energy transfer by a sensitizer of higher E_T such as acetophenone (302-310 kJ/mol)[37,38] to a mixture of the *cis-trans* isomers (28)/(29), which are structural congeners of (16), leads, on the other hand, to a sole product (30) in 93% yield.[46]

A number of cases have been found where the ODPM rearrangement also occurs on direct excitation. Notably, the first example of an ODPM reaction belongs to this category: (31) rearranges without sensitizer to (32) in 7% yield[47] (for further examples of ODPM reactions on direct excitation, see the following Sections B and C).

Scheme 7

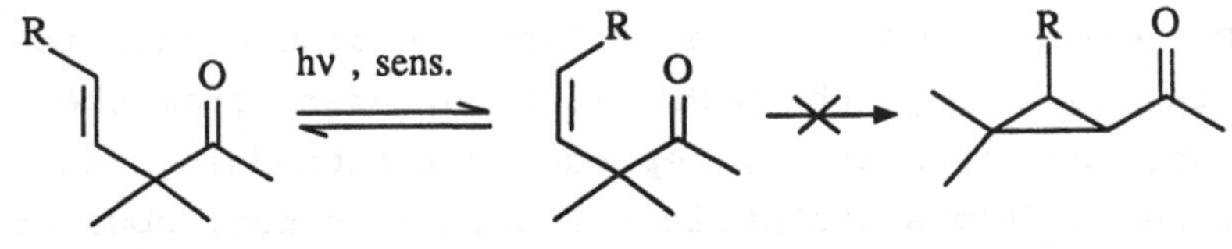

R=H, D

    (25)   (26)   (27)

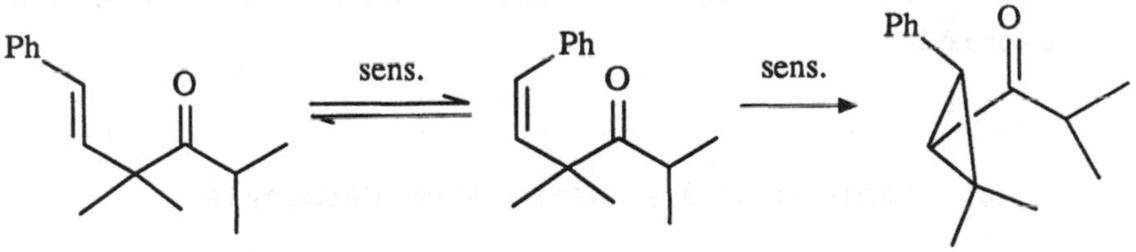

    (28)   (29)   (30)

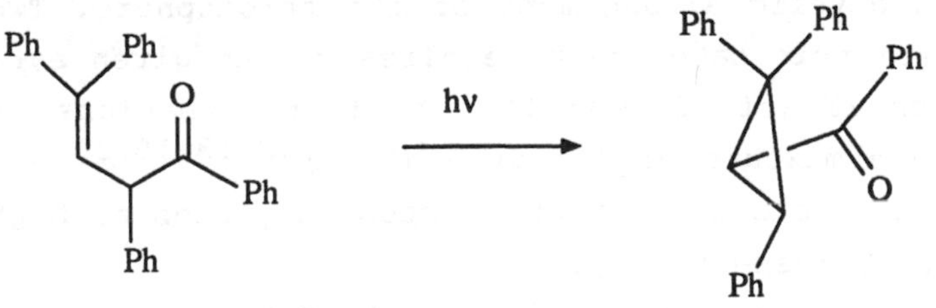

    (31)   (32)

Introduction of a ^{14}C label has shown that the
rearrangement to (32) proceeds by migration of the benzoyl
group, a partial chromophore that is known to facilitate
intersystem crossing to the triplet. A closely related but
more 'flexible' analog of (31), bearing a methyl ketone

instead of the benzoyl group, undergoes again exclusively
cis–trans isomerization from the triplet state.[48]

The generally encountered reluctance of the acyclic
substrates to undergo the ODPM reaction, apart from the
few exceptions cited above, represents a methodological
gap, at least from a synthetic viewpoint; it can, however,
be overcome by resorting to the respective imine
derivatives of such β,γ-unsaturated aldehydes and ketones.
These again are photochemically reactive and cyclize in
high yields when irradiated in the presence of a triplet
sensitizer: they undergo the azadi-π-methane rearrangement
(cf. Section VII).

B.   Semicyclic β,γ-Unsaturated Carbonyls

In cases where either the C=C, the C=O or both
moieties of a β,γ-unsaturated enone are part of a cyclic
system, the ODPM reactivity is noticeably enhanced and
less substituent-dependent as compared to the previously
discussed acyclic arrangement of the chromophore. Two
examples of this category have already been given earlier
in Section II (cf. Scheme 1), the transformations of
(3a-d) to a mixture of (4a-d) + (5a-d)[12,13,16,30] and
(6c,e,f) to (7c,e,f)[21] which proceed in yields as high as
65% and 84%, respectively.

Efficiency and limitation of the ODPM reaction are
strongly coupled to the degree of flexibility of the
alkene moiety. The energy dissipation from the triplet
state via twisting of the C=C bond is indicated by a
comparison of the following examples. When the C=C bond
is geometrically constrained, for example being part of
a 5-membered ring, the rearrangement proceeds smoothly as
documented by the representative reactions of (3a-d) and

(6f), cited above, as well as (20) -> (21) (cf. Scheme
4),[35] (33) -> (34),[45] and (35) -> (36) (cf. Scheme
8).[35,49] On the other hand, no rearrangement occurs when
the vinyl group is embedded in a 6-membered ring such
as in (37).[45] In this connection it should be pointed out
that strong evidence of the existence of alkene twisting
has recently been gained in a closely related field, the
photochemistry of α,β-unsaturated cyclohexenones.[50]

Substrates in which the methane carbon is not fully
substituted as for example in (35) (R = H) are similarly
ODPM unreactive, irrespective of the ring size of the
alkene moiety; β,γ -> α,β isomerization of the C=C bond
appears to constitute a major path of chemical energy
dissipation.[51] One exception in this category is, however,
the earlier cited (3d) in which the alkene is part of a
styrene moiety.[16]

When the C=C bond is part of an α,β-unsaturated
ketone moiety such as in (38) (cf. Scheme 9) direct, i.e.
π,π* excitation triggers the ODPM reaction: a stereo-
isomeric mixture of (39) and the 1,3-shift product (40)
are formed in 93% and 6%, respectively, at 91% conversion
of starting material.[52] (40) and (39) are suggested to be
generated from the singlet and triplet state,
respectively.[52] This finding of 1,2-acyl migration via the
singlet state seems to be characteritic for substrates
which have an extended chromophore in common, i.e.
additional conjugation of the vinyl group to either phenyl
or C=O sites. Notably, examples of this type are also
encountered with bicyclic (Scheme 3), acyclic (Scheme 7)
and bridged chromophores (Scheme 13).

Mechanistically of interest, albeit not of synthetic
value, is one specific representative of the otherwise
ODPM-resistant β,γ-unsaturated aldehyde class. (41)
exhibits on direct excitation at least moderate ODPM

Scheme 8

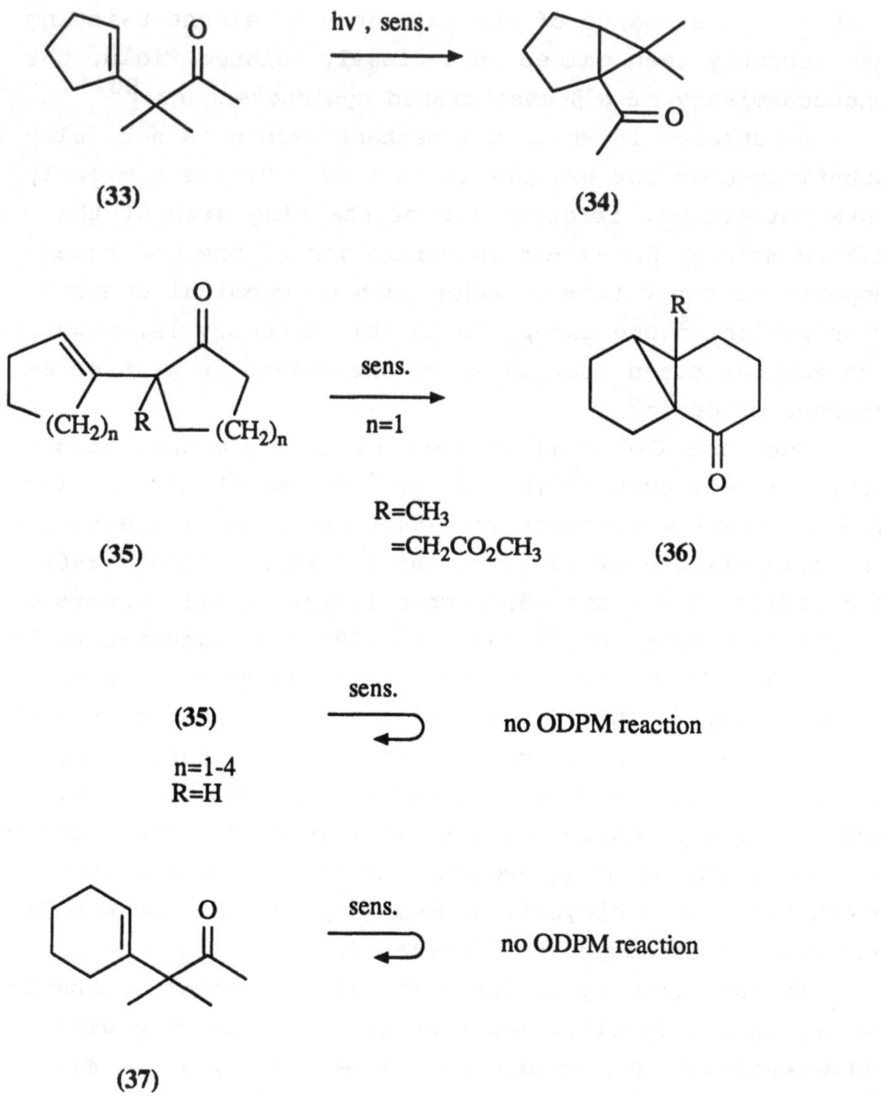

(33)   hv , sens.   (34)

(35)   sens.   n=1

R=CH$_3$
=CH$_2$CO$_2$CH$_3$ (36)

(35)   sens.   no ODPM reaction

n=1-4
R=H

(37)   sens.   no ODPM reaction

Scheme 9

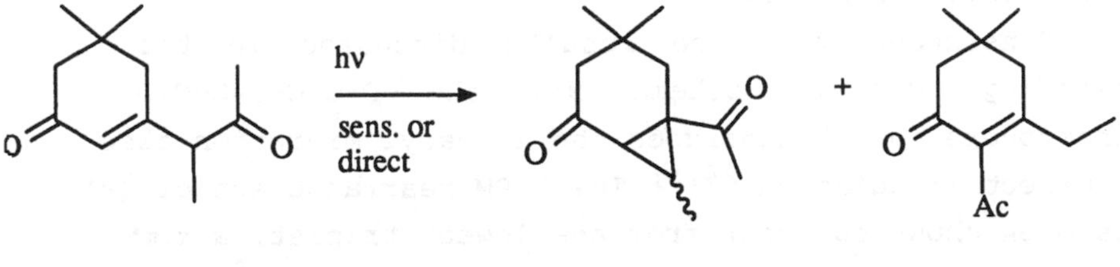

(38)   (39)   (40)

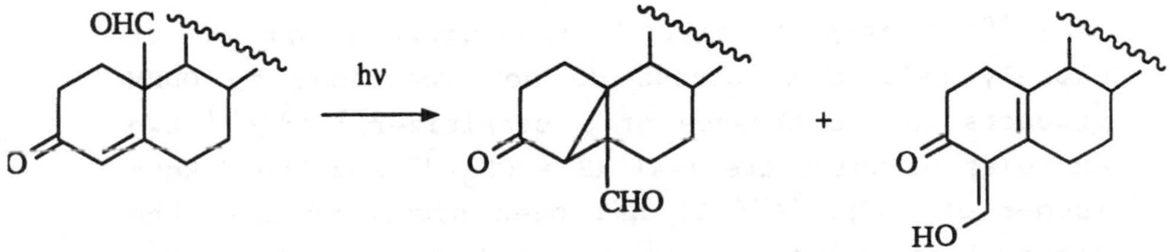

(41)   (42)   (43)

reactivity giving rise to the formation of (42) besides
the major reactions to wit decarbonylation and 1,3-acyl
migration to (43).[53]

C.   Mono-, Bi- and Spirocyclic β,γ-Unsaturated
                    Ketones

     Three representative examples of conversions of
bicyclic β,γ-enones should suffice to illustrate the
synthetic potential of this class of compounds: the
efficient transformation of (8) into (11),[24] the reaction

of an analog of (8) lacking the C-7 carbonyl,[54,55] and ultimately the preparation of (14) and (15) from (12)[25] (cf. Scheme 3).

Consistent with the results discussed in the preceding section (Scheme 9), the O=C-C-C=C-C=O chromophore in (8) undergoes the 1,2-acyl migration also on direct irradiation.[24,56] The ODPM rearrangement of (8) has been shown to occur from the lowest triplet, a π,π^* state in the energy range typical of α,β-enones.[56] Two unusual findings concern ODPM reactions encountered on direct excitation of structurally simple β,γ-enone chromophores as in (44) (R = H, Scheme 10) and in the steroidal analog (46) which rearrange to (45)[57,58] and (47),[58,59] respectively. Intriguingly, a number of closely related compounds do not rearrange to ODPM products in the absence of a sensitizer,[54,55,60] two examples of which are (44) (R = CH$_3$)[58] and the 2-keto isomer of (46).[58,59] It has been suggested that the dihedral angle between the carbonyl and the vinyl group in the enones controls the efficiency by which the $^3(\pi,\pi^*)$ state is populated on direct irradiation, hence the reaction selectivity of 1,2- vs. 1,3-acyl migration.[1,61]

The photochemical rearrangement of 2,4-cyclohexadienones represents a reaction of potential synthetic value.[62] For efficacious 1,2-acyl migration, only a relatively small number of substrates are suitable. These must be highly substituted, as for example (48). Whereas on direct excitation in methanol cleavage to the isomeric ketenes (49) ($\Phi \geq 0.42$) predominates,[63] the highly stereoselective 1,2-acyl shift to the bicyclohexenone (50) is found either in trifluoroethanol or when the dienone is adsorbed on silica gel.[64] The reaction to (50) is followed by a reversible phototransformation to the cross-conjugated dienone (51)[65,66] and accompanied by

Scheme 10

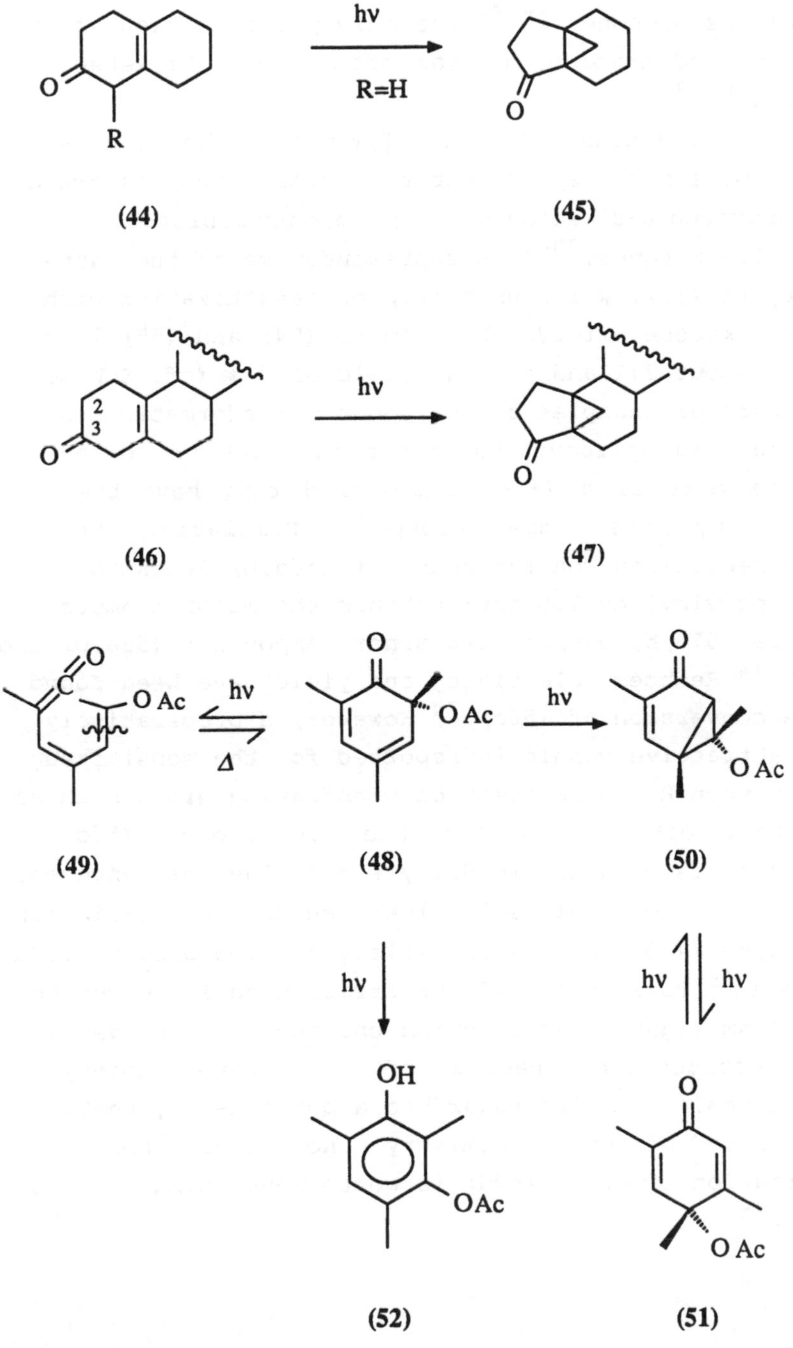

(44)
(45)

(46)
(47)

(49)
(48)
(50)

(52)
(51)

aromatization to (52) to a minor extent.[67] Such reactivity
has also been verified for tetra- and pentamethylated
2,4-cyclohexadienones.[68,69] The only photoreaction of the
hexamethylated homolog, on the other hand, is ketene
formation.[68,70,71]

In a third class of cyclic β,γ-enones the methane
carbon represents a spiro center; typical examples are a
β,γ-unsaturated δ-diketone[72] and β,γ;δ,ε-unsaturated
spirocyclic ketones.[73-76] A representative of the latter
category is (53), which upon triplet sensitization with
Michler's ketone affords the isomers (54) and (55) in a
ratio of about 2:1 and a total yield of 82% (cf. Scheme
11).[73] Further examples of analogous transformations of
homoconjugated spirocyclobutanones in which the diene
moiety is part of a 7- or 8-membered ring have been
reported by the same group.[73] Similarly, the
acetone-sensitized rearrangement of (56a,b) leads to a
mixture of vinyl cyclopropyl ketones the major isomers of
which are (57a,b) besides two minor components (58a,b) and
(59a,b).[74] Reduced selectivity and yield have been found
for the conversion of (56c).[74] However, a preparatively
highly attractive result is reported for the sensitized
reaction when R^1 is an electron withdrawing group such as
the methyl ester in (56d): the *trans* isomer (57d)
exclusively is obtained in 95% yield.[75] The rearrangement
of (56d) is also highly selective when direct irradiation
is employed (> 340-nm light): solely the *trans* product (57d)
is formed.[75] Puzzlingly, if the irradiation is conducted
with 254-nm light other reaction channels are activated,
i.e. electrocyclic opening of the diene moiety
predominates.[74,75] With regard to a potential synthetic
exploitation of this chemistry the clean thermal
isomerization (57d) -> (59d) is worth mentioning in this
context.[75]

Scheme 11

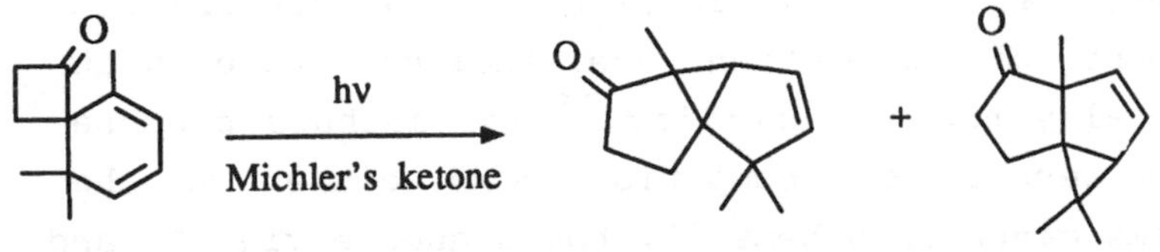

(53)   (54) (58 %)   (55) (24 %)

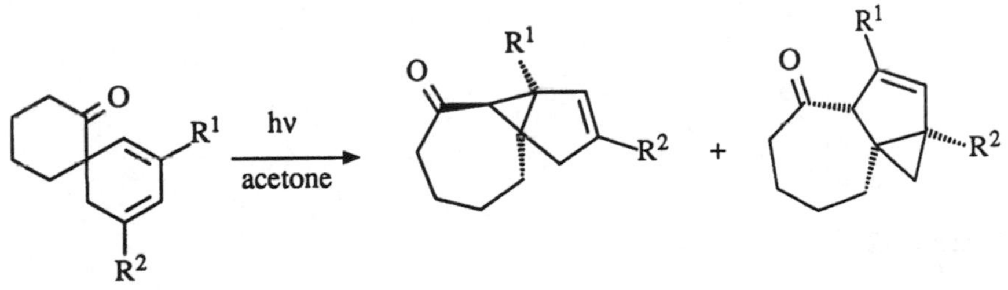

(56)   (57)   (58)

(a)	R¹=CH₃ , R²=H	55 %	15 %
(b)	R¹=H, R²=CH₃	41 %	14 %
(c)	R¹=R²=H	17 %	28 %
(d)	R¹=CO₂CH₃ , R²=H	95 %	—

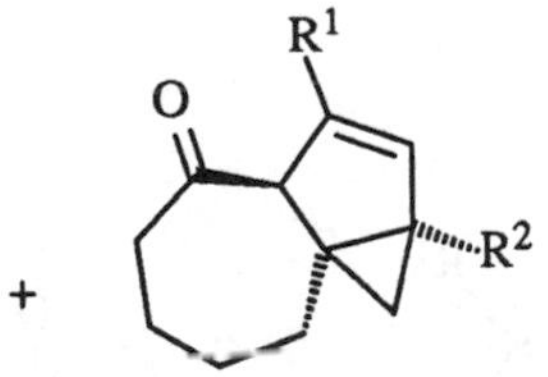

(59)

(a)	4 %
(b)	5 %
(c)	6 %
(d)	—

The photochemical results with (56a-d), i.e. the striking wavelength dependence of the reactions, are interpreted as being indicative of the mechanistic involvement of more than one excited state with exceptionally high selectivity.[76] The authors propose that the results are compatible with either of the mechanisms given in Scheme 12, the sequence via (A) and (B) and/or via the concerted variants (C) and (D); the data available so far do not allow for differentiation between the two processes.[74,76]

Scheme 12

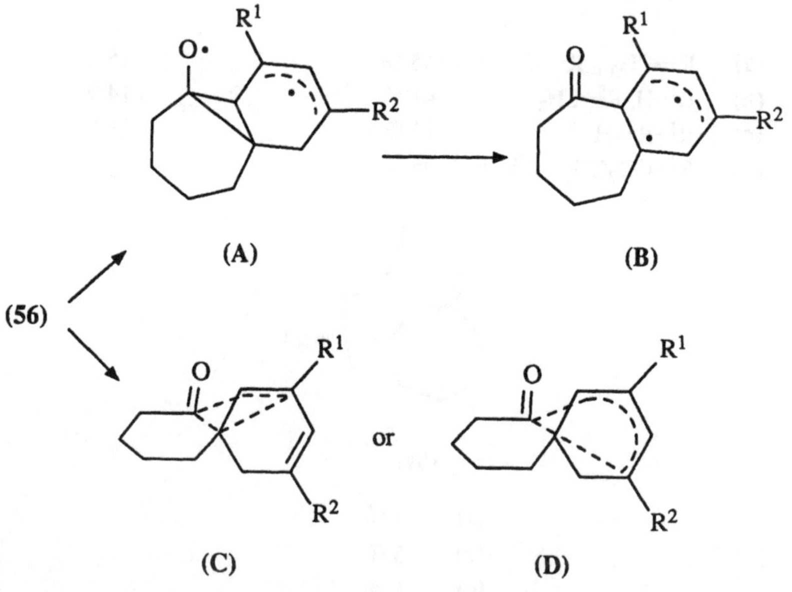

(A)   (B)

(56)

(C)   or   (D)

D. β,γ-Unsaturated Ketones as Part of Bridged Cyclic Substrates

1. GENERAL

Since the efficiency of the ODPM reaction is strongly coupled with the degree of flexibility of the C=C moiety of the β,γ-enone, the conformationally rigid [2.2.1]- and especially the [2.2.2]-bridged skeletons are generally excellent candidates for this rearrangement. Synthetically very useful examples of bicyclo[2.2.2]octenone transformations via the ODPM path have already been given in Scheme 5; for the extension of the list further analogous conversions are cited in Scheme 13 (these schemes take care of more recent examples only; for older entries, see refs. 1-4).

The acetone-sensitized rearrangement of bicyclo[2.2.1]heptenone (60a) affords mainly the 1,2-acyl shift product (61a) accompanied by a small amount of the 1,3-shift isomer (62a),[32,89] the latter being convertible to (61a) photochemically by acetone sensitization (cf. Scheme 14).[89] In accord with the results discussed in Section II, formation of (62a) can be suppressed if acetophenone ($\lambda_{irr.}$ >340 nm) is employed as the sensitizer. Identical photoreactivity has been found for the two higher substituted congeners (60b)[90] and the spiro-2-norbornenone (60c).[31,32]

In a mechanistic study, biradical (63), the assumed precursor of (61a), has been generated by benzophenone-sensitized nitrogen extrusion from the parent ketoazoalkane.[91] Three reaction channels originate from (63): formation of (60a) as the main product besides the cyclopropyl ketone (61a) in 28% and some (62a). This study has revealed that (60a) and (62a) both give (61a)

Scheme 13

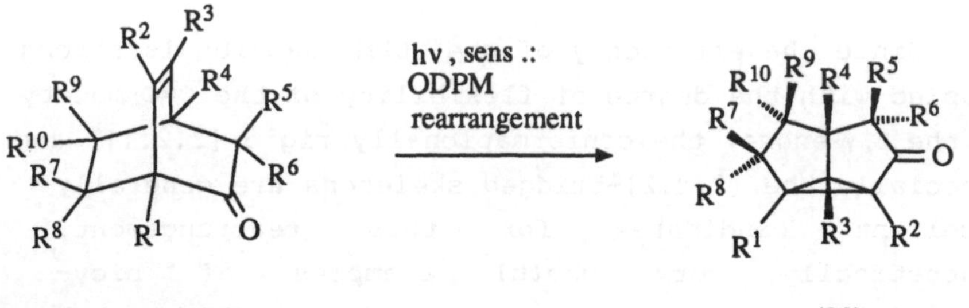

(22)   (23)

(a)-(d)   cf. Scheme 5

(e) $R^{1,4,7-10}$=H , $R^{2,3}$=CO$_2$CH$_3$, $R^{5,6}$=CH$_3$ (76% yield) (refs. 77,78)

(f) $R^{1,2,4,7-10}$=H , R^3=CO$_2$CH$_3$, $R^{5,6}$=OCH$_3$ (ref. 79)

(g) $R^{1-6,9-10}$=H , R^7=CH$_3$, R^8=CH$_2$OH   (91% yield) (ref. 80)

(h) $R^{1-6,8,10}$=H , $R^{7,9}$=CH=CH (66% yield) or CBr=CH (80% yield) (ref. 81)

(i) $R^{1-3,5-8}$=H , $R^{4,9}$=cycle , R^{10}=CH$_3$ (structure cf. Scheme 26 , 70% yield)
      (refs. 82-84)

(j) R^{1-7}=H , $R^{8,10}$=cycle , R^9=CH$_3$ (structure cf. Scheme 26 , 72% yield)
      (refs. 82-84)

(k) R^1=CH$_3$, R^2=H , $R^{3,4}$=cycle , R^{5-10}=H (structure cf. Scheme 23, 50% yield)
      (ref. 85)

(l) $R^{1,4-8}$=H , $R^{2,3}$=C(CH$_3$)$_2$–(CH$_2$)$_2$–CO , $R^{9,10}$=CH$_3$ (direct or sens. irrad. ,
      78% yield) (ref. 86)

(m) $R^{1,4,7-10}$=H , $R^{2,3,5 \text{ or } 6}$=CO$_2$CH$_3$, $R^{5 \text{ or } 6}$=CH$_3$ or CH$_2$Ph (51-65% yield) (ref. 87)

(n) $R^{1-4,7-9}$=H , R^5=CO$_2$CH$_3$, $R^{6,10}$=4-, 5-, and 6-membered cycles (66-89% yield)
      (ref. 88)

Note : (22a,i,j) and (23 a,i,j) are available in enantiomerically pure form ; all other
       compounds are racemic

Scheme 14

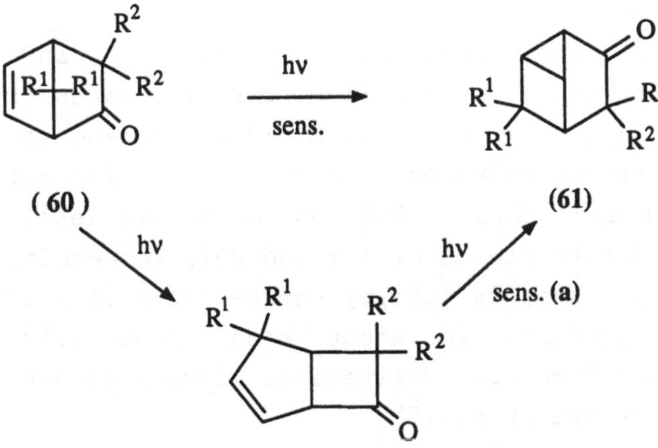

(60)

hv
sens.

(61)

hv

hv
sens. (a)

(62)

(a) $R^1=R^2=H$
(b) $R^1=H$, $R^2=OCH_3$
(c) $R^1=CH_2-CH_2$, $R^2=H$

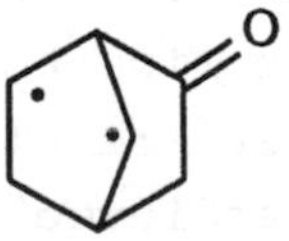

(63)

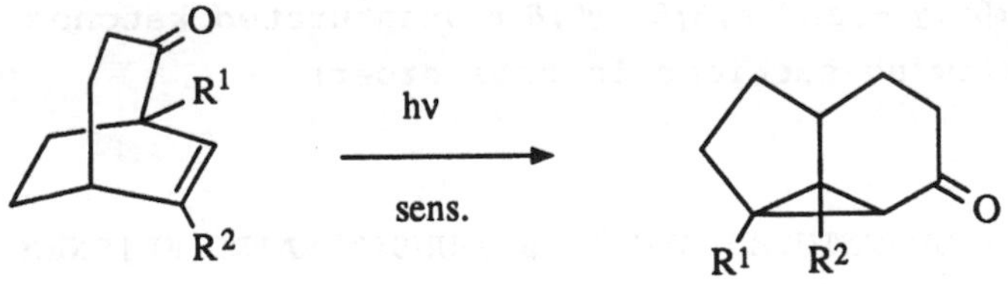

(64)

hv
sens.

(65)

(a) $R^1=R^2=H$
(b) $R^1=CH_3$, $R^2=H$
(c) $R^1=OCH_3$, $R^2=H$ } no reaction
(d) $R^1=OCH_3$, $R^2=CH_3$

on triplet-sensitization and that the triplet biradical
(63), formed via the 1,2-acyl shift from (60a) or (62a),
is an intermediate.

While both the bicyclo[2.2.1]heptenones and
bicyclo[2.2.2]octenones are excellent ODPM reactants the
higher homologs, the bicyclo[3.2.2]nonenones
(64),[92-94] are rather reluctant regarding the 1,2-shift.
In acetone solutions ($\lambda_{irr.}$ >290 nm) (64a) and (64b)
rearrange in moderate yields of 18% and 23%, respectively,
to the cyclopropyl ketones (65a,b) besides formation of
the 1,3-shift products in about equal amounts.[94]
Surprisingly, the 1-methoxy derivatives (64c,d) do not
undergo the 1,2-shift at all.[94]

Specific constellations of substituents and vinyl
group(s) in the enone can strongly influence or even alter
the course of the parent ODPM mechanism to give rise to
the formation of modified product skeletons that are of
great synthetic value. Such variations can be achieved
with the following classes of substrates: 1-methoxy-sub-
stituted β,γ-unsaturated ketones, β,γ-unsaturated ε-diketones, β,γ;β',γ'- and α,β;β',γ';δ,ε-unsaturated ketones
(see the following sections in this order).

2. 1-METHOXY-SUBSTITUTED CYCLIC β,γ-UNSATURATED KETONES

The rearrangement of the 1-methoxy-substituted
bicyclo[2.2.2]octenones (66a-d) follows a mechanistic
variant of the ODPM reaction, which includes two
consecutive single-step phototransformations, i.e. the
ODPM path to (67a-d), again triplet-sensitized, and the
conversion to the final 1,4-diketones (68a-d) (cf. Scheme
15).[5-11,95-98] The second step involves cleavage of the
transient cyclopropane in (67a-d) accompanied by loss of

Scheme 15

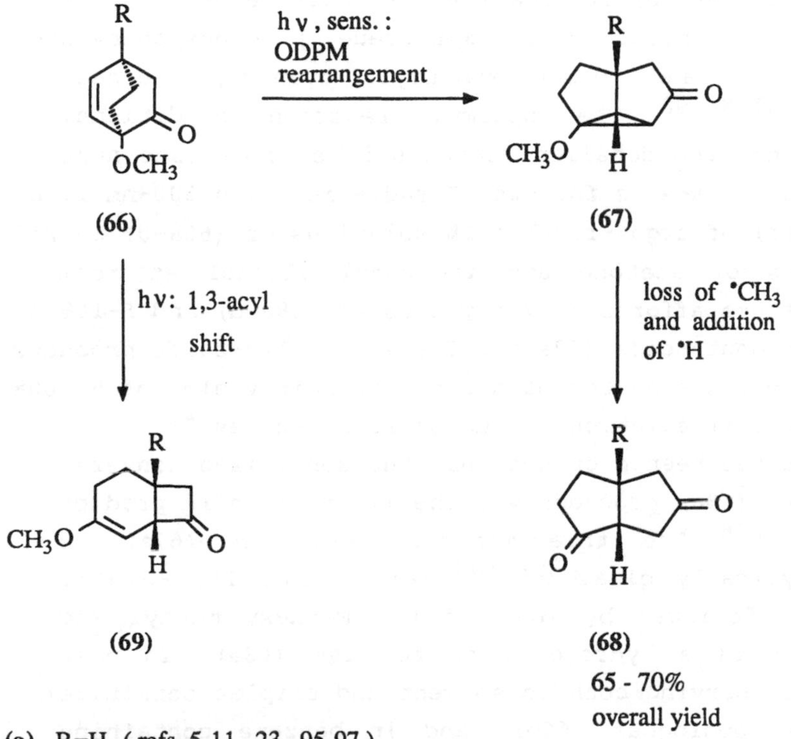

(66)   **(67)**

h ν , sens. :
ODPM
rearrangement

h ν: 1,3-acyl
shift

loss of ·CH₃
and addition
of ·H

(69)   **(68)**
65 - 70%
overall yield

(a) R=H (refs. 5-11 , 23 , 95-97)

(b) R=CH₂CH₂OCH₃ (refs. 41 , 97)

(c) R=CH₂CH(OCH₃)₂ (refs. 41 , 97)

(d) R=CH₃ (ref. 98)

the methoxy methyl, most likely via methyl radicals, and transfer of a hydrogen atom from the solvent.

Notably, diquinane products of the type (68) are of particular synthetic interest. Firstly, their rings are conveniently functionalized by keto groups that can be

chemically distinguished and allow further individual
elaboration of the functionality pattern into
polyquinanes (cf. Section VI). Secondly, the starting
materials (66a-d) are readily available, each in about 30%
yield, from anisol, p-methoxytoluene, p-methoxyphenylace-
tic acid and p-methoxybenzyl cyanide, respectively.[41,95-98] The optimum reaction conditions,
resulting from detailed mechanistic studies (see next
paragraph), are as follows. Irradiation with 300-nm light
(Rayonet) of argon-flushed 1% solutions of (66a-d) in 2:1
mixtures of acetone and isopropyl alcohol at room
temperature affords 65-70% yields of (68a-d) and 5-10% of
the cyclobutanones (69a-d). The latter 1,3-shift products
are accessible as the main components in yields of 50-60%
on direct irradiation of the starting enones.[41]

It has been proposed that the sensitized conversion
(66a) -> (68a) proceeds via the 1,2-acyl shift product
(67a).[95,96] The three-membered ring in (67a) is
photolytically cleaved[99,100] [cf. Scheme 16: -> (B),
R = H], followed by loss of the methoxy methyl and
addition of a hydrogen atom to give (68a). In neat
acetone, serving both as solvent and triplet sensitizer
of the homologous (66b), and in benzene containing
acetophenone the formation of a transient in low
concentration has actually been observed.[41,97,98] The
analytical data and chemical behaviour are clearly in
favor of the proposed tricyclic structure (67b).[41,97]
However, the donor/acceptor substitution of the
three-membered ring[101,102] renders the compound too
labile to allow its isolation. (67b) reacts not only
photochemically to (68b) but also undergoes this
transformation in protic media, presumably via
intermediate (D). The same study[97] has uncovered further
mechanistic details concerning the light-induced

Scheme 16

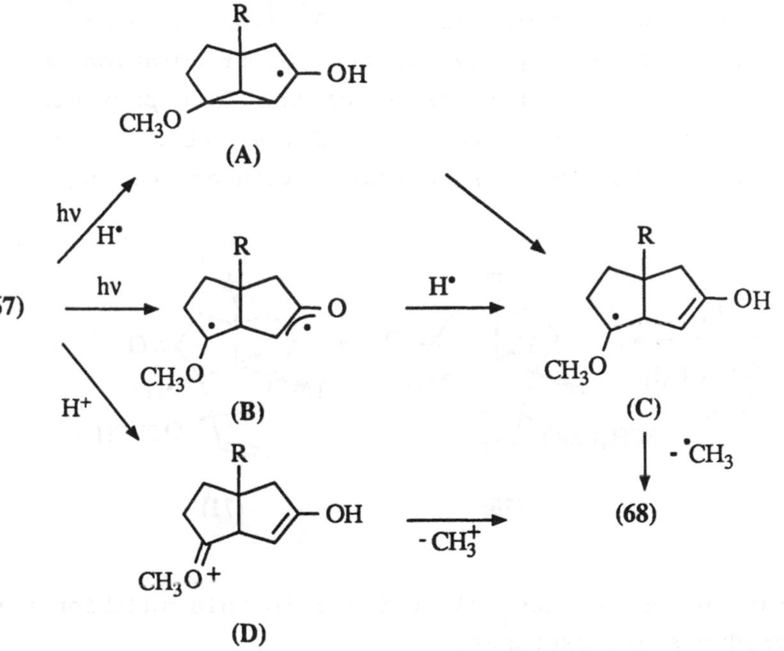

formation of (68b). An H/D isotope effect of 1.7 shows that the hydrogen transfer from the solvent (acetone vs. D_6-acetone) is rate determining. Furthermore, significant amounts of C_2H_6, but neither CH_3OH nor C_2H_4, have been detected by mass spectroscopy demonstrating that the methoxy group is homolytically cleaved and that that heterolytic cleavage and carbene elimination processes are excluded. Consequently, either hydrogen abstraction by the keto group [(67b) -> (A)] or the previously proposed[95,96] mechanism, i. e. regioselective β-cleavage[99] of the cyclopropyl ketone after n -> π* excitation [(67b) -> (B)], or both, are reasonable mechanistic interpretations. In case of the transient (A) cyclopropane cleavage to (C) would be expected to proceed rather readily.[103,104] A potential intermediacy of (B) has on the other hand been made plausible by a related experiment in which tricyclo[3.3.0.0^{2,8}]octan-3-one (23a) adds to 2-trimethylsilyloxybutadiene on irradiation at 300 nm in benzene to form three of the four possible products (70) and (71) in a 1 : 2 : 4 ratio.[97] The formation of (70) and (71) increases with decreasing

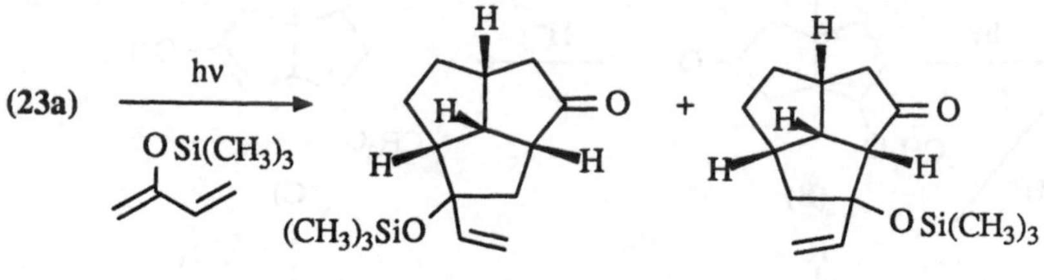

(70)   **(71)**

temperature and reaches 10% at 5 °C; in this addition the main products are oxetanes.

The ODPM reactivity of higher homologs in the series of bridged 1-methoxy compounds, i. e. bicyclo[3.2.2]non-enones, has also been investigated.[94] In contrast to the nor-methoxy analogs (64a,b) (cf. Scheme 14), which undergo the 1,2-migration in moderate yield, exclusively a 1,3-shift is observed for the derivatives (64c,d) [note: the substituents Y and Z in structure (4) of ref. 94 are erroneously interchanged].

3. β,γ-UNSATURATED ε-DIKETONES

The ODPM rearrangement has been very successfully probed with extended chromophores such as β,γ-unsaturated ε-diketones (cf. Scheme 17). The ε-diketones (72),[105,106]

Scheme 17

                                       φ

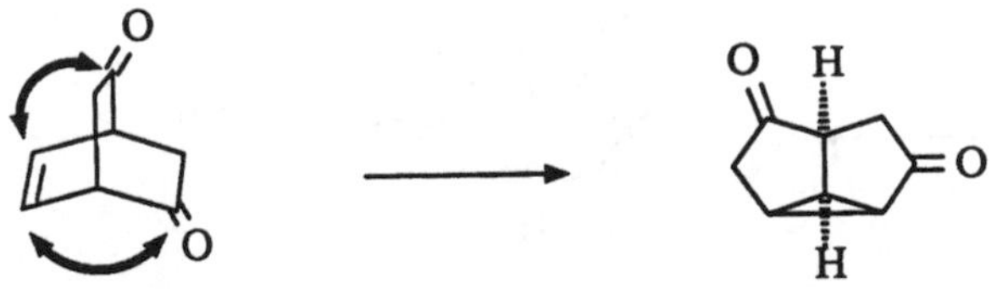

(72) → (73)   0.93

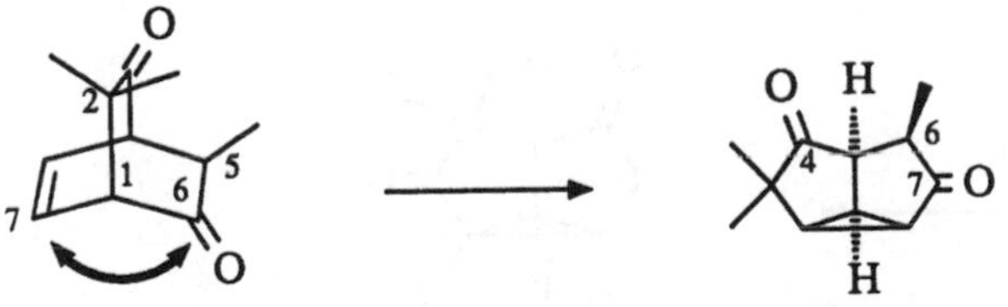

(74a) → (75a)   0.36

(74b) → (75b)   0.17

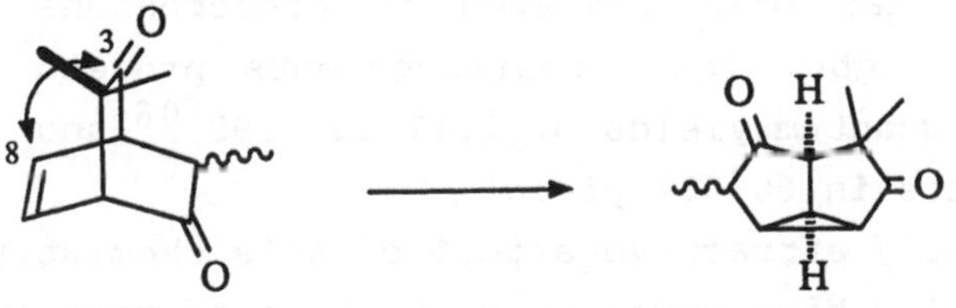

(74a,b) → (76a,b)

Scheme 17 (continued)

(77)   (78)   0.34

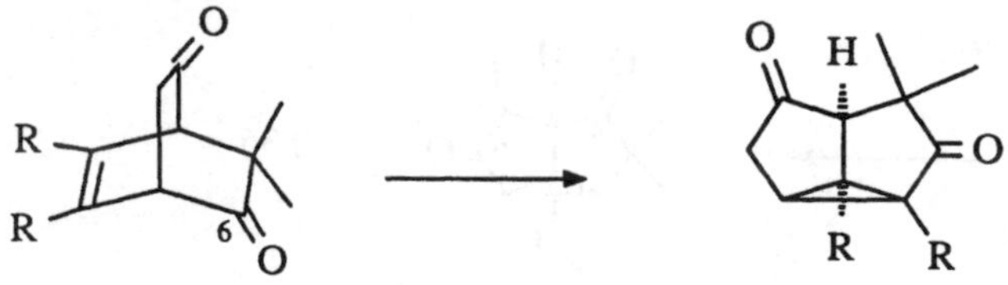

(79)   R=CO$_2$CH$_3$ (80)

(74),[105-107]   (77)[105,106]   and   (79)[108]   afford   the
tricyclo[3.3.0.0^{2,8}]octane-4,7-diones (73), (75), (76),
(78) and (80), respectively, upon triplet sensitization.
The rearrangements of (72), (74) and (77) have been
carried out in acetone at $\lambda_{irr.}$ = 300 nm and the conversion
of (79) to (80) has been achieved in acetophenone
employing 350-nm light. These rearrangements proceed
efficiently with quantum yields of 0.17 to 0.93[106] and
render the products in 60-95% yield.

      A preparatively attractive aspect of this chemistry
is the exceptionally high enone concentration of more than
20%, as shown with (74), at which the transformations can
still be run without noticeable formation of side

products. In the case of the trimethyl derivative (74) a regioselective photochemical rearrangement, favoring the formation of (75) over (76), has been encountered. The course of the reaction seems to be sterically controlled by the methyls. In situ photochemical equilibration of the C-5 epimers (74a,b) allows preferential rearrangement involving the C-6 [-> (75a,b)] rather than the C-3 [-> (76a,b)] keto group.[106,107]

It has been argued that bridging in the primary photochemical step between C-6 and C-7 of (74a), on the way to (75a), should be the sterically least hindered ODPM path since the C-5 methyl occupies the "*exo*" position with respect to the bonding sites.[106] In contrast, the corresponding bridging in (74b) should be impeded by the secondary methyl now occupying the "*endo*" position. The same argument should apply to the bridging process between C-3 and C-8 of (74a,b) which leads to (76a,b). Quantum yield measurements indeed support the notion of steric hindrance exerted by the "*endo*" methyls; decreasing values are found in the order of the reactions (72) -> (73), (74a) -> (75a) and (74b) -> (75b). The efficiency of the tetramethyl homolog (77) [-> (78)], that has the option of two identical bridgings yielding a single product, is greater than for (74b) -> (75b) which competes with the process (74a,b) -> (76a,b).

In conflict with this coherent picture of steric control of the ODPM rearrangement of the ϵ-diketones (74a,b) is the course of the transformation of (79).[108] This compound is reported to yield cleanly the 1,4-diketone (80) which is the isomer derived from bridging between the vinyl and C-6 in (79). In the previously described examples this reaction path has been shown to be unfavorable - a mechanistic riddle which has yet to be solved!

4. β,γ;β',γ'- AND α,β;β',γ';δ,ε-UNSATURATED KETONES

Besides parent ODPM arrangements mixed chromophores have also been investigated in which DPM-reactive sites can compete with the ODPM reaction [cf. also the general representations (E) and (F) in Scheme 6].

The acetophenone-sensitized rearrangement of the heavily substituted barrelenones (81) [β,γ;β',γ'-unsaturated ketones, type (F) in Scheme 6] leads to the corresponding unsaturated cyclopropyl ketones (82) via a highly regioselective DPM process, rather than the optional ODPM variant, in 42 - 73% yields.[109,110] These results contrast with the observations in a

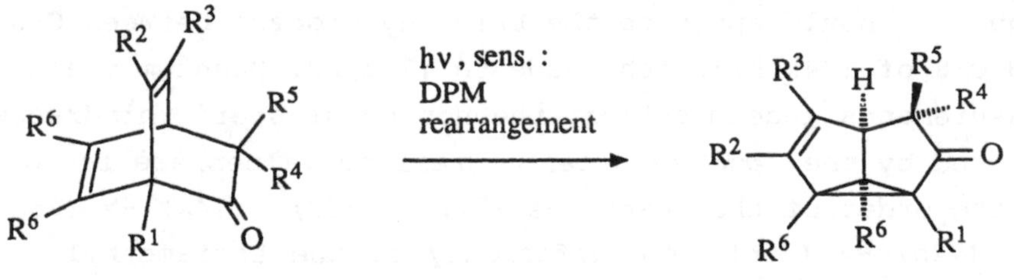

hν, sens. :
DPM
rearrangement

(81)   (82)

R^1	R^2	R^3	R^4	R^5	R^6
H	H	Me	O—CH$_2$	CO$_2$Me
H	H	Bu^t	O—CH$_2$	CO$_2$Me
H	H	Bu^t	CH$_2$—O	CO$_2$Me
H	Me	Me	O—CH$_2$	CO$_2$Me
Me	H	H	O—CH$_2$	CO$_2$Me
Me	H	H	OH	Me	CO$_2$Me
Me	H	Me	O—CH$_2$	CO$_2$Me
Me	H	Me	OH	Me	CO$_2$Me
Me	H	Bu^t	O—CH$_2$	CO$_2$Me

benzobarrelenone series[111,112] but are in full agreement
with the following mechanistic investigation.[113] The
question as to which of the two possible mechanisms is
operative has been answered by the acetone-sensitized
reaction of the labelled compound (83) in which the two
products (85) and (86) are formed. They must derive from
a DPM process[114] rather than from the alternative ODPM
path(s). The latter rearrangement would require formation
of a mixture of the isomers (84) and (85) the former of

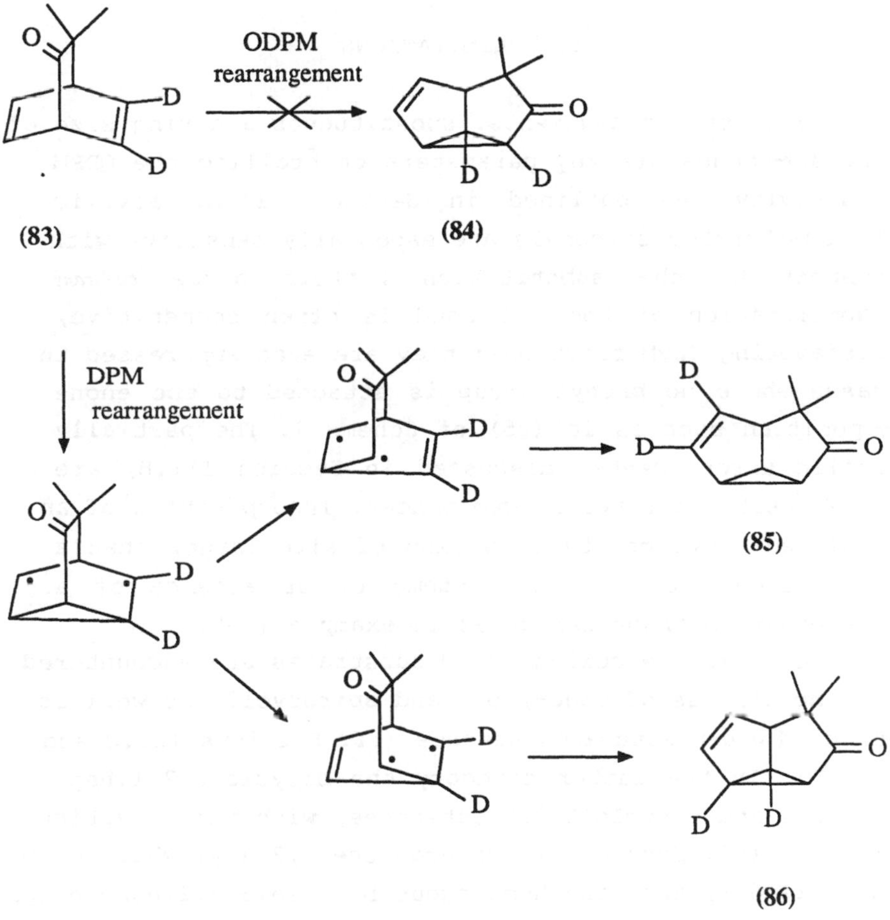

(83)   (84)

(85)

(86)

which has however not been detected. If the barrelenone
(83) is excited directly neither of the two reaction
channels is active. In summary, these results[109-113]
reveal relative bridging 'rates' in bicyclo[2.2.2]octadi-
enone-type substrates in the following order: vinyl-vinyl
> keto-vinyl > benzo-vinyl bridging. Similarly,
$\alpha, \beta; \beta', \gamma'; \delta, \varepsilon$-unsaturated bicyclic ketones [type (E) in
Scheme 6] react in a DPM fashion as shown for the
rearrangement of steroidal dienones.[115,116]

IV. LIMITATIONS

Skeletal arrangements, substituents and ring size of
the β, γ-enones are key parameters controlling the ODPM
reactivity. As outlined in Section III.A, acyclic
β, γ-unsaturated carbonyls are especially sensitive with
respect to the substitution pattern since *cis-trans*
isomerization of the C=C bond is often competitive,
disfavoring ODPM reactions; they are even suppressed in
cases where no phenyl group is attached to the enone
γ-position such as in (25) of Scheme 7. The partially
cyclic arrangements, discussed in Section III.B, are
ODPM-inert under two circumstances: incorporation of an
isolated vinyl moiety in a ring of size larger than a
5-membered one [(37) in Scheme 8] or absence of a
quaternary methane carbon as in example (35).
Generally excellent ODPM substrates are encountered
in the classes of mono-, bi- and spirocyclic as well as
bridged α, β-unsaturated ketones (cf. Sections III.C and
III.D). In the latter category the bicyclo[2.2.1]hep-
tenones and bicyclo[2.2.2]octenones, with the exceptions
(22o-s) (cf. Scheme 18), undergo the 1,2-acyl shift much
more readily than the homologous bicyclo[3.2.2]nonenones

Scheme 18

ODPM reactive : unreactive substrates :

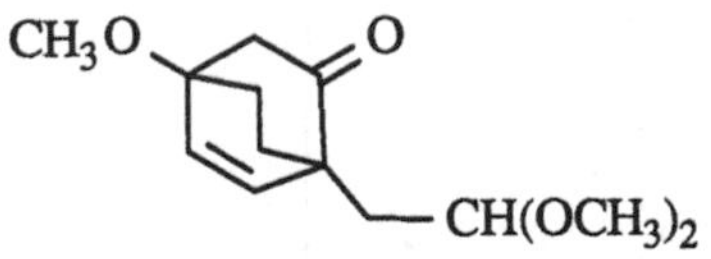

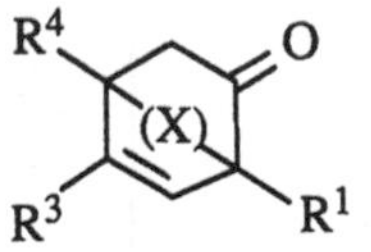

CH_3O ... O

—$CH(OCH_3)_2$

R^4 ... O

(X)

R^3 ... R^1

(22d) (22o-s)

(o) $X=CH_2CH_2$, $R^1=CH_3$, $R^3=H$, $R^4=OCH_3$
    (refs. 8 , 41)

(p) $X=CH_2CH_2$, $R^1=CO_2CH_3$, $R^{3,4}=H$ (ref. 5)

(q) $X=CH_2CH_2$, $R^1=CO_2H$, $R^{3,4}=H$ (ref. 5)

(r) $X=CH_2CH_2$, $R^1=CO_2Na$, $R^{3,4}=H$ (ref. 5)

(s) $X=CH_2$, $R^{1,4}=H$, $R^3=OCH_3$ (ref. 117)

(64) in Scheme 14. The enones of type (64a,b) rearrange in moderate yields to (65a,b), but most interestingly, the 1-methoxy derivatives (64c,d) do not give any ODPM product, a result which is hard to explain at present. The switch of reactivity from the [2.2.1] and [2.2.2] skeletons to the [3.2.2] homologs, however, could reasonably be explained by enhanced skeletal flexibility (see the discussion of the acyclic series) or unfavorable orbital overlap or both.

At this point we propose a model which allows an empirical prediction of reactivity for this class of compounds.[7] In Scheme 19 the substituents A (= acceptor: CO_2CH_3, CO_2H or CO_2^-) and D (= donor: OCH_3, CH_3 or H) are encircled in those cases where the β,γ-enones have been

Scheme 19

I: II:

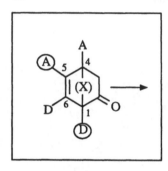   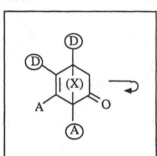

A :   acceptor

D :   donor

X : $(CH_2)_n$; n=1, 2

subjected to the conditions of the ODPM reaction. Notably,
this proposal is valid only for bicyclo[2.2.1]heptenones
and bicyclo[2.2.2]octenones. The types of compounds which
correspond to the patterns (I) or (II) are reactive or
unreactive, respectively. A comparison of the circled
substituents suggests a complementarity of A/D and D/A
patterns in (I) and (II), respectively.

Four arrangements with single A- or D-type
substituents have so far been established experimentally:
an acceptor at C-1 [(22p-r), cf. Scheme 18] or C-5 [(22f),
cf. Scheme 13] as well as a donor at C-1 [(66a), cf.
Scheme 15] or C-5 [(22s), cf. Scheme 18]. This empirical
rule may possibly be extended to substrates with two
substituents of very similar electronic donor (or
acceptor) properties as seen from the reactions of (66b-d)
to (68b-d) (Scheme 15) and the following two examples.

With the compounds (22d) (Schemes 5 and 18) and (22o)
(Scheme 18) it has been possible to demonstrate that the
borderline of reactivity is determined by a subtle
interplay of electronic effects. However, in the absence
of calculations it is yet difficult to rationalize how
small changes of the functionality pattern, such as
shortening the acetal chain of (22d) to R^1 = CH_3 of (22o),
render a photoreactive substrate [(22d) -> (23d): 90%
yield] photochemically inert.[8,41] This loss of reactivity
could be explained by a switch of relative donor capacity,
i.e. a change of D_1 (22d) to predominant D_4 character of
(22o). The significant influence of substituents on the
reactivity which are not directly attached to the
chromophore but occupy the bridgeheads, i.e. the α- and/or
δ-position(s) of the enone, is of particular interest in
this context and still needs further exploration. The
inhibition of reaction in the cases of (22p-r) could,
however, be due to the electron density at C-1 not being
adequate enough for the bond breaking in the intermediate
(87) of the ODPM rearrangement.[5]

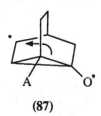

(87)

Bicyclo[3.2.2]nonenones [(64c,d): no reaction, cf.
Scheme 14, Section III.D.2] and 'mixed' bicyclo[2.2.2]octenones [(22e) -> (23e), cf. Scheme 13,
Section III.D.2] need not fit into the above reactivity
scheme.

V.   PREPARATION OF ENANTIOMERICALLY PURE ODPM PRODUCTS

An impressive number of enantiomerically pure ODPM products as starting materials for synthetic "ventures" have been made available in recent years. Resolution techniques prior to the photorearrangement (this section) as well as the use of enantiomerically pure starting materials in general (see the examples in Scheme 26) provide ready access to optically pure photoproducts. In contrast, enantioselection via sensitization has so far turned out to be of little preparative value. The latter possibility has been probed with a racemate of (22a) the rearrangement of which has been sensitized by the hexa-hydrofluorenone (-)-(88) (cf. Scheme 20).[39] The enantiomeric excess of (-)-(23a) after 7-44% conversion of (+)-(22a) reaches 4.5% in benzene and in ethyl acetate at ambient temperature and 10% in ethyl acetate at low temperature. Interestingly, the same degree of induction can be achieved when racemic (88) is employed for the rearrangement of (22a) in optically active solvent, i.e. in diethyl (+)-tartrate.[39]

The ODPM transformation of bridged β,γ-unsaturated

Scheme 20:

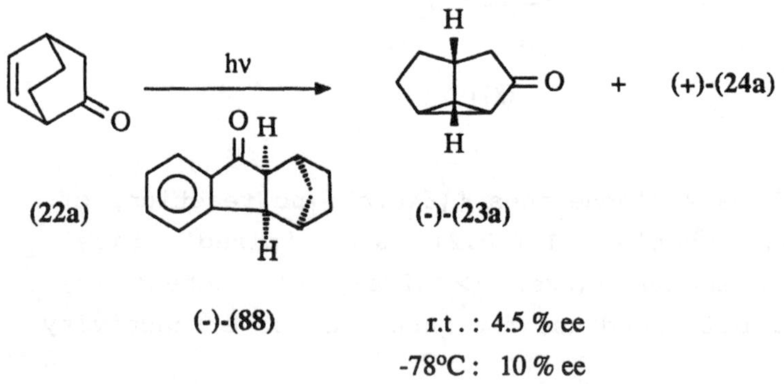

(22a)

(-)-(88)

(-)-(23a)

+   (+)-(24a)

r.t. : 4.5 % ee
-78°C : 10 % ee

Scheme 21

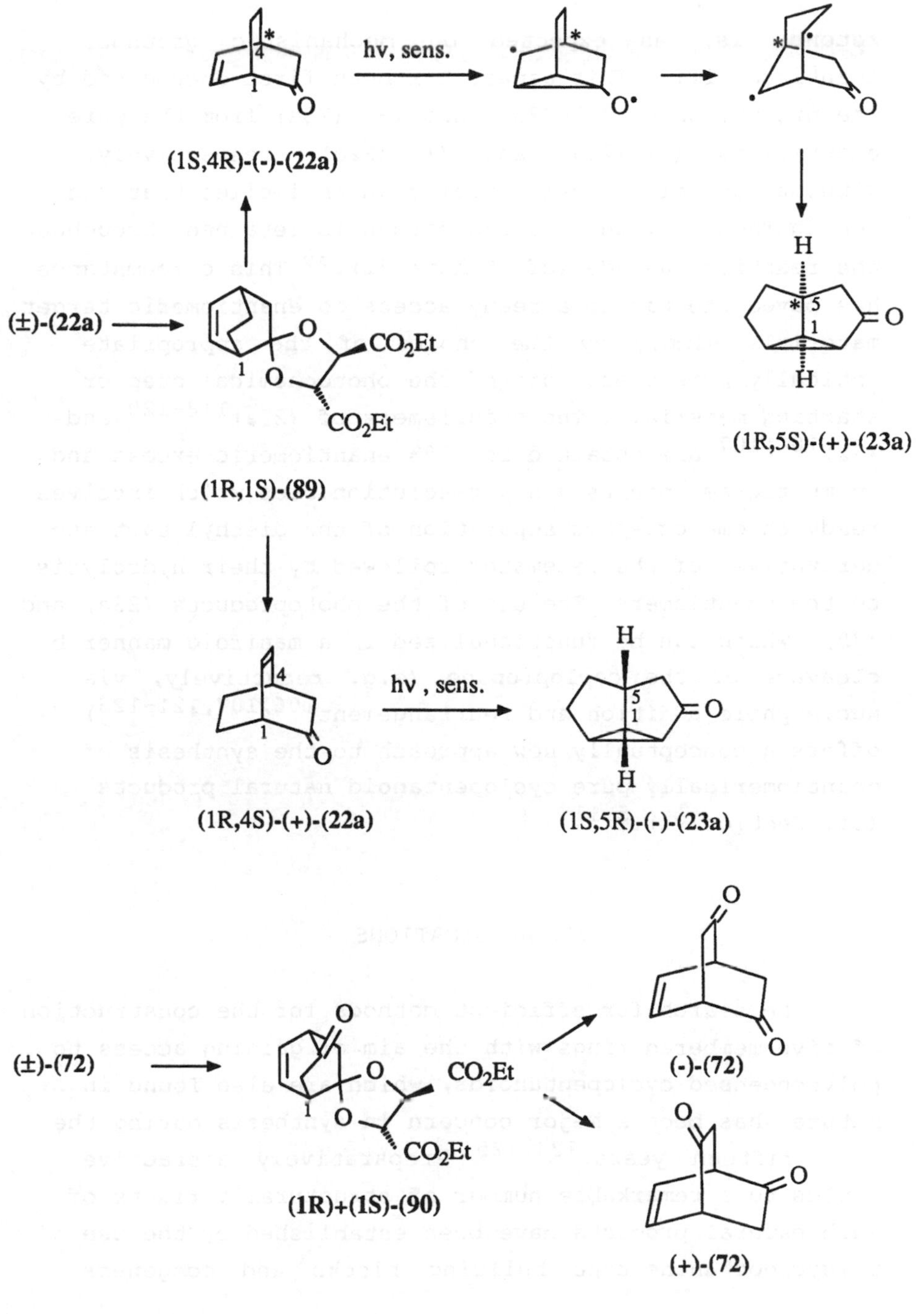

(1S,4R)-(-)-(22a)

(±)-(22a) →

(1R,1S)-(89)

(1R,5S)-(+)-(23a)

(1R,4S)-(+)-(22a)

hv, sens.

(1S,5R)-(-)-(23a)

(±)-(72) →

(1R)+(1S)-(90)

(-)-(72)

(+)-(72)

ketones is, as expected on mechanistic grounds, enantiospecific. This aspect has been first documented by the production of (+)-(23a) and (-)-(23a) from the pure enantiomers (-)-(22a) and (+)-(22a), respectively, without loss of optical activity which implies that the configuration of the starred carbon is retained throughout the reaction cascade (cf. Scheme 21).[39] This circumstance has paved the way to a ready access to enantiomeric target materials simply by the choice of the appropriate optically pure precursors of the photochemical step or starting materials. The enantiomers of (22a)[118-120] and (72)[106,107] are obtained in >98% enantiomeric excess and in multigram amounts via a resolution step which involves ready chromatographic separation of the diethyl tartrate derivatives of the racemates followed by their hydrolysis to the enantiomers. The use of the photoproducts (23a) and (75), which can be functionalized in a manifold manner by cleavage of the cyclopropane (e.g. reductively, via nucleophile addition and rearrangement[106,107,121-123]) offers a conceptually new approach to the synthesis of enantiomerically pure cyclopentanoid natural products (cf. Section VI).[5-11]

VI. APPLICATIONS

The search for efficient methods for the construction of five-membered rings with the aim of gaining access to polycondensed cyclopentanoids, which are also found in nature, has been a major concern in synthesis during the past fifteen years.[124,125] Preparatively attractive routes to a remarkable number of structural variants of such natural products have been established by the use of tricyclooctanone-type building blocks and congeners

Scheme 22

I   (Section III . D . 1)

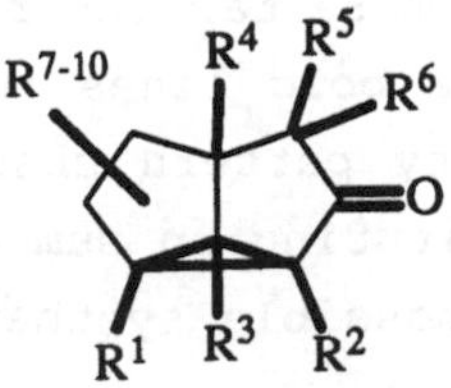

II   (Section III . D . 2)

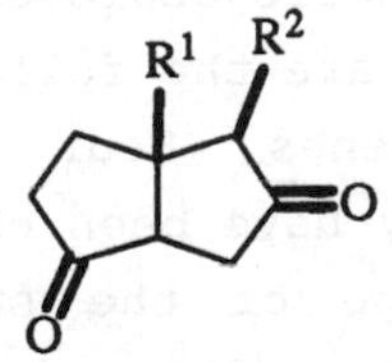

III   (Section III . D . 3)

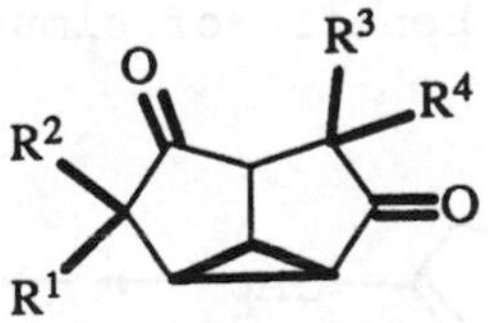

IV

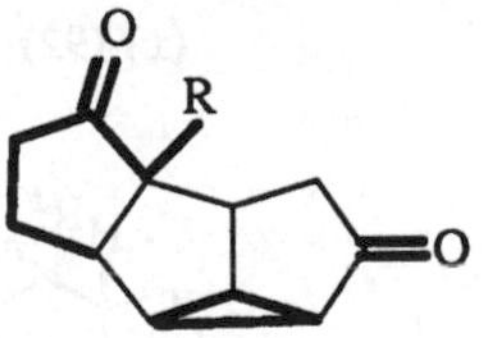

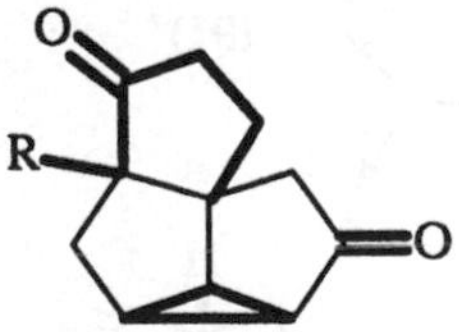

(this Section)

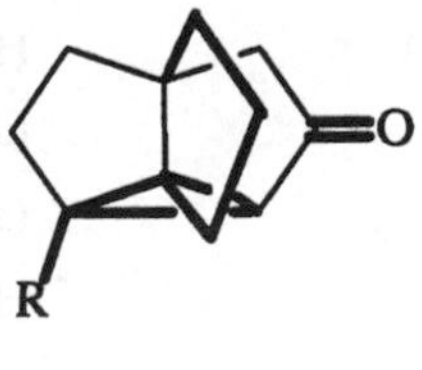

thereof, all products of high-yield ODPM transformations.
In summary, four basic structural arrangements as shown
in Scheme 22 have so far proven to be of broad utility in
synthesis, a parent type (I) with a minimal set of
functionalities and three more target-oriented and more
complex units (II-IV). The bold lines in the formulae
emphasize the functionality pattern characteristic of
each representative constituting in sum a fully flexible
and complementary set of possible synthetic modifications
at the diquinane core.

Among the first synthetic achievements, based on the
use of tricyclooctanones, are the following preparations
of mono- and sesquiterpenes. Four members of the
monoterpene iridoid family have been obtained by a single
approach. The disadvantage of the stereo*non*selective
reduction of (23b) affording (91) and (93) in a 1:1 ratio
is compensated for by the benefit of simultaneous access

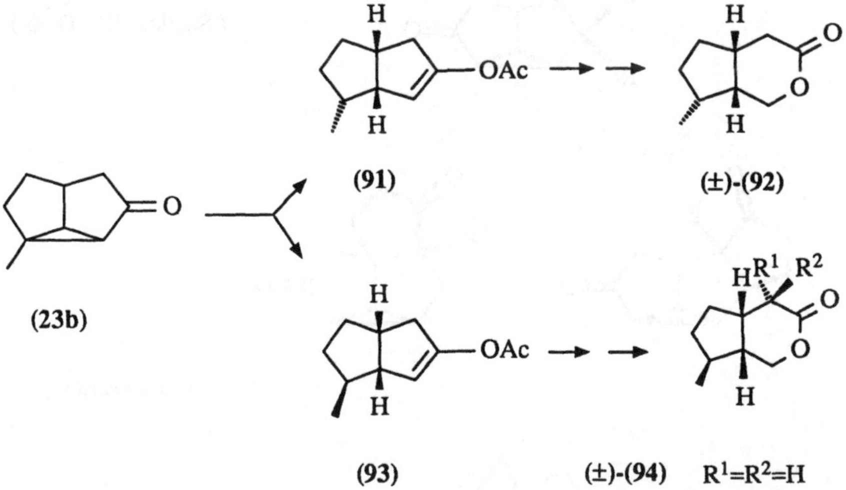

(±)-(94)	$R^1=R^2=H$
(±)-(95)	$R^1=R^2=CH_2$
(±)-(96)	$R^1=CH_3$, $R^2=H$
(±)-(97)	$R^1=H$, $R^2=CH_3$

to four natural products, viz. (±)-boschnialactone (92) and the epimer (94) which is a precursor to (±)-allodolicholactone (95), (±)-iridomyrmecin (96) and (±)-isoiridomyrmecin (97).[5] The sesquiterpene cedrol and more highly oxygenated cedranoids, in racemic form, have been obtained from the Stork-Clarke β-diketone (98) which in turn can be assembled in a novel way via the 76%-yield rearrangement of (22e) to (23e) (cf. Scheme 13).[77,126] In 10 steps and in a 15% overall yield a synthesis of (±)-forsythide aglucone dimethyl ester has very recently been accomplished by similarly exploiting the potential of ODPM photochemistry for the key step.[127]

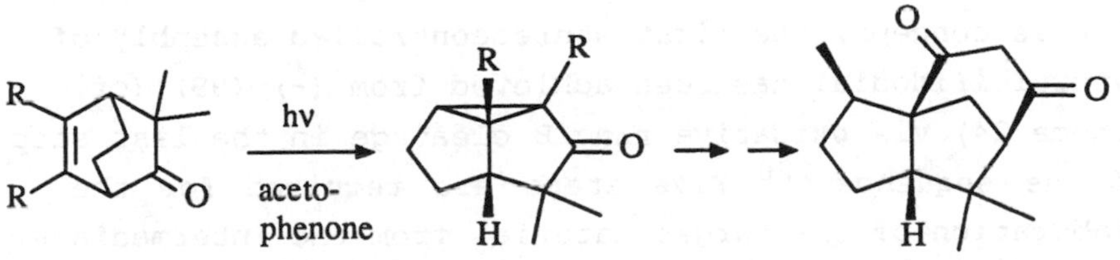

| (22e) | (23e) | (±)-(98) |

R=CO₂CH₃

    A very economic preparation of racemic modhephene (cf. Scheme 23), a unique natural product with [3.3.3]propellane structure, has been prepared from (23k) which in turn is obtained in 50% yield upon irradiation of (22k) in acetone solution (cf. Scheme 13).[85]

    An enantioselective access to a number of different structural variants of the cyclopentanoids is established by the use of the tricyclooctanone building blocks (-)-(23a) and (+)-(23a) (cf. Schemes 24 and 25) both of which are available in optically pure form (see Section V). These photoproducts offer an alternative to previous

Scheme 23

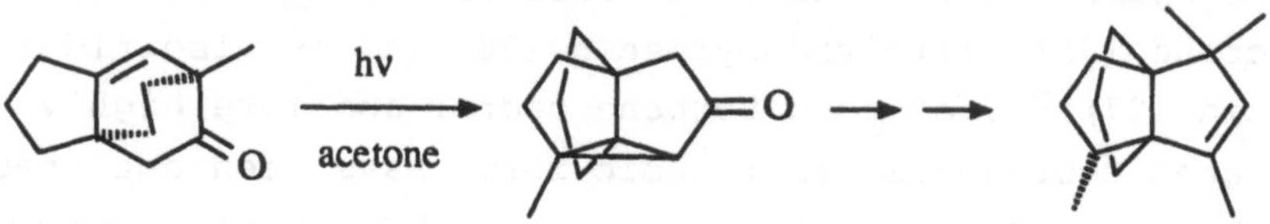

(22k)   (23k)   (±)-modhephene

individually designed target-oriented syntheses. The
enantiomers (23a) exhibit structural features allowing
for a multitude of synthetic transformations of
predictable regio- and stereoselectivity.[121-123] As part
of this concept, the first stereocontrolled assembly of
natural iridodial has been achieved from (-)-(99) (cf.
Scheme 24) via oxidative ring B cleavage in the last step
of the sequence.[120] Five steps are required for the
elaboration of the target material from the intermediate
(-)-(99) which is the product of exclusive *exo* methylation
of (-)-(23a). Rearrangement of the cyclopropane moiety of
(23a) constitutes the basis for extensive ring A
functionalizations, including oxidative enlargement, on
the way to (+)-loganin aglucone 6-acetate.[118] This
synthesis is distinguished by a higher overall yield of
7%, starting from 1,3-cyclohexadiene and acrylonitrile,
than previous efforts and by the lack of separation
problems.

The cleavage of all rings of (23a) has led to a
stereocontrolled access to (103) which serves as a segment
in the construction of the 16-membered lactone carbamycin
B.[128] The numbered carbons in the target derive from the
respective centers of the intermediates (102) and (103),
the first of which has been prepared from (23a) in a
high-yield sequence of consecutive cyclopropane

Scheme 24

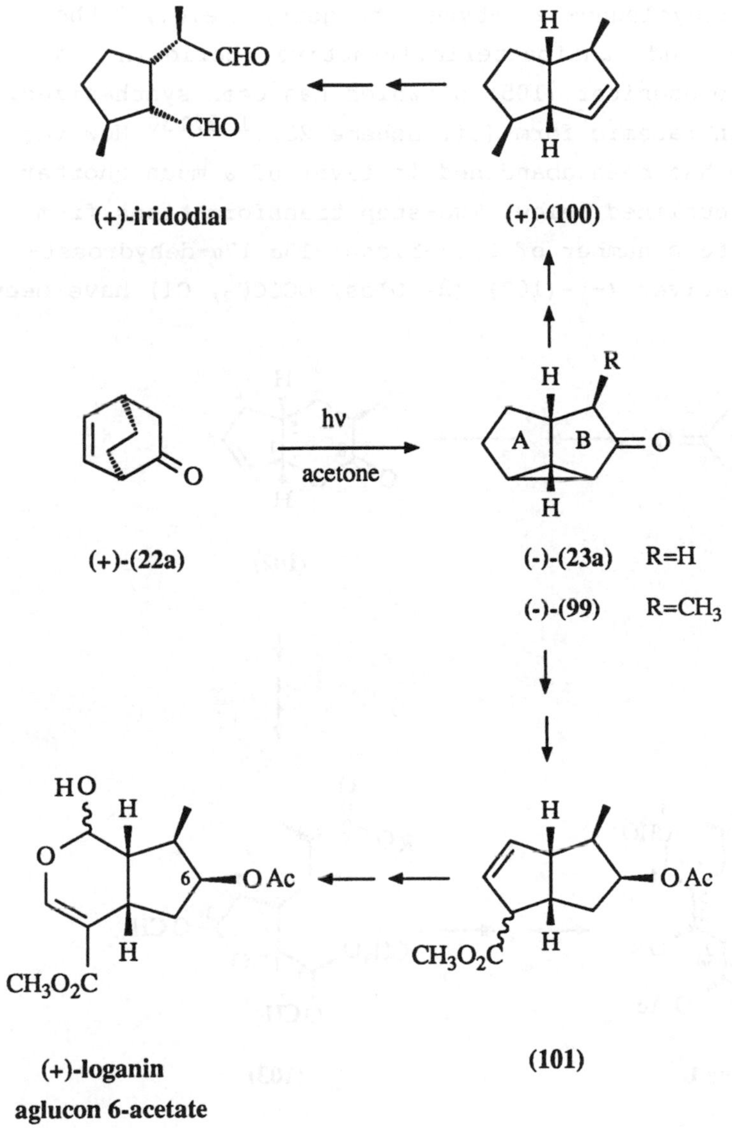
CHO
CHO
H
H
(+)-iridodial
(+)-(100)
hν
acetone
R
H
A
B
O
O
H
(+)-(22a)
(-)-(23a) R=H
(-)-(99) R=CH3
HO
H
O
6
OAc
H
CH3O2C
H
OAc
H
CH3O2C
(+)-loganin
aglucon 6-acetate
(101)

rearrangement,   Baeyer-Villiger   oxidation,   and
methylation.

The enantiomer (+)-(23a) has been considered
configurationally suited as starting material to approach
cis,anti,cis tricycloundecane-type targets, e.g., the
antitumor- and antibacterially-active coriolin, a
potential progenitor (105) of which has been synthesized,
at first in racemic form (cf. Scheme 25).[129,130] However,
this route has been abandoned in favor of a much shorter
synthesis outlined below. Two-step transformations from
(+)-(23a) to a number of C,18-bisnor-13α,17α-dehydroest-
rone derivatives (-)-(107) (R= OTos, OCOCF$_3$, Cl) have been

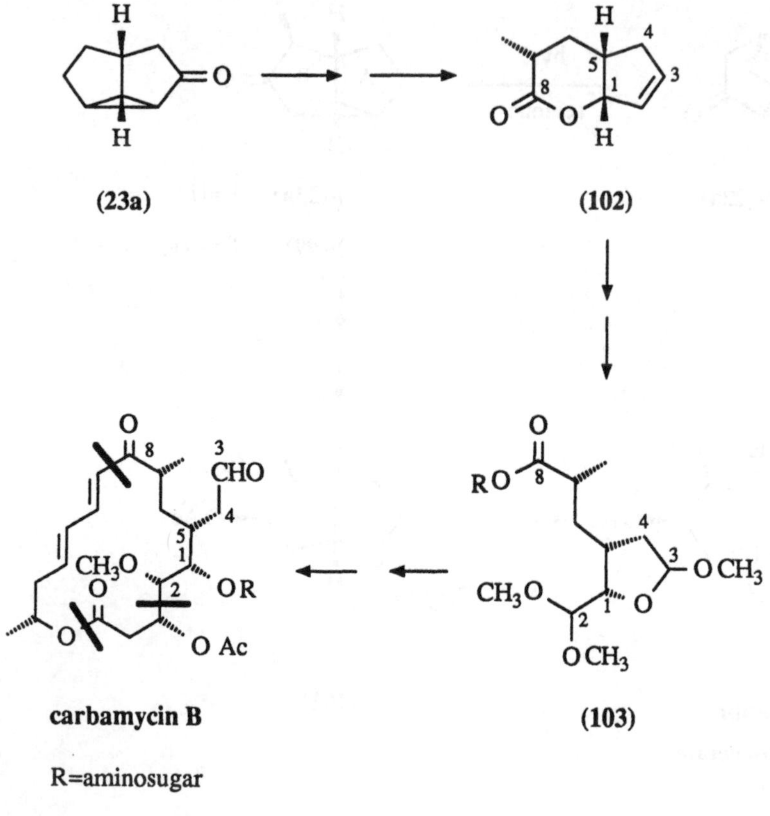

(23a)   (102)

carbamycin B   (103)

R=aminosugar

Scheme 25

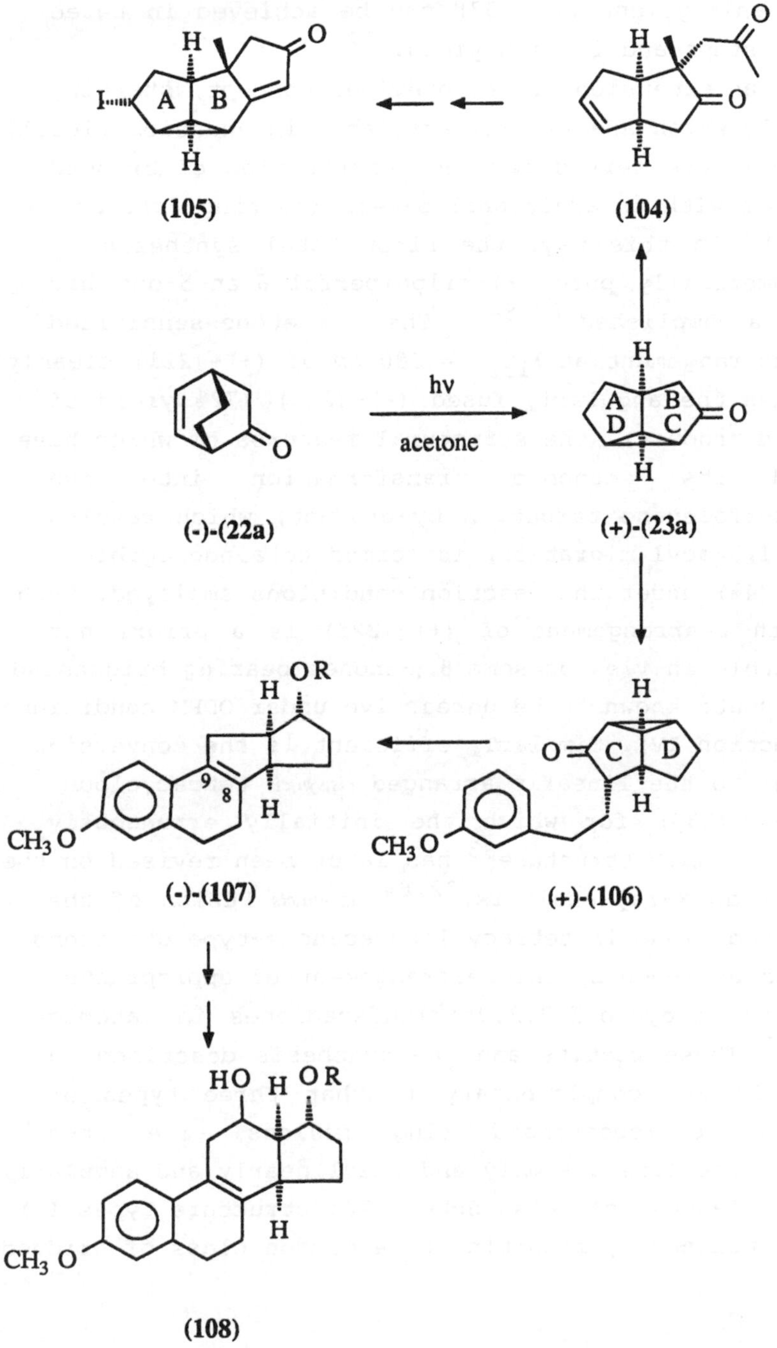

(105) ← ← (104)

(-)-(22a)   →hv, acetone→   (+)-(23a)

(-)-(107) ← (+)-(106)

(108)

elaborated;[123,131] these are intermediates from which
ring C enlargement to (108) can be achieved in three
further steps and in high yield.[132]

In an extension of the previous concept, starting
materials which are readily available in enantiomerically
pure form have served for the construction of bridged
β,γ-enones with an additional 5-membered ring (cf. Scheme
26).[82,83] In this way, the first total synthesis of
enantiomerically pure (-)-silphiperfol-6-en-5-one has
been accomplished.[82,84] The acetone-sensitized
photorearrangement at $\lambda_{irr.}$ = 300 nm of (+)-(22i) cleanly
furnishes the angularly fused (-)-(23i) (72% yield of
purified product), the structural features of which have
allowed its economic transformation into the
silphiperfolenone target. A by-product, which results
from a 1,3-acyl migration, is formed to a negligible
extent (4%) under the reaction conditions employed. Such
a smooth rearrangement of (+)-(22i) is a priori not
predictable in view of some β,γ-enones bearing bridgehead
substituents known to be unreactive under ODPM conditions
(see Section IV). Similarly efficient is the conversion
of (22j) to the linearly arranged *cis,syn,cis* tetracycloun-
decanone (23j) for which the initially erroneously
assigned *cis,anti,cis* structure[83] has later been revised on the
basis of an X-ray analysis.[82,84] *cis,anti,cis* Fusion of the
5-membered rings in tetracycloundecanone-type skeletons
has been achieved by the rearrangement of appropriately
assembled tricyclo[5.2.2.0^{2,6}]undecadiones in racemic
form.[133] These results and the synthesis described in
Scheme 23 are complementary in that three types of
important cyclopentanoid ring fusions, i.e. the
[3.3.3]propellane assembly and the linearly and angularly
fused skeletons (cf. also Scheme 22: structure types IV),
can be attained by resorting to a common class of bridged

Scheme 26

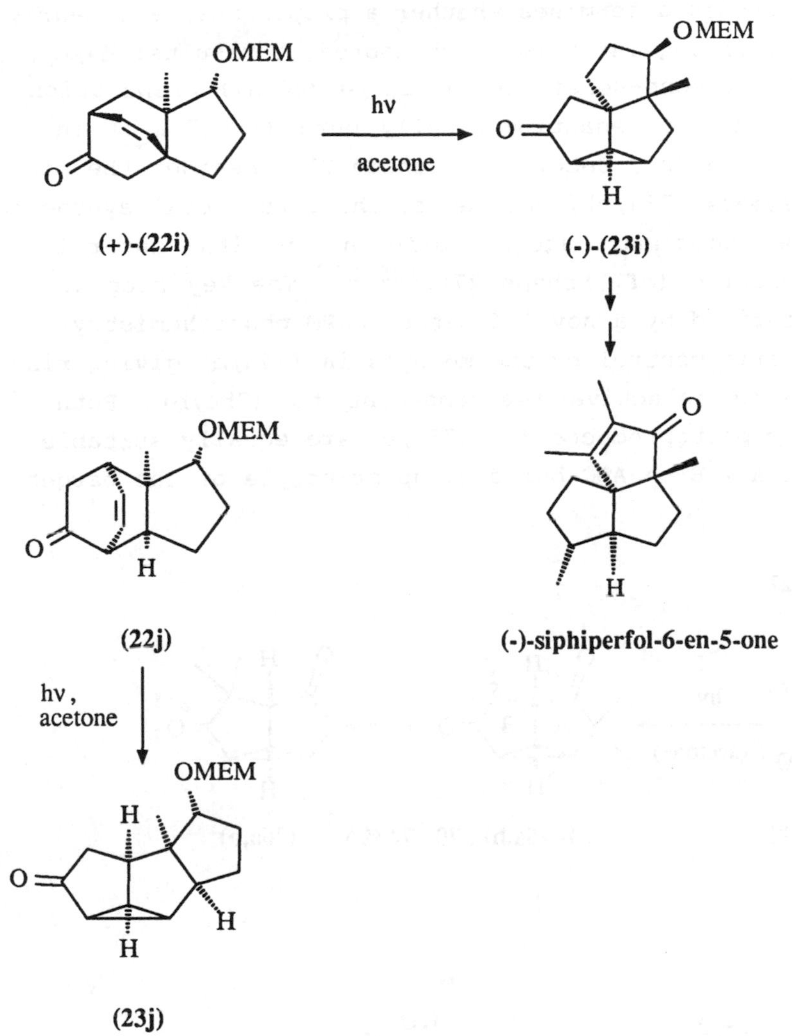
OMEM
hν
acetone
O
(+)-(22i)
OMEM
O
H
(-)-(23i)
OMEM
O
H
(22j)
O
H
(-)-siphiperfol-6-en-5-one
hν,
acetone
OMEM
H
O
H
H
(23j)

β,γ-enones (22i,j,k). These enones are accessible via thermal [4+2] cycloadditions in an efficient regio- and stereocontrolled manner. Whereas the stereochemical result of the addition of the etheno bridge is responsible for the ultimate *syn/anti* relation of the rings in the

photoproducts, the regiochemical location of the C=C bond
in the enones determines whether a propellane, a linearly
fused or an angularly fused photoproduct results.

The acetone-sensitized ($\lambda_{irr.}$ = 300 nm) preparation
of a mixture of enantiomerically pure (-)-(75a,b) in
70-74% yield (cf. Section V, Scheme 21), besides the minor
regioisomers (76a,b), has led to the first total synthesis
of the antitumor agent coriolin in its natural
configuration (cf. Scheme 27).[105-107] The key step is
characterized by a novel facet of ODPM photochemistry,
viz. steric control by the methyls in (74a,b) giving rise
to a site selective rearrangement to (75a,b). Both
epimeric photoproducts (-)-(75a,b) are equally suitable
for the A + B -> ABC building-up principle of the target

Scheme 27

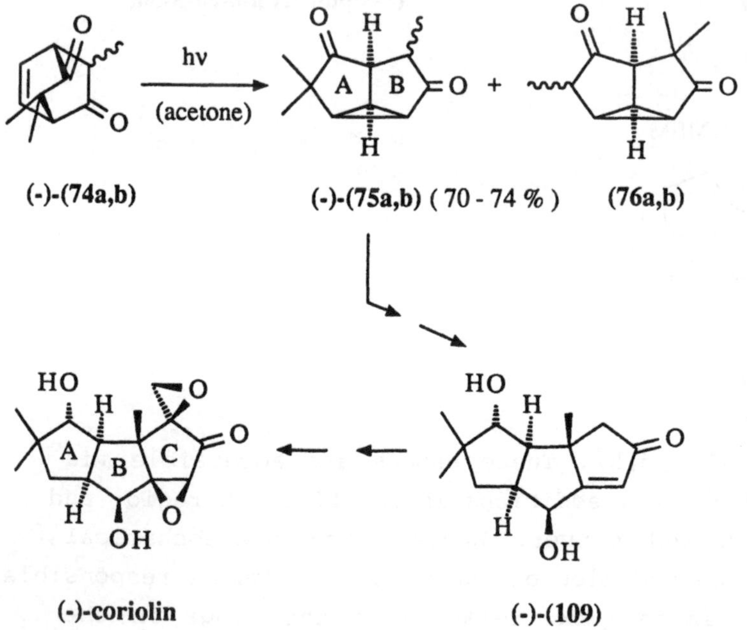

(-)-(74a,b)   (-)-(75a,b) (70 - 74 %)   (76a,b)

(-)-coriolin   (-)-(109)

via (-)-(109). Concurrent with the ODPM reaction and
favoring the most efficient ODPM channel, a second and
fast photoreaction results in a 5 : 1 equilibration of the
epimers (74a) and (74b). Although the mechanism of this
process has not been fully identified, it has at least
been shown that Norrish type I clevage and reclosure is
not responsible for the epimerization.[106] It has been
mentioned before and should be stressed again at this
point that the transformation of (74a,b) can be run at
preparatively attractive high substrate concentrations of
more than 20% without noticeable side reactions (1,3-acyl
shift). This synthesis of (-)-coriolin comprises fourteen
steps from readily available starting materials; the
sequence compares favorably with the shortest syntheses
of the racemate achieved earlier.

    Similarly, the photorearrangement of a β,γ-un-
saturated ε-diketone (79) (cf. Scheme 17) has been probed
in view of its potential application as a key step for the
synthesis of cedranoids.[108] Although the tricyclo-
octanedione (80) is cleanly accessible the scheme has been
abandoned in view of difficulties in assembling the
ultimately required functionalities; routine β,γ-enone
photochemistry has been employed instead.

    In a final example use has been made of a mechanistic
variant of the ODPM reaction, the conversion of a
1-methoxy-substituted bicyclo[2.2.2]octenone (110) to the
bicyclo[3.3.0]octadione (111). The highly substituted
β,γ-enone (110) is accessible via hydroxymethylation of
(66c), a compound which itself is photoreactive (cf.
Scheme 15). (110) rearranges in two consecutive
single-step photoreactions to (111), a photoproduct which
upon purification on silica gel reacts smoothly to a
potential progenitor (112) (58% overall yield) of
pentalenolactone G (cf. Scheme 28).[9,10,41,134] This

Scheme 28

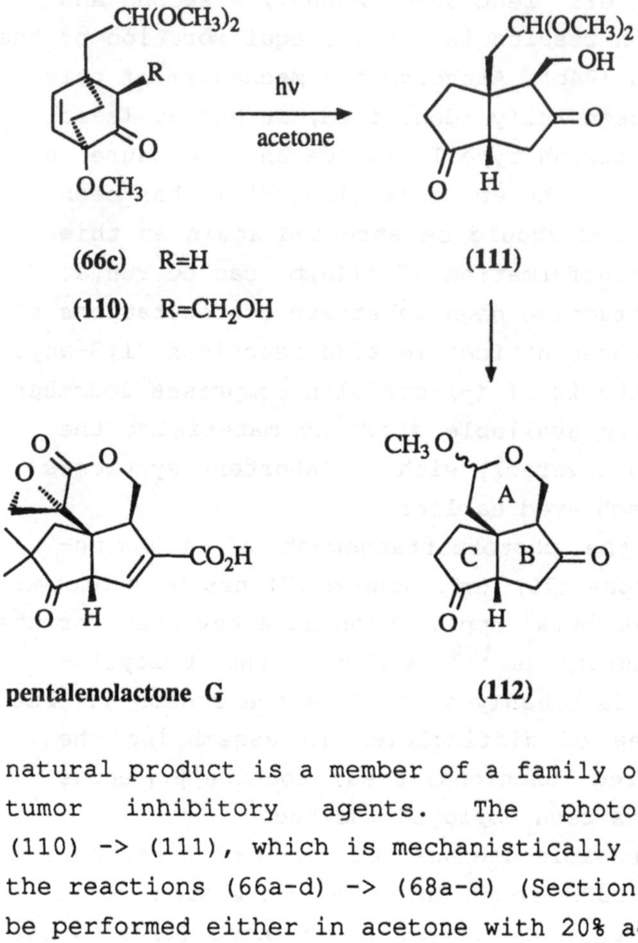
CH(OCH3)2
R
O
OCH3
hv
acetone
CH(OCH3)2
OH
O
O
H
(66c) R=H
(110) R=CH2OH
(111)
O
O
O
CO2H
O
H
CH3O
O
A
C
B
O
O
H
pentalenolactone G
(112)

natural product is a member of a family of antibiotic and tumor inhibitory agents. The photorearrangement (110) -> (111), which is mechanistically in accord with the reactions (66a-d) -> (68a-d) (Section III.D.2), can be performed either in acetone with 20% acetophenone added ($\lambda_{irr.} \geq 340$ nm) or by following the procedure described in Section III.D.2.[41]

VII. AZADI-π-METHANE REACTION

As outlined in Section III.A, acyclic β,γ-unsaturated aldehydes are ODPM-unreactive and the majority of the analogous ketones, apart from a few exceptions, undergo the ODPM rearrangement extremely sluggishly, if at all; these substrates prefer to *cis-trans* isomerize upon photonic excitation.

Readily accessible imine derivatives of such substrates, however, fill this gap in the ODPM reaction spectrum and cyclize in high yields when irradiated in presence of triplet sensitizers. They undergo the so-called azadi-π-methane rearrangement, typical examples of which are shown for (113) to (114) (R^1 = alkyl or phenyl).[135,136]

This reaction dependends on the type of nitrogen substitution. It is proposed that the reactivity may correlate with the ionization potential of the imine.[137] Oxidized derivatives, i.e. the oxime (113) (R^3 = OH)[138] and the oxime ether (113) (R^3 = OCH_3),[139] fail to undergo the azadi-π-methane reaction. Instead, isomerizations around the C=N and the C=C bond are the

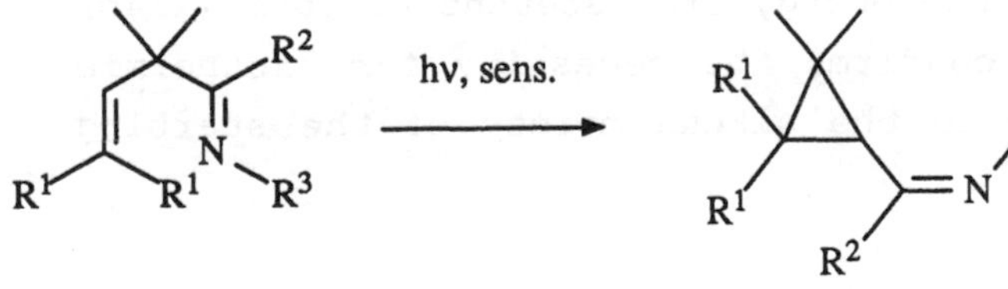

(113)   (114)

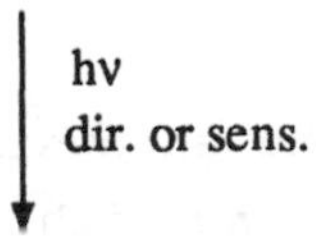

hv
dir. or sens.

(a) R^1=Ph; R^2=H, CH_3 ;
     R^3=OAc

(b) R^1=CH_3; R^2=H;
     R^3=OAc

no reaction

R^1=Ph; R^2=H, CH_3 ;
R^3=OH or OCH_3

main paths of deactivation. Only for one oxime, a cyclic
and additionally constrained derivative, has
azadi-π-methane reactivity been reported.[140]

    In order to increase the ionization potential of the
oxime group, oxime acetates have been prepared and
investigated recently.[141] The acetophenone-sensitized
rearrangement of (113a) turns out to be highly efficient
providing the cyclopropane (114a) in 79% yield, in
addition to the nitrile derivative which derives from
elimination of acetic acid to the extent of 11%. The
azadi-π-methane reaction also proceeds in benzene; direct
excitation, on the other hand, fails to trigger this
process. The authors suggest that initial energy transfer
to the 1,1-diphenyl alkene moiety may be responsible for
the success of cyclization. This notion is substantiated
in the same work[141] by the lack of rearrangement of (113b)
in the presence of acetophenone. The sole product formed
under these conditions is an oxetane. The azadi-π-methane
reaction path, on the other hand, is again activated when
(113b) receives the energy from acetone to form (114b),
a result which confirms the necessity for exothermic
energy transfer to the alkene moiety of the starting
material.

ACKNOWLEDGMENTS

    The author gratefully acknowledges the enthusiastic
collaboration of his group and, in particular with respect
to this chapter, the fruitful contributions of the
postdoctoral research associates, graduate students and
technical assistants whose names appear as coauthors in
the references. Special thanks are extended to Professor

Kurt Schaffner who has given a strong impetus to the author to take care of the ODPM reaction from a synthetic viewpoint. Furthermore, the critical reading of the manuscript and the comments by K. S. and Mr. Henry Gruen have valuably contributed to the final shape of this chapter.

VIII. REFERENCES

1. W. G. Dauben, G. Lodder, and J. Ipaktschi, Top. Curr. Chem., 54, 73 (1975).

2. K. Schaffner, Tetrahedron, 32, 641 (1976).

3. K. N. Houk, Chem. Rev., 76, 1 (1976).

4. D. I. Schuster, Rearrangements in Ground and Excited States, vol. 3, (P. de Mayo, ed.), Academic Press, New York, p. 167 (1980).

5. M. Demuth and K. Schaffner, Angew. Chem., 94, 809 (1982); Angew. Chem. Int. Ed. Engl., 21, 820 (1982).

6. M. Demuth, Chimia, 38, 257 (1984).

7. M. Demuth, Habilitationsschrift, University of Essen (1985).

8. K. Schaffner and M. Demuth, Modern Synthetic Methods, vol. 4, (R. Scheffold, ed.), Springer, Berlin, p. 61 (1986).

9. M. Demuth, Modern Synthetic Methods, vol. 4, (R. Scheffold, ed.), Springer, Berlin, p. 89 (1986).

10. M. Demuth, Pure Appl. Chem., 58, 1233 (1986).

11. M. Demuth and G. Mikhail, Synthesis, 145 (1989).

12. D. E. Sadler, J. Wendler, G. Olbrich, and K. Schaffner, J. Am. Chem. Soc., 106, 2064 (1984).

13. B. Reimann, D. E. Sadler, and K. Schaffner, J. Am. Chem. Soc., 108, 5527 (1986).

14. K. N. Houk, D. J. Northington, and R. E. Duke, J. Am. Chem. Soc., 94, 6233 (1972).

15. G. Marsh, D. R. Kearns, and K. Schaffner, J. Am. Chem. Soc., 93, 3129 (1971).

16. H.-U. Gonzenbach, I.-M. Tegmo-Larsson, J.-P. Grosclaude, and K. Schaffner, Helv. Chim. Acta., 60, 1091 (1977).

17. M. J. Mirbach, A. Henne, and K. Schaffner, J. Am. Chem. Soc., 100, 7127 (1978).

18. P. S. Engel, M. A. Schexnayder, and W. V. Phillips, Tetrahedron Lett., 1157 (1975).

19. A. Henne, N. P. Y. Siew, and K. Schaffner, Helv. Chim. Acta, 62, 1952 (1979).

20. A. Henne, N. P. Y. Siew, and K. Schaffner, J. Am. Chem. Soc., 101, 3671 (1979).

21. M. J. C. M. Koppes and H. Cerfontain, Rec. Trav. Chim. Pays-Bas, 107, 549 (1988).

22. T. J. Eckersley and N. A. J. Rogers, <u>Tetrahedron</u>,
     <u>40</u>, 3759 (1984).

23. T. J. Eckersley, S. D. Parker, and N. A. J. Rogers,
     <u>Tetrahedron</u>, <u>40</u>, 3749 (1984).

24. S. Domb and K. Schaffner, <u>Helv. Chim. Acta</u>, <u>53</u>, 677
     (1970).

25. B. Winter and K. Schaffner, <u>J. Am. Chem. Soc.</u>, <u>98</u>,
     2022 (1976).

26. W. G. Dauben, G. Lodder, and J. D. Robbins, <u>J. Am.
     Chem. Soc.</u>, <u>98</u>, 3030 (1976).

27. W. G. Dauben, G. Lodder, and J. D. Robbins, <u>Nouv.
     J. Chim.</u>, <u>1</u>, 243 (1977).

28. J. I. Seeman and H. Ziffer, <u>Tetrahedron Lett.</u>, 4409
     and 4413 (1973).

29. R. S. Givens and W. F. Oettle, <u>J. Am. Chem. Soc.</u>,
     <u>93</u>, 3963 (1971).

30. E. Baggiolini, K. Schaffner, and O. Jeger, <u>J. Chem.
     Soc. Chem. Commun.</u>, 1103 (1969).

31. J. Ipaktschi, <u>Chem. Ber.</u>, <u>105</u>, 1840 (1972).

32. J. Ipaktschi, <u>Tetrahedron Lett.</u>, 2153 (1969).

33. K. N. Houk and D. J. Northington, <u>J. Am. Chem. Soc.</u>,
     <u>94</u>, 1387 (1972).

34.   H. U. Gonzenbach, Ph.D. Dissertation, Eidgenös-
      sische Technische Hochschule, Zürich (1973).

35.   R. L. Coffin, R. S. Givens, and R. G. Carlson, J.
      Am. Chem. Soc., 96, 7554 (1974).

36.   I.-M. Tegmo-Larsson, Ph.D. Dissertation, University
      of Geneva (1976).

37.   W. Amrein, I.-M. Larsson, and K. Schaffner, Helv.
      Chim. Acta, 57, 2519 (1974).

38.   P. J. Wagner and T. Nakahira, J. Am. Chem. Soc., 95,
      8474 (1973).

39.   M. Demuth, P. R. Raghavan, C. Carter, K. Nakano, and
      K. Schaffner, Helv. Chim. Acta, 63, 2434 (1980).

40.   R. S. Givens, W. F. Oettle, R. L. Coffin, and R. G.
      Carlson, J. Am. Chem. Soc., 93, 3957 (1971).

41.   B. Wietfeld, Ph.D. Dissertation, Max-Planck-Insti-
      tut   für   Strahlenchemie,   Mülheim   and   Ruhr
      University, Bochum (1984).

42.   K. G. Hancock and R. O. Grider, Tetrahedron Lett.,
      4281 (1971).

43.   K. G. Hancock and R. O. Grider, Tetrahedron Lett.,
      1367 (1972).

44.   K. G. Hancock and R. O. Grider, J. Am. Chem. Soc.,
      96, 1158 (1974).

45. P. S. Engel and M. A. Schexnayder, *J. Am. Chem. Soc.*, 94, 9252 (1972).

46. W. G. Dauben, M. S. Kellogg, J. I. Seeman, and W. A. Spitzer, *J. Am. Chem. Soc.*, 92, 1786 (1970).

47. L. P. Tenney, D. W. Boykin, Jr., and R. E. Lutz, *J. Am. Chem. Soc.*, 88, 1835 (1966).

48. D. O. Cowan and A. A. Baum, *J. Am. Chem. Soc.*, 93, 1153 (1971).

49. R. G. Carlson, R. L. Coffin, W. W. Cox, and R. S. Givens, *J. Chem. Soc. Chem. Commun.*, 501 (1973).

50. M. Mintas, D. I. Schuster, and P. G. Williard, *J. Am. Chem. Soc.*, 110, 2305 (1988).

51. R. C. Cookson and N. R. Rogers, *J. Chem. Soc. Chem. Commun.*, 809 (1972).

52. H. Eichenberger, K. Tsutsumi, G. de Weck, and H. R. Wolf, *Helv. Chim. Acta*, 63, 1499 (1980).

53. E. Pfenninger, D. E. Poel, C. Berse, H. Wehrli, K. Schaffner, and O. Jeger, *Helv. Chim. Acta*, 51, 772 (1968).

54. H. Sato, N. Furutachi, and K. Nakanishi, *J. Am. Chem. Soc.*, 94, 2150 (1972).

55. H. Sato, K. Nakanishi, J. Hayashi, and Y. Nakadaira, *Tetrahedron*, 29, 275 (1973).

56. S. Domb, G. Bozzato, J. A. Saboz, and K. Schaffner,
     Helv. Chim. Acta, 52, 2436 (1969).

57. J. R. Williams and H. Ziffer, J. Chem. Soc. Chem.
     Commun., 194 (1967).

58. J. R. Williams and H. Ziffer, Tetrahedron, 24, 6725
     (1968).

59. J. R. Williams and H. Ziffer, J. Chem. Soc. Chem.
     Commun., 469 (1967).

60. P. S. Engel, M. A. Schexnayder, H. Ziffer, and J.
     I. Seeman, J. Am. Chem. Soc., 96, 924 (1974).

61. J. R. Williams and G. M. Sarkisian, J. Chem. Soc.
     Chem. Commun., 1564 (1971).

62. K. Schaffner and M. Demuth, Rearrangements in Ground
     and Excited States, vol. 3, (P. de Mayo, ed.),
     Academic Press, New York, p. 281 (1980).

63. G. Quinkert, Angew. Chem., 87, 851 (1975); Angew.
     Chem. Int. Ed. Engl. 14, 790 (1975).

64. M. R. Morris and A. J. Waring, J. Chem. Soc. C, 3269
     (1971).

65. T. R. Rodgers and H. Hart, Tetrahedron Lett., 4845
     (1969).

66. H. Hart and D. W. Swatton, J. Am. Chem. Soc., 89,
     1874 (1967).

67. D. H. R. Barton and G. Quinkert, <u>J. Chem. Soc.</u>, 1 (1960).

68. J. Griffiths and H. Hart, <u>J. Am. Chem. Soc.</u>, <u>90</u>, 3297 (1968).

69. J. Griffiths and H. Hart, <u>J. Am. Chem. Soc.</u>, <u>90</u>, 5296 (1968).

70. A. J. Waring, M. R. Morris, and M. M. Islam, <u>J. Chem. Soc. C</u>, 3274 (1971).

71. A. D. Dickinson, A. T. Hardy, and H. Hart, <u>Org. Photochem. Synth.</u>, <u>2</u>, 62 (1976).

72. M. Kimura and S. Morosawa, <u>J. Am. Chem. Soc.</u>, <u>103</u>, 2433 (1981).

73. T. A. Lyle, H. B. Mereyala, A. Pascual, and B. Frei, <u>Helv. Chim. Acta</u>, <u>67</u>, 774 (1984).

74. J. Zizuashvili, S. Abramson, U. Shmueli, and B. Fuchs, <u>J. Chem. Soc. Chem. Commun.</u>, 1375 (1982).

75. J. Oren, L. Schleifer, U. Shmueli, and B. Fuchs, <u>Tetrahedron Lett.</u>, <u>25</u>, 981 (1984).

76. J. Oren and B. Fuchs, <u>J. Am. Chem. Soc.</u>, <u>108</u>, 4881 (1986).

77. P. Yates, and K. E. Stevens, <u>Tetrahedron</u>, <u>37</u>, 4401 (1981).

78. P. Yates, and K. E. Stevens, Can. J. Chem., 60, 825
     (1982).

79. J.-T. Chen, J.-T. Huang, C.-C. Liao, and C.-P. Wei,
     Proceedings of the VIIth IUPAC Conf. on Org.
     Synthesis (1988).

80. L. A. Paquette, C. S. Ra, and T. W. Silvestri,
     Tetrahedron, 45, 3099 (1989).

81. G. Mehta and A. Srikrishna, Tetrahedron Lett., 3187
     (1979).

82. W. Hinsken, Ph.D. Dissertation, Max-Planck-Institut
     für Strahlenchemie, Mülheim and Ruhr University,
     Bochum (1986).

83. M. Demuth and W. Hinsken, Angew. Chem., 97, 974
     (1985); Angew. Chem. Int. Ed. Engl., 24, 973 (1985).

84. M. Demuth and W. Hinsken, Helv. Chim. Acta, 71, 569
     (1988).

85. G. Mehta and D. Subrahmanyan, J. Chem. Soc. Chem.
     Commun., 768 (1985).

86. R. Kilger, W. Körner, and P. Margaretha, Helv. Chim.
     Acta, 67, 1493 (1984).

87. A. G. Schultz, J. P. Dittami, F. P. Lavieri, C.
     Salowey, P. Sundararaman, and M. B. Szymula, J. Org.
     Chem., 49, 4429 (1984).

88. A. G. Schultz, F. P. Lavieri, and T. E. Snead, J. Org. Chem., 50, 3086 (1985).

89. P. S. Engel and M. A. Schexnayder, Tetrahedron Lett., 1153 (1975).

90. H.-D. Scharf and W. Küsters, Chem. Ber., 104, 3016 (1971).

91. W. Adam, O. De Lucchi, K. Hill, E.-M. Peters, K. Peters, and H. G. von Schnering, Chem. Ber., 118, 3070 (1985).

92. L. A. Paquette, R. P. Henzel, and R. F. Eizenber, J. Org. Chem., 38, 3257 (1973).

93. T. Uyehara, Y. Kabasawa, T. Kato, and T. Furuta, Tetrahedron Lett., 26, 2343 (1985).

94. T. Uyehara, Y. Kabasawa, T. Furuta, and T. Kato, Bull. Chem. Soc. Jpn., 59, 539 (1986).

95. S. D. Parker and N. A. J. Rogers, Tetrahedron Lett., 4389 (1976).

96. T. J. Eckersley and N. A. J. Rogers, Tetrahedron, 40, 3759 (1984).

97. M. Demuth, B. Wietfeld, B. Pandey, and K. Schaffner, Angew. Chem., 97, 777 (1985); Angew. Chem. Int. Ed. Engl., 24, 763 (1985).

98. M. Demuth and P. Ritterskamp, unpublished result.

99. L. D. Hess, J. L. Jacobson, K. Schaffner, and J. N. Pitts, Jr., J. Am. Chem. Soc., 89, 3684 (1967).

100. W. G. Dauben and G. W. Shaffer, Tetrahedron Lett., 4415 (1967).

101. A. de Meijere, Angew. Chem., 91, 867 (1979); Angew. Chem. Int. Ed. Engl., 18, 809 (1979).

102. H. N. C. Wong, M.-Y. Hon, C.-W. Tse Y.-C. Yip, J. Tanko, and T. Hudlicky, Chem. Rev., 89, 165 (1989).

103. D. C. Neckers, A. P. Schaap, and J. Hardy, J. Am. Chem. Soc., 88, 1265 (1966).

104. D. G. Marsh, J. N. Pitts, Jr., and K. Schaffner, J. Am. Chem. Soc., 93, 333 (1971).

105. E. Weigt, Ph.D. Dissertation, Max-Planck-Institut für Strahlenchemie, Mülheim and Ruhr University, Bochum (1985).

106. M. Demuth, P. Ritterskamp, E. Weigt, and K. Schaffner, J. Am. Chem. Soc., 108, 4149 (1986).

107. M. Demuth, P. Ritterskamp, and K. Schaffner, Helv. Chim. Acta, 67, 2023 (1984).

108. P. Yates, D. J. Burnell, V. J. Freer, and J. F. Sawyer, Can. J. Chem., 65, 69 (1987).

109. H.-D. Becker and B. Ruge, Angew. Chem., 87, 782 (1975); Angew. Chem. Int. Engl., 14, 761 (1975).

110. H.-D. Becker and B. Ruge, J. Org. Chem., 45, 2189
      (1980).

111. H. Hart and R. K. Murray, Jr., Tetrahedron Lett.,
      379 (1969).

112. R. S. Givens and W. F. Oettle, J. Am. Chem. Soc.,
      93, 3963 (1971).

113. R. T. Luibrand, B. M. Broline, K. A. Charles, and
      R. W. Drues, J. Org. Chem., 46, 1874 (1981).

114. For comprehensive accounts on the 'Zimmerman
      reaction', see: H. E. Zimmerman, Rearrangements in
      Ground and Excited States, vol. 3, (P. de Mayo,
      ed.), Academic Press, New York, p. 131 (1980), and
      this volume.

115. B. Nann, D. Gravel, R. Schorta, H. Wehrli, K.
      Schaffner, and O. Jeger, Helv. Chim. Acta, 46, 2473
      (1963).

116. B. Nann, H. Wehrli, K. Schaffner, and O. Jeger,
      Helv. Chim. Acta, 48, 1680 (1965).

117. W. Kirmse, Ruhr University Bochum, private communi-
      cation.

118. M. Demuth, S. Chandrasekhar, and K. Schaffner, J.
      Am. Chem. Soc., 106, 1092 (1984).

119. M. Demuth and K. Schaffner, patent appl.: Fed. Rep.
      of Germany P 3046,106.4 (1980); U. S. patent
      4,415,756 (1983); Canadian patent 1199332 (1986).

120.   P. Ritterskamp, M. Demuth, and K. Schaffner, J. Org.
       Chem., 49, 1155 (1984).

121.   M. Demuth, G. Mikhail, and M. V. George, Helv. Chim.
       Acta, 64, 2759 (1981).

122.   M. Demuth and G. Mikhail, Tetrahedron, 39, 991
       (1983).

123.   G. Mikhail and M. Demuth, Helv. Chim. Acta, 66, 2362
       (1983).

124.   L. A. Paquette, Topics Curr. Chem., 119, 1 (1984).

125.   L. A. Paquette and A. M. Doherty, Reactivity and
       Structure - Concepts in Organic Chemistry, 26, 1
       (1987).

126.   K. E. Stevens and P. Yates, J. Chem. Soc. Chem.
       Commun., 990 (1980).

127.   C. C. Liao and C. P. Wei, Tetrahedron Lett., in
       press.

128.   M. Demuth, P. Dalmases, P. R. Kanjilal, and G.
       Mikhail, manuscript in preparation.

129.   M. Demuth, A. Cánovas, E. Weigt, C. Krüger, and
       Y.-H. Tsay, Angew. Chem., 95, 747 (1983); Angew.
       Chem. Int. Ed. Engl., 22, 721 (1983).

130.   M. Demuth, A. Cánovas, E. Weigt, C. Krüger, and
       Y.-H. Tsay, Angew. Chem. Suppl., 1053 (1983).

131.   G. Mikhail, Ph.D. Dissertation, Max-Planck-Institut

für Strahlenchemie, Mülheim and Instituto Quimico de Sarriá, Barcelona (1984).

132. T. Planas, Ph.D. Dissertation, Max-Planck-Institut für Strahlenchemie, Mülheim and Instituto Quimico de Sarriá, Barcelona (1987).

133. V. K. Singh, P. T. Deota, and B. N. S. Raju, Synth. Commun., 17, 115 (1987).

134. M. Demuth, B. Wietfeld, and A. Pascual, unpublished results.

135. D. Armesto, J. A. F. Martin, R. Perez-Ossorio, and W. M. Horspool, Tetrahedron Lett., 23, 2149 (1982).

136. D. Armesto, W. M. Horspool, J. A. F. Martin, and R. Perez-Ossorio, J. Chem. Res. (S), 46 (1986).

137. D. Armesto, W. M. Horspool, F. Langa, and R. Perez-Ossorio, J. Chem. Soc., Perkin Trans. II, 1039 (1987).

138. D. Armesto, W. M. Horspool, F. Langa, J. A. F. Martin, and R. Perez-Ossorio, J. Chem. Soc., Perkin Trans. I, 743 (1987).

139. A. C. Pratt and Q. Abdul-Majid, J. Chem. Soc., Perkin Trans. I, 359 (1987).

140. M. Nitta, I. Kasahara, and T. Kobayashi, Bull. Chem. Soc. Jpn., 54, 1275 (1981).

141. D. Armesto, W. M. Horspool, F. Langa, and Y. Peñas, Abstr. XII IUPAC Symp. on Photochem., 678 (1988).

3

The Photochemistry of Carbenium Ions and Related Species

RONALD F. CHILDS

McMaster University
Hamilton, Ontario, Canada

GARY B. SHAW

University of Alberta
Edmonton, Alberta, Canada

I   INTRODUCTION

Remarkable strides have been made in recent years in the study of carbenium ions.[1,2] While a variety of reasons lie behind this expansion of knowledge, it has been particularly helped by the development of non-nucleophilic super acid media which has allowed for the widespread preparation and characterization of carbocations as long lived species.[3] Indeed, techniques for the isolation of carbenium ions have developed sufficiently that many of these species are now readily available, some even being sold commercially. Wide ranging studies, including structure determinations by x-ray crystallography[4], have been conducted. The stage has now been reached that carbenium ions can no longer be considered as fleeting intermediates or, even if prepared as stable entities, as chemical curiosities, but rather as one of the standard forms in which carbon compounds exist. Perhaps no better indication of the "arrival" of carbenium ions is the recognition that they occur as natural products. The iminium salts of retinal which form the basis of the visual pigment chromophores are prime examples of "natural" carbenium ions.[5] Other important examples are the anthocyanin pigments which, among other things, are responsible for the color of red wine.[2]

Carbenium ions exhibit a rich chemistry. Many ground state reactions have been studied in great detail including proton or electrophile loss, nucleophile additions, isomerizations, cycloadditions, fragmentations, redox reactions, complex formation, etc.. However, with the exception of the visual pigment chromophore and the associated retinylidene iminium salts, there have been far fewer systematic examinations of the chemistry of their excited states.[6] The present review explores what is currently known about the photoreactions of carbenium ions. As will be shown, some of these reactions have considerable potential in terms of the synthesis of both carbenium ions themselves as well as their neutral derivatives. It will also be clear from this review that there is considerable scope for additional work in this area.

A.   Scope of Review

There are two distinctly different aspects of the photochemistry of carbenium ions. The first of these, which forms the substance of this review, is concerned with the chemistry which occurs as a result of the electronic excitation of a carbenium ion.[6] The second aspect of the photochemistry of carbenium ions that is currently receiving considerable attention is the use of light to generate carbocations. These reactions which either involve the formation and oxidation of radicals[7], the photo-heterolysis of R-X[8], the photo-protonation of C=C bonds[9], or the heterolysis of radical cations[10], represent important methods for the generation of carbenium ions. However, these reactions are not reviewed here as in general the carbenium ions are produced and react in their ground states.

The material presented in this review will be restricted to the photochemistry of carbocations in condensed phases. The range of carbenium ions covered includes not only the parent carbenium ions but also the heteroatom substituted derivatives of carbenium ions. Oxygen and nitrogen containing derivatives of carbenium ions are important chemical entities.[11,12] Depending on the electronegativity of the heteroatom substituent, these substituted cations have varying fractions of the positive charge located on the carbon framework, Scheme 1, and an extra degree of complexity associated with the partial double bond between the heteroatom and the carbenium center. The photochemistry of onium salts where the positive charge is located on a hetero or halogen atom and cannot formally be delocalized onto the carbon framework, is not considered here.[13]

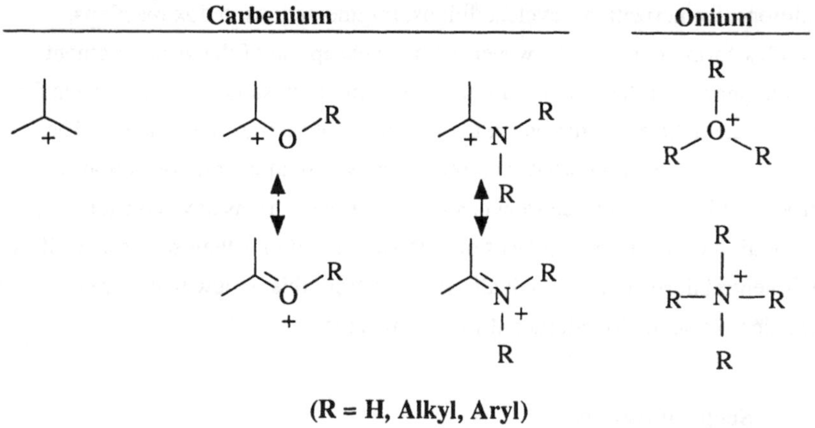

Carbenium			Onium

(R = H, Alkyl, Aryl)

Scheme 1.

B.   Early Studies

Interest in the photoreactions of carbenium ions largely stems from the late 1960's. Prior to this period there are scattered reports of the photochemistry of these species. For example, Dauben and co-workers in 1960 noted that trityl salts isomerized to give low yields of 9-phenylfluorene on standing as solids under normal laboratory illumination.[14] However, it was not until 1968 that van Tamelen and colleagues reported the first systematic examination of the photoreactions of the trityl, tropylium and triphenylcyclopropenium cations.[15]

The early investigations by van Tamelen of the photochemistry of these stable carbocations were conducted in aqueous acid media.[15] As a consequence the photoproducts were not stable as carbocations and neutral products were isolated. This situation was altered later in 1968 by Childs and Winstein who used FSO_3H as a solvent to study the photo-isomerization of substituted benzenium ions to the corresponding bicyclo[3.1.0]hexenyl cations.[16] This work represented the first example where both the starting materials and photoproducts were carbocations. The techniques developed by Childs and Winstein were rapidly applied to cations such as the tropylium ion[17,18] and hydroxy-substituted cations derived by the protonation of carbonyl compounds.[17,19,20]

Studies on the photochemistry of nitrogen substituted carbenium ions also gathered momentum in the late 1960's with the reports of the occurrence of an isomerization about 11-cis double bond of retinal during irradiation of the rhodopsin.[21]

C.   Media Used for the Study of Carbenium Ion Photochemistry

Carbenium ions are ionic materials and generally relatively polar solvents are required in order to obtain solutions of these materials. Beyond this solubility limitation there are three factors which determine the choice of a solvent for the study of a photoreaction of a carbenium ion. First, the inherent reactivity of carbenium ions towards nucleophiles means that in general non-nucleophilic solvents must be used. Second, the general tendency for carbenium ions to undergo facile thermal isomerizations means that the temperature at which reactions are carried out may have to be low in order to retard these rearrangements and consequently the solvent should have a low melting point. Third, the solvent must meet the normal requirements of any solvent used in a photoreaction, particularly being transparent in the region where the substrate absorbs light.

The propensity for carbenium ions to react with nucleophiles or lose a proton varies depending on the stability of the cation. At one end of the scale are the very stable cations such as the aromatic tropylium or cyclopropenium cations, the nitrogen substituted carbenium ions or iminium salts, and extensively delocalized systems such as the trityl cations.[22] In all of these cases there is a wide choice of solvents available ranging from the polar protic to polar aprotic media. In aqueous environments it is frequently necessary to use acidic solutions in order that the cation, and not an alkene or alcohol in equilibrium with the cation, is the major species present in solution.

The stability of any potential products is an important factor to be borne in mind in selecting a medium for a photochemical experiment even with one of the more stable carbocations. Frequently the products of the photoisomerization of stable carbocations are more reactive than the starting materials and they can react with the medium used for the reaction. The photoisomerization of the tropylium cation provides a typical example of this effect. While the tropylium cation is stable in aqueous acid solutions and can undergo photoisomerization on irradiation in this medium, the products isolated are alcohol and ether derivatives of the bicyclo[3.2.0]heptadienyl cation.[15] On the other hand, irradiation of the same cation in FSO_3H leads to the formation of the 7-norbornadienyl cation.[17,18]

The choice of solvent becomes more restricted with the less stable cations. Non-nucleophilic, and frequently very strongly acidic solvents are required in order for the starting cations to be stable. Sulfuric and fluorosulfuric acids are the most frequently used acid media. Both of these solvents are readily available and, when pure, transparent in the ultraviolet region down to ca. 200nm. Sulfuric acid suffers from the disadvantages of having a higher viscosity, a much higher melting point, and a lower acidity than fluorosulfuric acid.[3] For this reason FSO_3H is often the solvent of choice for the study of the photoreactions of carbenium ions. The acidity of FSO_3H can be enhanced by the addition of SbF_5, however, the resulting acid media are much more viscous and this can affect the quantum efficiencies of the photoreactions of carbenium ions.[23]

Spectroscopic studies of carbenium ions dispersed in a variety of different types of glasses have been reported. Where an acidic glass is required these have been obtained by cooling concentrated sulfuric acid or mixtures of this acid with acetic acid or ethanol down to dry ice/acetone or liquid nitrogen temperatures.[24,25] Alternatively, mixtures of alkylsulphonic acids have been used.[26]

One factor which reduces the complexity of the choice of a suitable solvent is that in general carbenium ions are more stable with respect to the medium in their excited states and less susceptible to either proton loss or collapse with a nucleophile. For example, Feldman and Thame in a study of the equilibrium between the dibenzotropylium cation and the corresponding alcohol in the ground and singlet excited states estimated a pK_R+ of 22.7 in the singlet excited state as compared to -3.7 in the ground state.[27] Similarly it has been shown that the pK_a's of both the singlet and triplet states of a variety of protonated ketones are more positive by 5 to 6 H_o units than those of the corresponding ground states.[25,28] The net result of these changes in pK_a is that it is unusual for a carbocation which is stable in a particular medium in the ground state to react with this medium by either proton loss or nucleophilic collapse when it is in one of its electronic excited states. Indeed, it is the enhanced basicity of alkenes and other unsaturated compounds in their excited states which allows the for the photochemical generation of carbenium ions under mildly acidic conditions.[9]

D.   Electronic Spectra and Excited States of Carbenium Ions

The electronic absorption spectra of a large variety of carbenium ions have been reported.[29] In general the absorption and emission spectra of carbenium ions are substantially red shifted as compared to their neutral counterparts. For example, the allyl cations with 2π electrons exhibit absorption maximum at ca. 300

nm^{30} as compared to ca. 190 nm of the isoelectronic ethylenes.[31]

Red shifts are also observed on the formation of heteroatom substituted carbenium ions. This effect has been extensively studied with nitrogen substituted systems where the red shift associated with the protonation of imine derivatives of retinal is of considerable importance in terms of the visual pigment chromophore. Simple visual pigment model systems have been found to be blue shifted by about 50 nm from the absorption maximum of rhodopsin. Nakanishi and co-workers[32] have suggested that a suitably placed anion in the protein is responsible for inducing a further bathochromic shift in the natural chromophore.

The modifications to the absorption spectra of carbonyl compounds on oxygen protonation are of considerable importance to understanding the differences in the photochemistry of the neutral and protonated materials. The n,π^* band of a neutral carbonyl compound is substantially blue shifted on protonation as a result of the n-electrons being bonded to the OH proton.[25,24,33] On the other hand the π,π^* band is red shifted. As a result the lowest energy excited states of protonated carbonyl compounds are of π,π^* character. The probability of intersystem crossing from the singlet to triplet manifold is reduced in a protonated carbonyl compound as compared to the neutral molecule as a result of the effective removal of the n,π^* states from consideration in the former species.[24,25] Consequently the photochemistry of protonated carbonyl compounds frequently occurs from the singlet π,π^* excited state. Photoreactions characteristic of n,π^* states, such as hydrogen atom abstraction, are eliminated with the protonated species.[34]

Mention should be made of the effect of the counterion and solvent on the absorption spectra of carbenium ions. The role and positioning of the counterion with respect to the chromophore of a carbocation has been investigated in detail with the iminium salts of retinal and related derivatives.[35] In terms of carbenium ions lacking polar substituents such as -OH or -NHR which can form strong hydrogen bonds with the solvent, there appears to be little difference in the absorption spectra of the ions in the gas phase or polar solvents such as FSO_3H.[36] In less polar solvents or with cations in which there can be a specific interaction with the solvent[37], then the spectrum in solution is generally blue shifted from that in the gas phase. This is typically the case, for example, with protonated carbonyl compounds and likely reflects the specific hydrogen bonding of an anion to the OH proton which occurs in condensed phases.[36,4b] With iminium salts in general, it has been found that the absorption maximum increases linearly with the reciprocal of the anion radius in non-polar solvents.[35] This arises from the formation of a hydrogen-bonded ion pair between the cation and anion. In polar protic solvents such as methanol this effect is removed and the absorption maximum is constant through a series of counterions. Secondary interactions with the chromophore and an anion can also affect the absorption wavelength depending on the placement of the anion.[32]

In some instances charge transfer bands can be observed between a carbocation and a suitable anion or other species in solution.[38] In these cases the position of the charge transfer band is dependent on the polarity of the solvent used. The solvent dependence of the charge transfer band between a pyridinium cation and iodide forms the basis of Kosower's Z scale of solvent polarity.[39]

The general techniques used in the sensitization and quenching of triplet state reactions of organic molecules[40] have not been employed to any significant extent with carbenium ions. The typical range of sensitizers and quenchers used with conventional media and neutral molecules are usually not amenable for work in the strong acid solvents most often used with carbocations. In many instances sensitizer or quencher molecules will be protonated in these media and as a result there will be considerable modification of their chromophores and electronic states. The likelihood of triplet energy transfer between two cationic species becomes an issue, although in this latter regard Filipescue and co-workers have shown that triplet energy transfer takes place from protonated coumarin to europium(III) in strong acid solutions.[26]

II PHOTOREACTIONS OF METHYLIUM IONS

There are no reported photoreactions of alkyl substituted carbenium ions in condensed phases.[41] This is not surprising in that simple alkyl cations have uv spectra with absorption maxima less than 200 nm.[29] In contrast the photochemistry of aryl substituted methylium cations has been investigated in some detail.

A.   Aryl Substituted Methylium Cations

1.   <u>Benzyl cation.</u> Andrews and Keelan have reported that irradiation
of the benzyl cation in an argon matrix leads to the formation of the tropylium
cation, Scheme 2.[42] Little is known about the steps involved in this transformation
although spectroscopic evidence indicates that an intermediate absorbing at 325
nm might be involved. The close relationship between the benzyl and tropylium
cations in the gas phase is well established.[43]

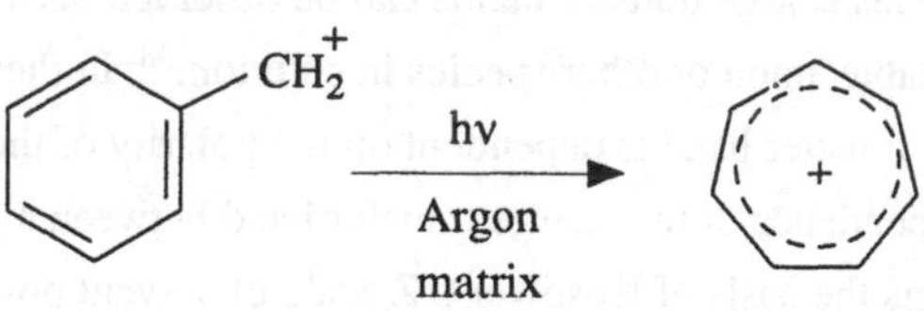

Scheme 2 .

        Cabell and Hogeveen in a preliminary report have indicated that the
pentamethylbenzyl cation is converted to hexamethylbicyclo[3.1.0]hexenyl cation
on irradiation in FSO_3H at -80°C.[44] No further details are available on this
reaction.

        2.   <u>Diarylmethyl cations.</u> de Mayo and colleagues have recently
studied the photochemistry of 1,1-dianisylethene in trifluoroacetic acid.[45] Under
the conditions used the cation (1), in equilibrium with the alkene, was the light
absorbing species. These workers showed that the key step involved in this
photoreaction is a single electron transfer (SET) from the alkene present in the acid
medium to an electronically excited cation. The radical cation/radical pair so
produced leads to the products observed in the photoreaction, Scheme 3. Single
electron transfer to an excited carbenium ion is a general photoreaction of
carbenium ions when a suitable electron donor is present in the reaction medium.

        3.   <u>Triphenylmethyl (trityl) cations.</u> The photoreactions of the
triphenylmethyl cation have been studied in a variety of media. A wide range of
different products and processes have been found depending on the solvent and
other substrates used in the reaction. The principle photoreactions observed are-

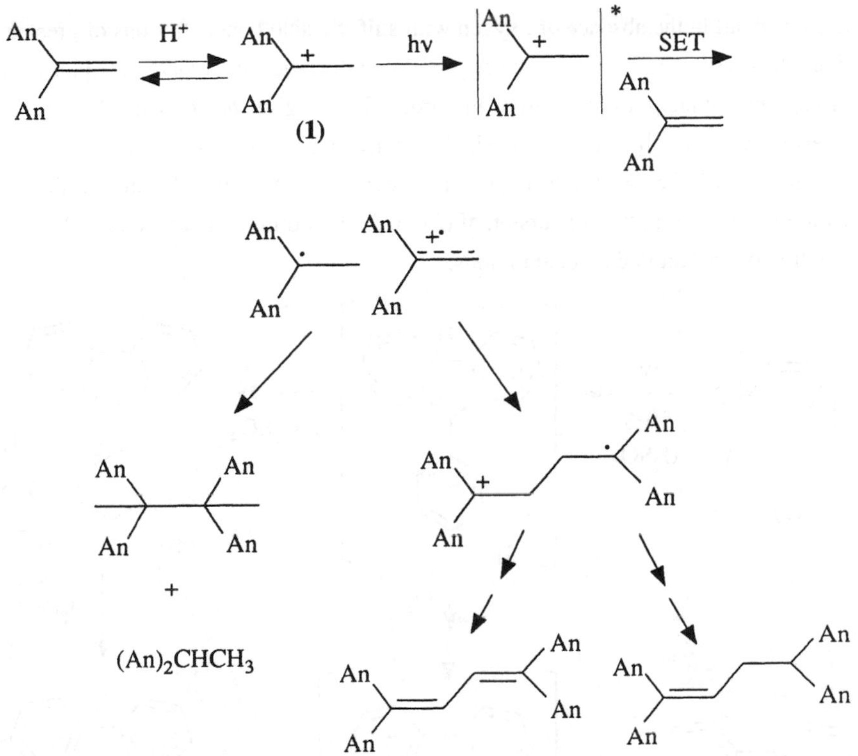

Scheme 3. An = [structure: para-substituted phenyl with O]

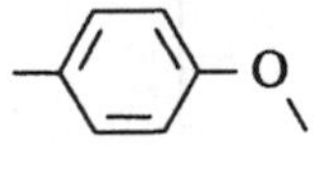

◊   Photoisomerization/oxidation reactions.

◊   Dimerization and related coupling reactions

◊   Electron transfer processes

    The initial work on the photochemistry of the triphenylmethyl cation was reported by van Tamelen and co workers in 1968.[15] Among other reactions, these workers showed that irradiation of solutions of the triphenylmethyl cation (2) in concentrated H_2SO_4 solutions led on work up to the formation of 9-phenylfluorenol, (3). Subsequent studies by these workers[46] and also Allen and Owen[47] showed that (4), the precursor of (3), is the sole product when the reaction is carried out in the absence of oxygen with sulfuric acid concentrations of greater than 88%. Allen and Owen have suggested that the reaction involves a triplet state of (2) which undergoes cyclization and proton loss to give the short-lived intermediate (5). This intermediate is then converted to (4) by reaction with the solvent. As will be evident throughout this review, the 5-C ring closure reaction which is involved in the conversion of (2) to (5) is a common photochemical reaction of unsaturated carbenium ions.

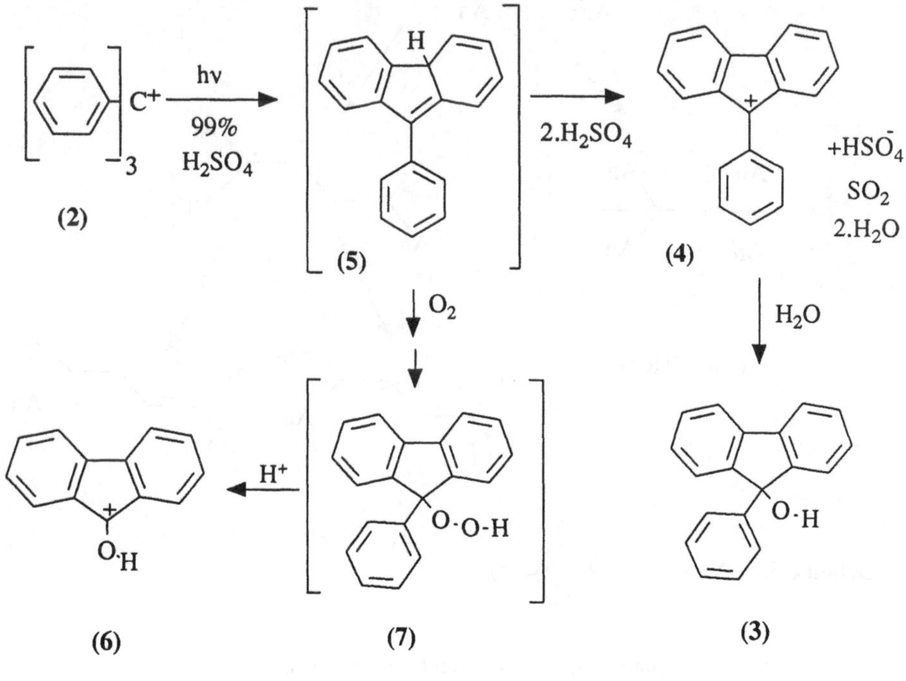

Scheme 4.

The introduction of oxygen into the reaction medium was found to retard the disappearance of (2)[47] and also leads to the formation of protonated fluorenone (6).[46,47] The effect of oxygen on the overall efficiency of the reaction would appear to be due in part to quenching of the triplet state of (2) as well as a further process giving (6) which occurs at higher oxygen concentrations. The formation of (6), which was shown to occur at the expense of (4), possibly occurs via an acid catalyzed rearrangement of the hydroperoxide (7), Scheme 4. Allen and Owen have examined the regio-selectivities of comparable photoisomerizations of substituted triphenylmethyl cations examined.[47]

Additional products are formed when (2) is irradiated in more dilute acid solutions. The nature of the products depends on the concentration of (2) used in the reaction. For example, Allen and Owen showed that irradiation of (2) at low concentrations (10^{-5}M) in 78% H_2SO_4 led to the formation of (8) as the major product.[47] The pathway leading to (8) is not known with any certainty although it has been suggested that attack by water on an excited state of (2) is involved. Such a process does not seem likely as re-ionization of any alcohol intermediate so derived would readily occur in the acidic media used for these reactions.

As the concentration of (2) is increased coupled products are formed. For example, van Tamelen and Cole showed that (9)-(11) were the major products formed on irradiation of a 10^{-3}M solution of (2) in 76% H_2SO_4.[46,48] These workers suggested that the reaction involves a hydrogen atom transfer from a ground state trityl cation to triplet trityl cation with the formation of a pair of radical cations.[46,48,49] Coupling of these radical cations would lead to a dication which could subsequently undergo a series of dark reactions involving intermolecular hydride shifts, etc. to give (9) - (11). Alternatively, Bethel and Clare, as a result of their studies on the use of trityl, xanthydryl or dibenzotropylium cations to catalyze the photo-oxidation of triphenylmethanes, have suggested that a more likely process involves a single electron transfer to (2) in its triplet state with formation of the triphenylmethyl radical.[50] Products (9)-(11) can then be derived from this radical using conventional chemistry.

(8)   (9) X=Y=H (10) X=H;Y=OH (11) X=Y=OH

In more dilute acid media, especially those containing organic co-solvents such as acetic acid, toluene, etc., a complex range of products is observed.[46,48,50] Once more most of these products would seem to derive from the trityl radical formed as a result of single electron transfer. For example, the products shown in Scheme 5 are obtained in good yield when (2) is irradiated in the presence of oxygen . It has been suggested that (12) is a secondary product derived from (13) on work up of the reaction. Compounds (13) and (14) would appear to derive from (15), the product of the addition of oxygen to the triplet state of (2).

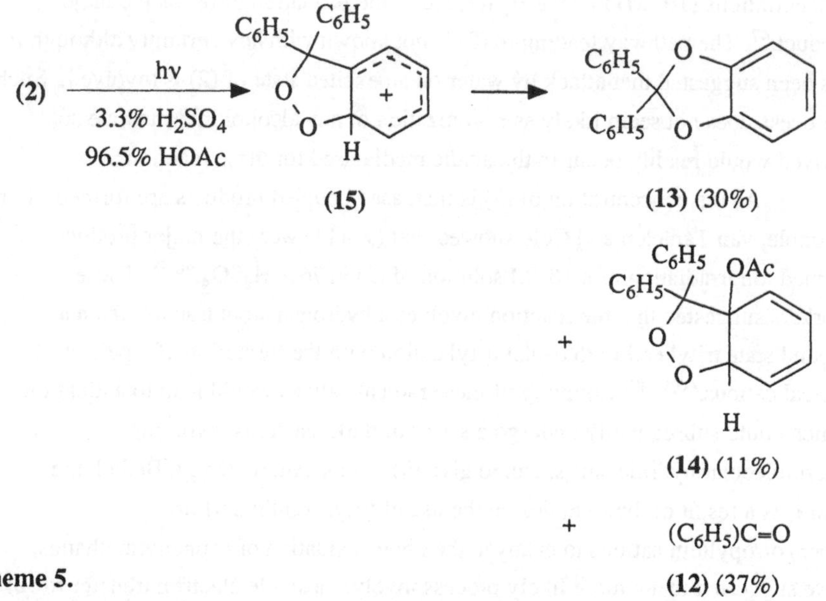

Scheme 5.

Single electron transfer processes to electronically excited trityl cations form the basis of the use of these cations as photocatalysts for the retro-cycloaddition reactions of cyclobutane rings[51], the initiation of polymerization reactions[52], and the catalysis of the addition of oxygen to dienes.[53]

B.   Oxygen Substituted Methylium Ions

A limited number of reports of photochemically induced reactions of oxygen substituted carbenium ions have been described. Asensio and Miranda and colleagues have recently reported that the irradiation of the protonated lactone (16H) in concentrated sulfuric acid leads to the formation of the dehydohalogenated cation (17H).[54] This photochemical dehydrohalogenation

stands in marked contrast to the behavior of (16) in methanol where the
hydroxy-ester (18) is obtained on irradiation. It was suggested by the authors that
the dehydrohalogenation of (16H) in sulfuric acid results from the "protection" of
the ester grouping to its normal photochemical reactions by oxygen protonation, so
preventing the ring opening reaction observed with (16). The authors propose that
irradiation of (16H) leads to a photoinduced homolytic cleavage of the C-Br bond
followed by single electron transfer[54,55] to give a transient dication which then
undergoes proton loss to give (17H).

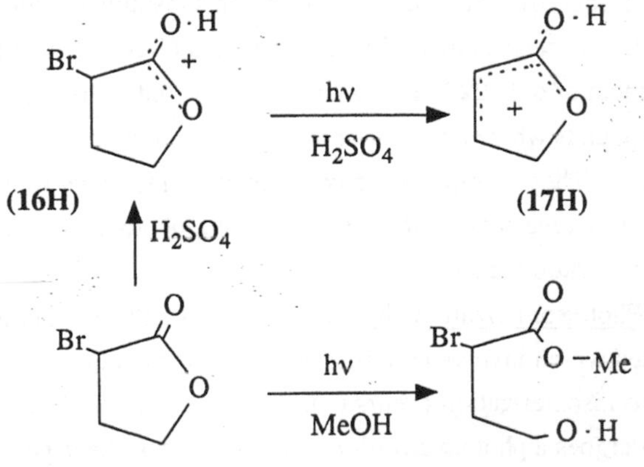

Scheme 6.

This photodehydrohalogenation reaction finds a parallel in the
photo-Friedel Crafts reaction observed by Ogata and colleagues.[56] Treatment of
benzene with ethyl chloroacetate and AlCl$_3$ at elevated temperatures in the dark
leads primarily to the formation of ethyl and various polyethyl benzenes. On the
other hand, irradiation of the same solution at room temperature yields ethyl
phenylacetate, Scheme 7. Coordination of the AlCl$_3$ to the carbonyl group of the
ester would seem to protect this group to photoreactions and instead
photochemically induced homolysis of the C-Cl bond followed by single electron
transfer to yield the $^+$CH$_2$CO$_2$Et/AlCl$_3$ cation/complex would appear to occur.

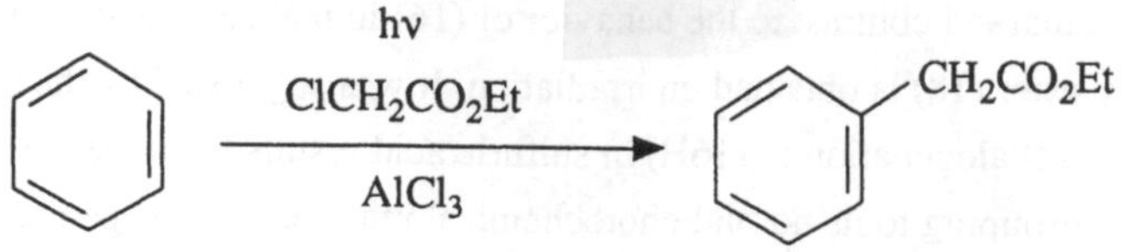

Scheme 7.

C.   Nitrogen Substituted Methylium Ions

The photochemistry of a variety of nitrogen substituted methylium cations or of iminium salts has been studied. Since an isolated $C=N^+$ fragment has an absorption maximum of only 220 nm[57] photochemical studies have generally been restricted to compounds where a further chromophoric group is conjugated to the $C=N^+$ functionality. These compounds have been shown to undergo three general classes of reactions; isomerization about the CN bond, photoaddition to the CN bond, and photocyclization reactions.

1.   <u>Photoisomerization about the CN bond</u>. Several photoreactions have been reported which involve an initial isomerization about the CN bond. Typically, *trans/cis* isomerization occurs to produce a *cis* iminium salt which subsequently undergoes a photocyclization. In this manner, the 9-phenanthridine (21) is formed in 38% yield, on irradiation of the *trans* isomer (20) in acid solution, Scheme 8.[58] A similar reaction sequence has been found for nitrogen substituted methylium cations where one of the phenyl rings has been substituted with a pyridine moiety.[59] Thus irradiation of trans-N-3-pyridyl-benzylidene iminium sulfate (22) leads to the formation of the photocyclized product (24). These photochemical transformations parallel those observed with stilbenes.[60]

2.   <u>Photoaddition</u>. A general type of photoreaction that simple iminium salts undergo is photoaddition via an electron transfer pathway. Since several reviews[61] on this subject have appeared recently a detailed discussion will not be presented here. Rather, the general concepts of photoaddition will be shown as well as several illustrative examples. As shown in Scheme 9, excitation of an iminium salt causes a π,π^* transition to give a singlet excited state. In the presence of an easily oxidized donor such as an alcohol, single electron transfer to

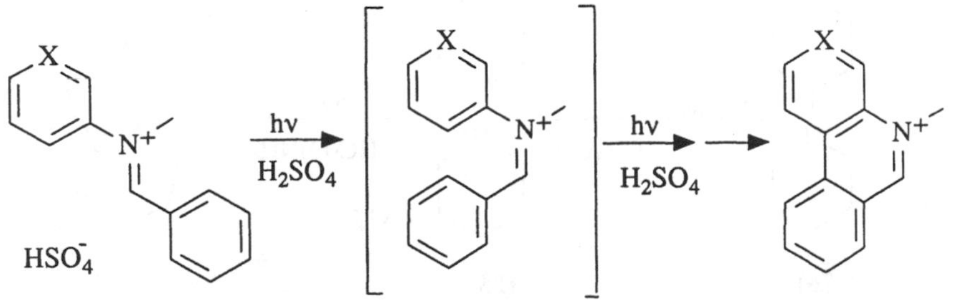

(20) X=CH   **(19) X=CH**   **(21) X=CH**
(22) X=N   **(23) X=N**   **(24) X=N**

Scheme 8.

the iminium salt occurs. This is followed by proton loss from the radical cation
derived from the alcohol and radical coupling to form an addition product.

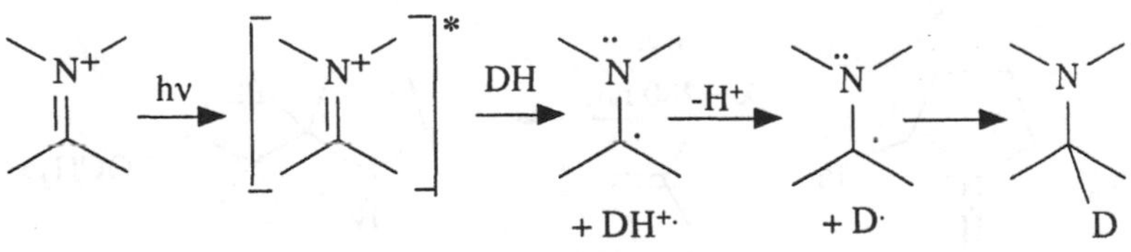

Scheme 9.

An example of this type of photoaddition reaction is provided by the
photosensitized addition of methanol to cyclohexylidene pyrrolidinium perchlorate
(25) which occurs in 74.5% yield to form the 1-hydroxymethyl adduct (26),
Scheme 10.[62] Interestingly, the photoaddition of formamide to (25) also occurs.[62]
However the initial product resulting from the addition of the formamyl radical to
C_1 of the cyclohexyl moiety undergoes a subsequent thermal ring closure, Scheme
10.

Stavinoha and Mariano have examined the addition of olefins to
2-phenyl-1-pyrrolinium perchlorate.[63] For example, irradiation of (27) in
methanol in the presence of isobutylene yields (28) in 81% yield, Scheme 11. This

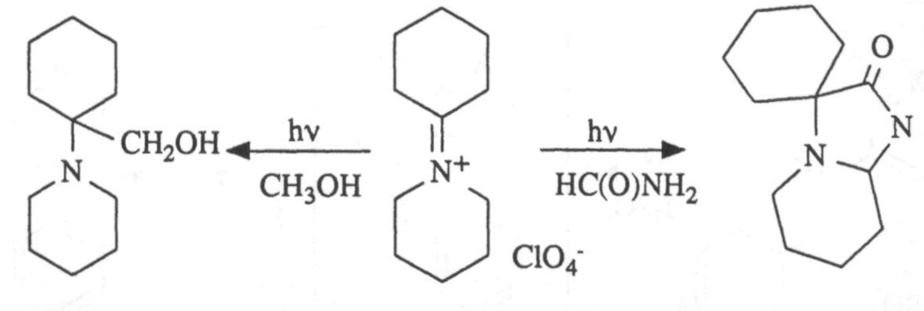

(26)   **(25)**

Scheme 10.

photoaddition has been rationalized on the basis of an electron transfer mechanism, Scheme 12. The regiochemistry of the product involves an anti-Markovnikov addition to the olefin with nucleophilic attack by the methanol occurring at the least substituted carbon of the intermediary isobutyl radical cation to give the more stable *t*-butyl radical. This then undergoes carbon-carbon bond formation with the 2-pyrrolinyl radical.

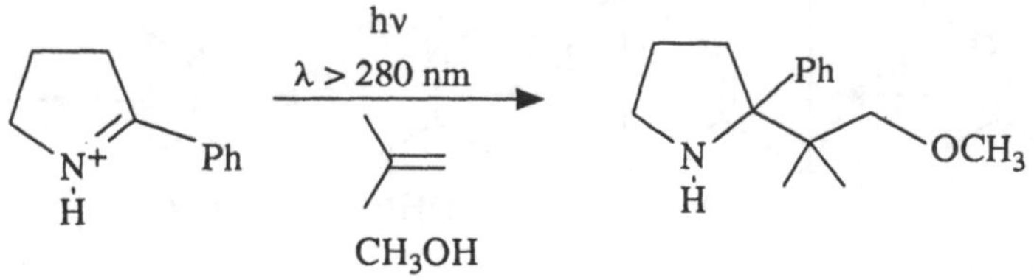

(27)   **(28)**

Scheme 11.

It should be noted that this photoaddition pathway is specific for electron-rich olefins.[63] With electron-deficient olefins such as acrylonitrile the unusual tricylic compound (29) is formed on irradiation of (27), Scheme 13.[63] Mariano and coworkers have suggested that the formation of (29) involves a π2+π2 photocycloaddition in the singlet state, followed by re-aromatization of the phenyl ring.

     3.   Photocyclization. Intramolecular photocyclization reactions of iminium salts provide a useful synthetic approach to complex heterocyclic

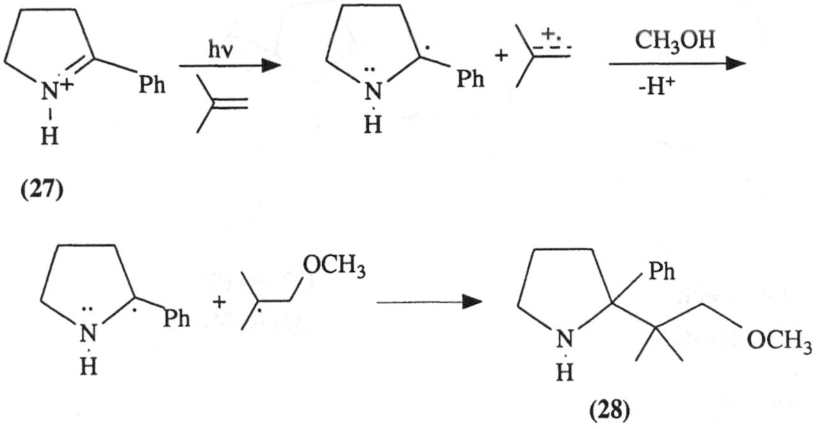

(27)

(28)

Scheme 12.

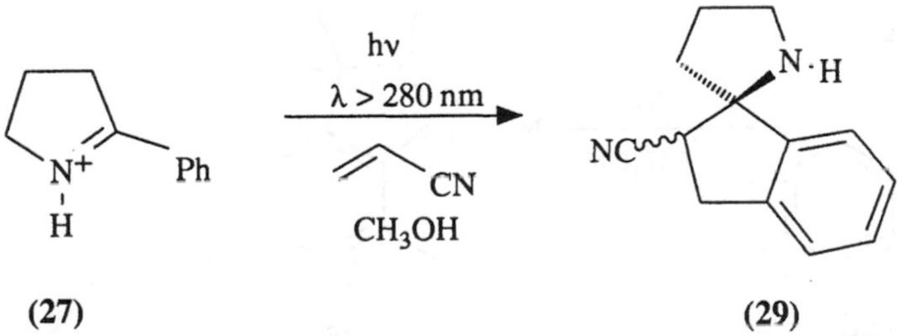

hv

λ > 280 nm

(27)   **(29)**

Scheme 13.

compounds. It has been found that a trimethylsilyl group has a marked influence on selectivity during the photocyclization process. Thus irradiation of the trimethylsilyl pyrrolinium salts (30) or (31) in oxygen-free acetonitrile produces the benzoindolizidine compounds (32) and (33) respectively, Scheme 14.[64] In the absence of the silyl group, formation of the benzoindolizidine is diminished and photofragmentation products or different cyclization products are noted. For example the iminium salt (34) undergoes a photocyclization reaction to yield the benzopyrrolizidine (35) in 55% yield, Scheme 15.

The mechanisms of these reactions have been probed using deuterium labelled compounds.[64,65,66] It has been found that the photocyclizations which

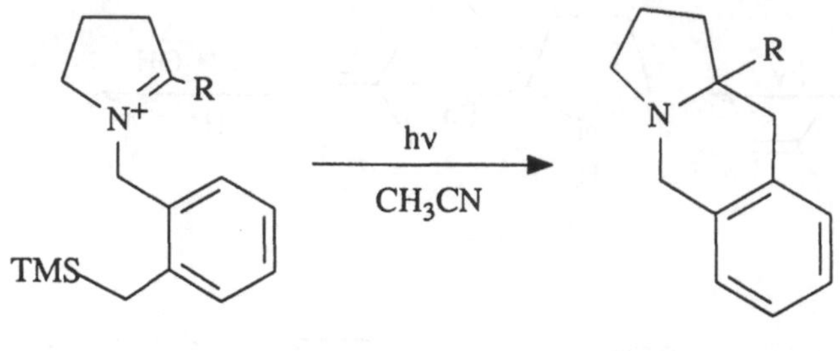

(30) R=Ph
(31) R=Me

(32) R=Ph
(33) R=Me

Scheme 14.

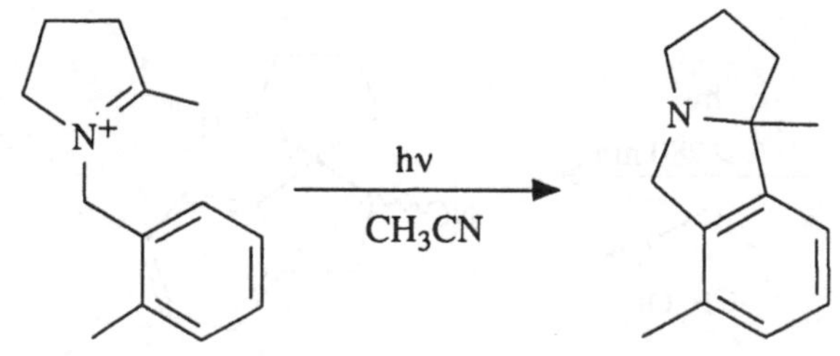

(34)   (35)

Scheme 15.

involve electron transfer proceed via either a diradical cation or a diradical coupling process. In the photocyclization of (31), excitation leads to a π,π* singlet state which undergoes electron transfer from the phenyl ring to the C=N⁺ bond, Scheme 16. This is followed by loss of the trimethyl silyl cation to yield a diradical. Radical coupling yields product (33). The photocyclization of (34), proceeds via radical coupling of the diradical cation and subsequent proton loss to yield the benzopyrrolizidine (35).

These types of novel photocyclizations have been used by Mariano and co-workers in for the synthesis of several alkaloids.[67-69] For example, the protoberberine (±)-xylopinine (36) has been synthesized using a photocyclization to close the C-ring in the final synthetic step, Scheme 17.[68]

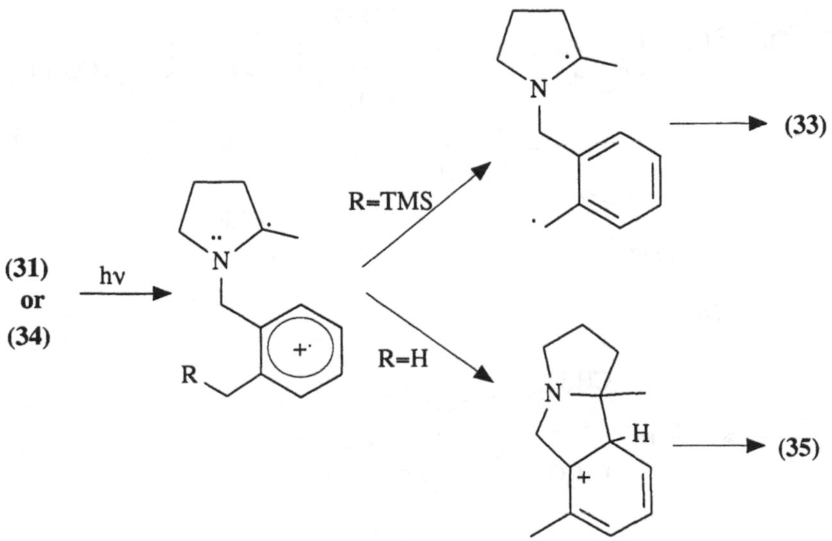

Scheme 16.

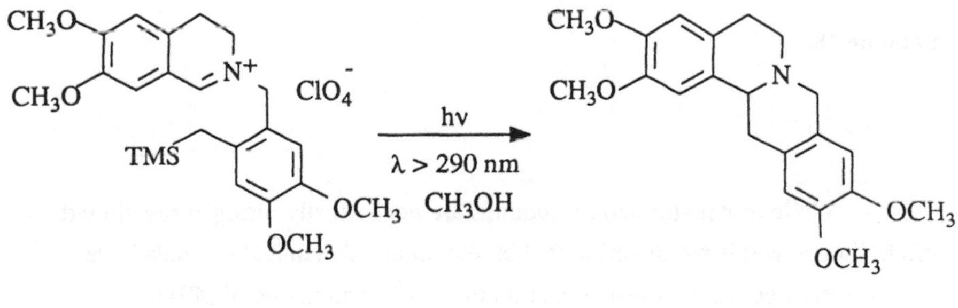

(36)

Scheme 17.

A similar strategy has been used for the synthesis of
15,16-dimethoxy-cis-erythrinan (37).[69] The key step is the cyclization of the
dihydroisoquinoline compound (38) to form a spirocyclic center between the B and
C rings. This was accomplished through irradiation of (38) in methanol to yield
(39) in 60% yield, Scheme 18. Several other synthetic manipulations subsequent
to this reaction are required in order to form the desired erythrinan.

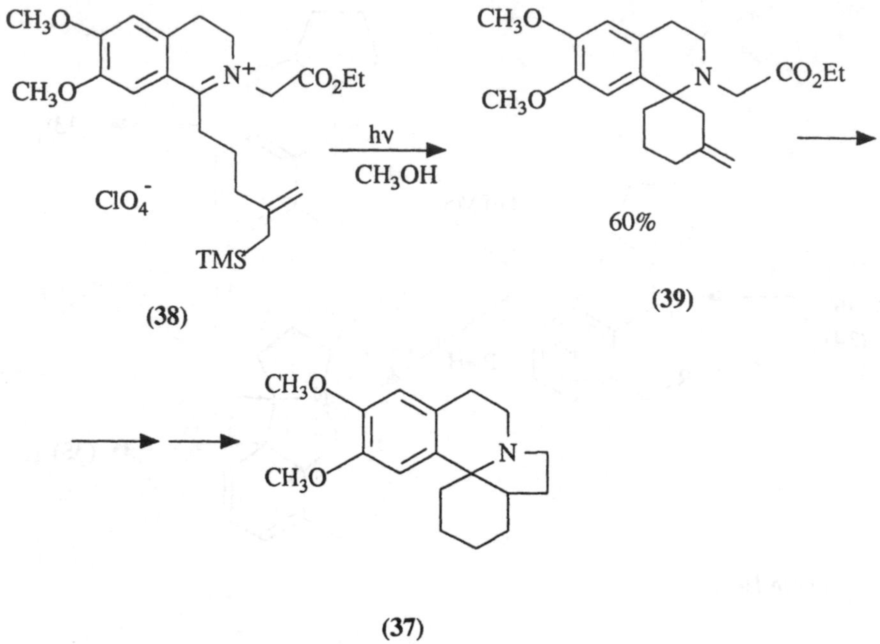

(38)   (39)

(37)

Scheme 18.

While protonated azo compounds are not formally nitrogen substituted methylium ions it is worth noting in this section that Adam and Miranda have recently reported the photoisomerization of (40).[70] Irradiation of (40) in concentrated sulfuric acid leads to the formation of the protonated pyridazine (41H).[70] The photochemical formation of this reaction product from what can be considered formally to be an α-amino-nitrenium ion, is in marked contrast to the products obtained either on irradiation of (40) in conventional solvents or from the thermally induced isomerization of (40) in H_2SO_4, Scheme 19. Details of the steps involved in the transformation of (40H) to (41H) are not known at this time. Corresponding bicyclic azo compounds lacking the gem-dimethyl group at C_7 appear to be photochemically stable in concentrated H_2SO_4. It is also interesting that yet a further product (42) can be obtained from (40) on irradiation in the presence of triphenylpyrylium tetrafluoroborate (TPT).[70] Under these conditions a single electron transfer occurs from (40) to the electronically excited pyrylium cation.[71]

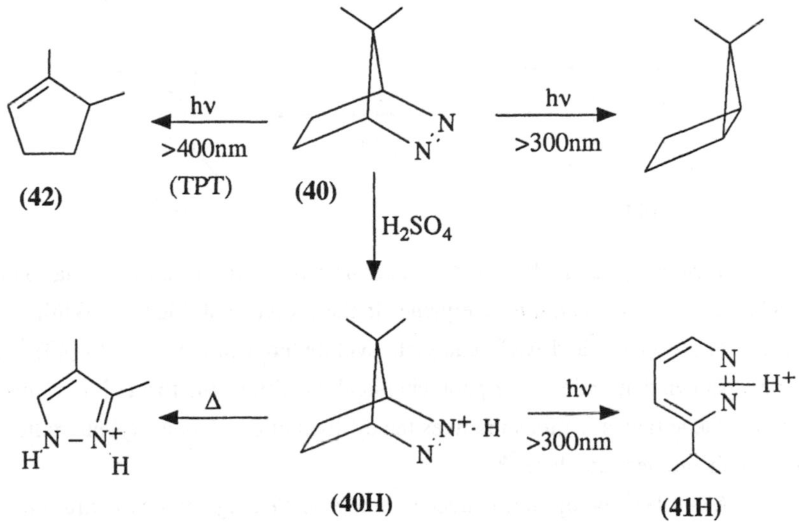

(42)   (TPT)   **(40)**

     (40H)   **(41H)**

Scheme 19.

III PHOTOREACTIONS OF ALLYL CATIONS

A.   Alkyl/Aryl Substituted Allyl Cations

No studies of the photochemistry of acyclic alkyl or aryl substituted allyl cations have been reported. However, two different types of photoreaction of cyclic, alkyl-substituted allylic cations have been described. The first reaction involves a photocleavage reaction and the formation of a radical cation. The second type of reaction involves a photoinduced electrocyclic ring opening of cyclopentenyl cations.

1. Photochemical formation of radical cations. Hogeveen and colleagues have irradiated solutions of the cyclobutenyl cation (43) at -80°C in an esr cavity and observed the formation of the cyclobutadienyl radical cation (44).[72] The

radical cation (44), produced by the homolytic cleavage of the $AlCl_3$ from (43), was identified by comparison of its esr spectrum with that previously reported by Maier.[73] A "dimeric" radical cation was formed from (44) on standing in the dark.

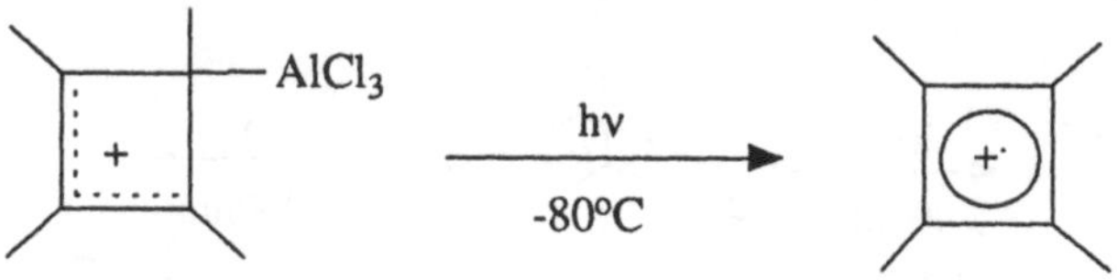

(43)   (44)

Various cyclobutadiene radical cations can be produced by irradiation of CH_2Cl_2 solutions of substituted acetylenes in the presence of $AlCl_3$.[74] While mixtures of acetylenes and $AlCl_3$ can yield cyclobutenyl cations such as (43)[75], it has been shown that under these photochemical conditions the formation of the cyclobutadiene radical cations involves the cycloaddition of an acetylene radical cation with another acetylene.[76]

A similar type of photoinduced homolytic cleavage of a carbenium ion has been described by Davies and co-workers.[77] Irradiation of a solution of hexamethylcyclopentadiene (45) in trifluoroacetic acid (TFA) led to the formation of the radical cation (46). As the diene (45) was photochemically stable in the absence of acid it was suggested that the formation of (46) involves the ground state protonation of (45) followed by a photocleavage of the C-H bond, Scheme 20.

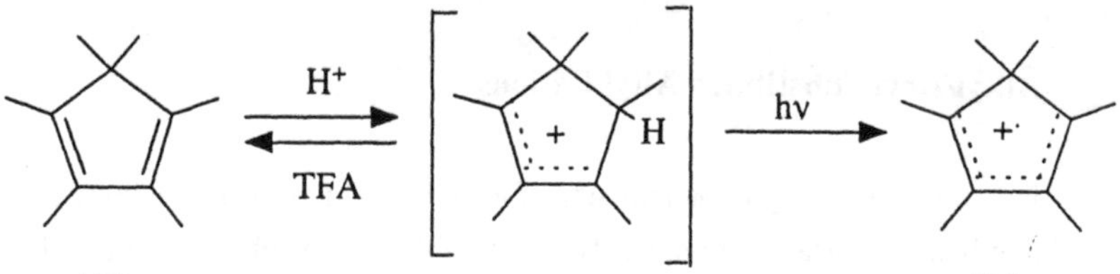

(45)   (46)

Scheme 20.

Photolysis of (47) in TFA produced the radical cation (48), however, in this case the reaction was shown to proceed via a homolytic cleavage of the C-H bond in (47) to form the cyclopentadienyl radical (49) which subsequently was protonated to form (48).[77]

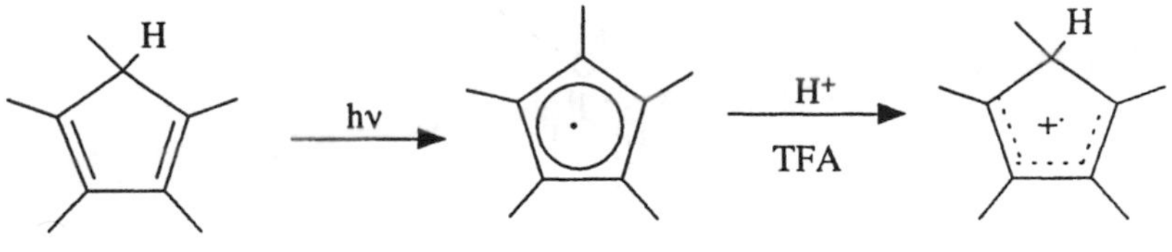

Scheme 21.

    2.   Ring opening reactions of cyclopentenyl cations. The
photochemically induced ring opening of cyclopentenyl cations has been observed
only with the bicyclo[3.1.0]hexenyl systems. Irradiation of bicyclo[3.1.0]hexenyl
cations causes them to undergo an electrocyclic ring opening reaction to form the
corresponding benzenium ions.[16] The reverse photochemical ring closure of a
benzenium ion to form the bicyclic cations also occurs and a photostationary state
is established between the two cations, Scheme 22.

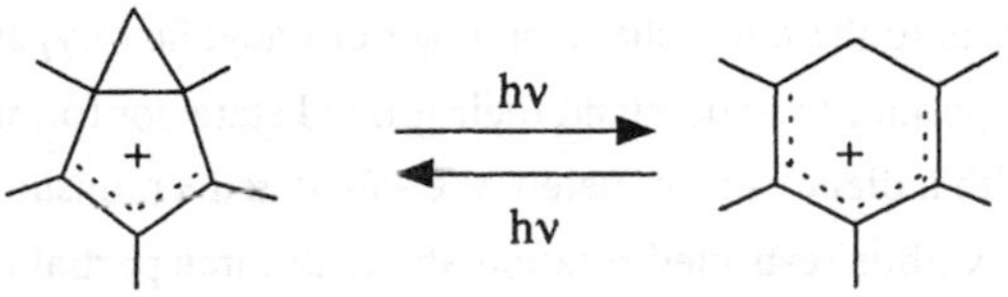

Scheme 22.

    The photochemical ring opening reactions of bicyclohexenyl cations are
of particular importance to the understanding of the photochemistry of protonated
phenols and will be discussed in more detail in a later section.

B.   Oxygen Substituted Allyl Cations

    There has been considerable work reported on the photoreactions of
oxygen substituted allyl cations. These relatively stable carbocations are obtained
by protonation, alkylation or reaction with a Lewis acid of α,β-unsaturated
aldehydes[78-82], acetals[80,83,84] ketones[78-82], and derivatives of carboxylic acids.[78-82]
    Oxygen substituted allyl cations undergo five different types of
photoreactions. Referring to the numbering scheme shown in (50) these are:

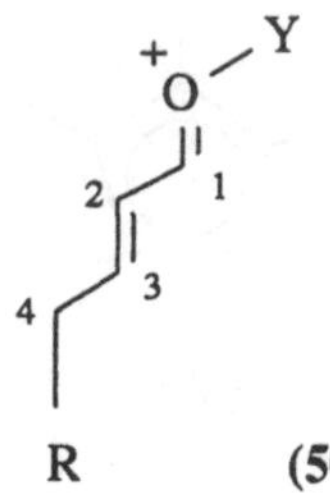

$$+O\!-\!Y$$

R   (50)

◊   cis/trans isomerization about the C_1O bond (when Y=OMe)
◊   cis/trans isomerization about the C_2C_3 bond
◊   migration of atom or group from C_4 to C_3
◊   cleavage of the C_4R bond
◊   cycloadditions to the C_2C_3 bond

Some of these reactions can be catalyzed by Lewis acids and offer considerable synthetic potential.

Before describing the photochemical reactions acyclic oxygen substituted allylic cations it is important to understand their ground state conformational chemistry. Just as with a diene, a protonated or O-alkylated α,β-unsaturated carbonyl compound exhibits restricted rotation about all three partial double bonds of the unsaturated framework of the cation, Scheme 23. The mechanisms and barriers to the ground state isomerizations about all the these bonds have been studied extensively.[79,80,83-86] In general it has been shown that the barriers to isomerization about the various partial double bonds increase in the order C_1C_2 (ca. 7-9 kcal/mol[86]) < C_1O (ca. 15-20 kcal/mol[80,83,84,85]) < C_2C_3 (ca. 20-25 kcal/mol[79,80,83]).

The barriers to the interconversion of the various conformations of methoxy and hydroxyallyl cations depend both on the substituents on the cation and the medium used. By working with FSO_3H solutions of the 1-methoxy cations at temperatures below -60°C, it is possible to sufficiently slow thermal isomerizations about the C_2C_3 and C_1O bonds that their photochemically induced stereomutations may be examined. Exchange reactions of the OH proton with the acid medium make it difficult to study isomerism about the C_1O bond of the

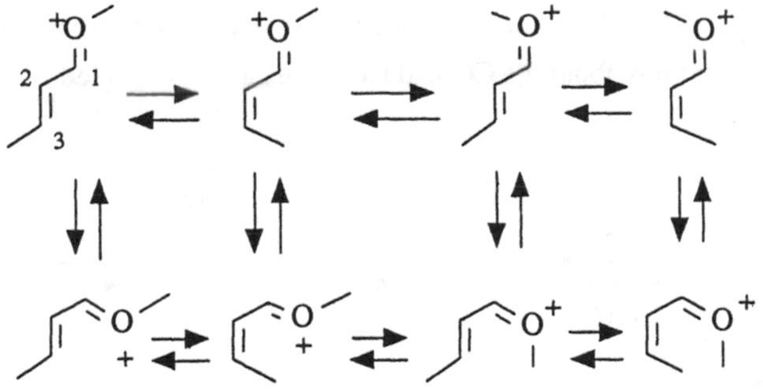

Scheme 23.

protonated species even at low temperatures.

　　　1.　　Photoisomerization about the C_1O bond. Hagar at McMaster showed that a thermodynamic equilibrium mixture of the two ions (51), 96%, and (52), 4%, is obtained on dissolution of the dimethyl acetal of acrolein in FSO_3H.[80] These cations each adopt an *s-trans* conformation about the C_1C_2 bonds. Irradiation at low temperatures of the thermodynamic mixture of (51) and (52) at 254 nm led to *cis/trans* isomerization about the C_1O bonds, Scheme 24. A photostationary state was established consisting of 73% (51) and 27% (52). The quantum efficiency of the conversion of (51) to (52) was shown to be 0.15 at -68°C.[23]

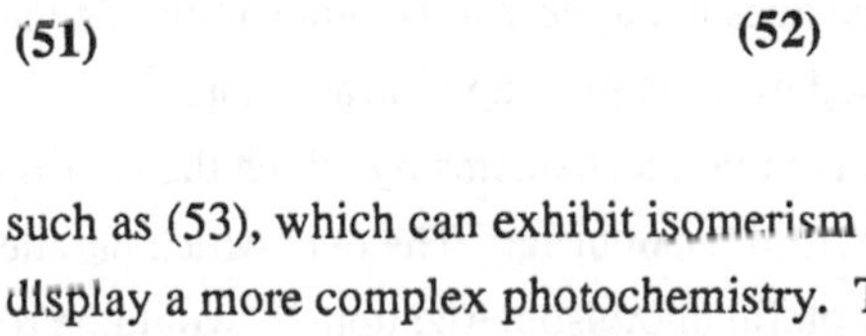

Scheme 24.

　　　Cations such as (53), which can exhibit isomerism about both the C_1O and C_2C_3 bonds, display a more complex photochemistry. Thus the irradiation of a FSO_3H solution of (53) at -70°C with 254 nm light leads to isomerization about both the C_1O and C_2C_3 bonds, Scheme 25.[80] It would appear that the photoisomerizations about the CO and CC bonds are not coupled.

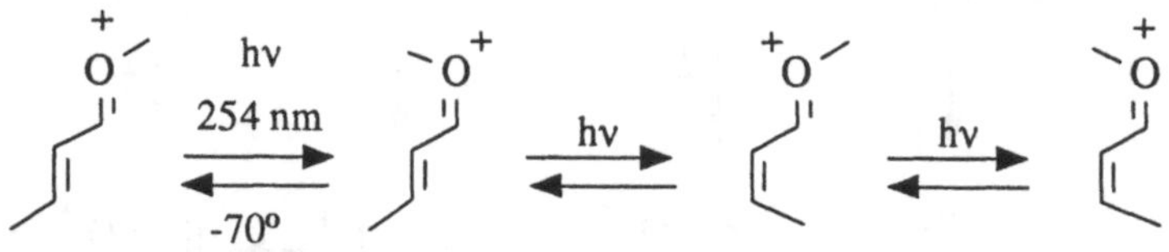

(53)

Scheme 25.

2.   Photoisomerization about the C_2C_3 bond is observed with all the protonated acyclic α,β-unsaturated carbonyl compounds whose photochemistry has been examined. Indeed for most of these cations this is the only reaction which is observed as any photoisomerizations which might be occurring about the C-O bonds are not detectable as a result of rapid exchange of the OH protons with the acid medium.

The photoisomerizations of protonated pent-3-en-2-one provide a typical example of this type of photoreaction. The photochemistry of this cation has been extensively studied at McMaster and the principle results are summarized in Table 1.[79] Protonated *trans*-pentenone (54H) in FSO_3H undergoes an efficient photo-induced isomerization about the C_2C_3 bond to form the corresponding *cis* isomer (55H) on irradiation at either 25° or -68°C. The presence of oxygen in the acid solutions has no observable effect on the efficiencies of these reactions and it was suggested that the reactions involve singlet excited states. The reverse photoisomerization of (55H) to (54H) also occurs and a photostationary state is established between the two isomers (38% (54H); 62% (55H)). The composition of this photostationary state does not appear to be temperature dependent. No other products were detected even on prolonged irradiation.[23]

In principle there are two mechanisms by which the *cis/trans* isomerization of (54H) and (55H) can occur. The one, which parallels that found with a wide range of other olefin photoisomerizations[87], would be rotation about the C_3C_4 bond in the singlet π,π^* excited states of (54H) and (55H) to form a common twisted intermediate, crossing to the ground state and relaxation to either

Table 1. Quantum efficiencies of *cis/trans* isomerization
of protonated pentenones

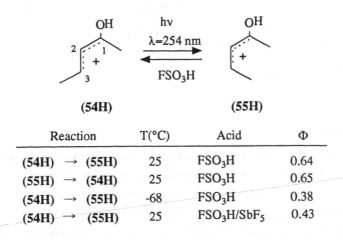

Reaction	T(°C)	Acid	Φ
(54H) → (55H)	25	FSO$_3$H	0.64
(55H) → (54H)	25	FSO$_3$H	0.65
(54H) → (55H)	-68	FSO$_3$H	0.38
(54H) → (55H)	25	FSO$_3$H/SbF$_5$	0.43

isomer. Alternatively, another possibility which could account for isomerization
about both the C$_2$C$_3$ and C$_1$O bonds is an electrocyclic ring closure of (54H) to
form a protonated oxete followed by a reverse thermal ring opening, Scheme 26.
A similar photochemical ring closure to form an oxete has been found to occur
from the π,π^* excited states of some highly substituted enones.[88] The formation of
a protonated oxete would require photoexcitation of a *s-cis* conformation of the
starting protonated pentenones. Mullen and co-workers have shown that the
preferred conformation of (54H) has a *s-trans* configuration about the C$_2$C$_3$ bond
(75%) with only some 25% of (54H) being in the *s-cis* form.[86] The high quantum
efficiencies of the *cis/trans* isomerization of these protonated pentenones would
seem to rule out the possibility that it is a minor conformer which is
photochemically active and as such the protonated oxete pathway would seem
improbable.

The photo-interconversions of (54H) and (55H) parallel the comparable
photoisomerizations of the parent ketones, (54) and (55). However, in the latter
case a further isomerization is observed which leads to the formation of the
deconjugated product (56).[89,90] Cations such as (56H) would not be observable in a

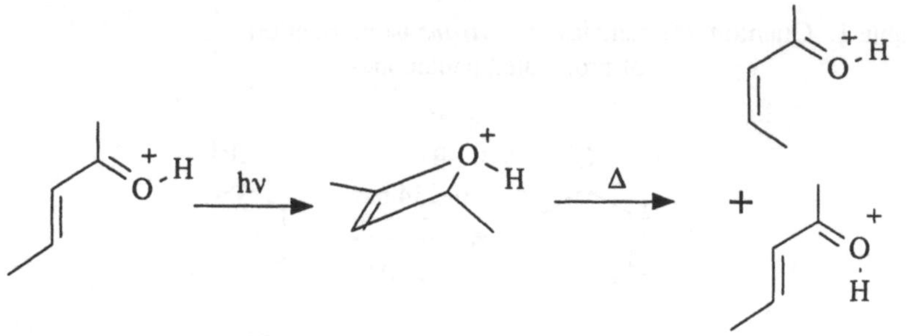

Scheme 26.

strong acid medium as they would rapidly isomerize to form either the conjugated
isomers (54H) and (55H) or undergo ring closure reactions. In order to rule out the
possibility of a γ-hydrogen abstraction reaction occurring in the
photoisomerizations of (54H) the reactions were examined in FSO_3D. No
deuterium incorporation, other than on oxygen, in either (54D) or (55D) was found
on irradiation of these cations ruling out the possibility of the occurrence of a
photo-deconjugation reaction.[79]

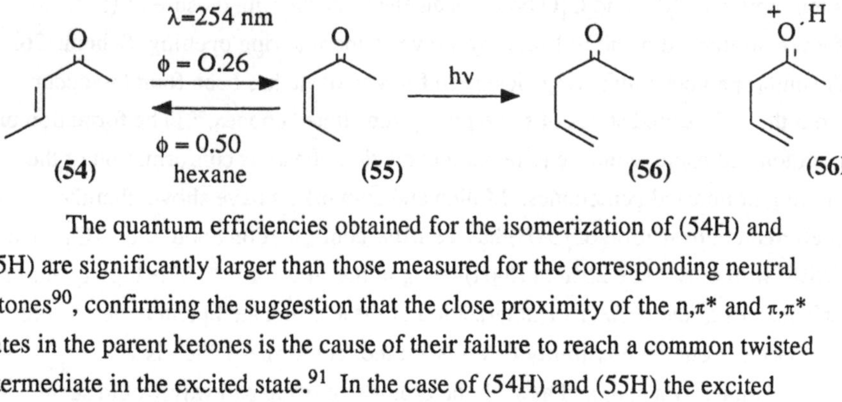

The quantum efficiencies obtained for the isomerization of (54H) and
(55H) are significantly larger than those measured for the corresponding neutral
ketones[90], confirming the suggestion that the close proximity of the n,π* and π,π*
states in the parent ketones is the cause of their failure to reach a common twisted
intermediate in the excited state.[91] In the case of (54H) and (55H) the excited
states involved are of π,π* character. The quantum efficiencies of the
interconversion of (54H) and (55H) in FSO_3H at 25°C sum to a value greater than
unity suggesting that the probability of a common twisted intermediate relaxing to
either (54H) or (55H) differs depending on the origin of this intermediate.[92]

The nature of the acid medium used as solvent (FSO_3H or FSO_3H/SbF_5 at

25°) and the temperature of the reaction (FSO$_3$H at 25° and -68°C) both affect the efficiency of the conversion of (54H) to (55H), Table 1.[23] Both of these effects were attributed to an increase in viscosity of the solvent associated with either adding SbF$_5$ or lowering the temperature. In this context it was noted that no photoisomerizations occurred on irradiation of (54H) in FSO$_3$H/SbF$_5$ at -68°C. At this temperature this acid mixture is very viscous.

Variation of the quantum efficiencies of *cis/trans* photoisomerizations with solvent viscosity has been previously noted for neutral systems.[93,94] The effects observed with (54H) and (55H) appear to be larger than those normally encountered. One possible reason for this large temperature dependence in the cations is a difference in charge distribution in the excited as compared to ground states of the cations and the necessity of re-ordering solvent as they undergo isomerization. Recent work on the crystal structures of protonated carbonyl compounds has shown that there is a consistent and specific pattern to the placement of anions around the protonated carbonyl group.[4b] Comparison of solution and solid state ^{13}C NMR spectra suggests that a similar ordering persists in solution.[4b]

Similar *cis/trans* photoisomerizations to those described for (54H) and (55H) have been reported for a range of other hydroxy-substituted allyl cations, Table 2. In many instances these photoreactions have been used to prepare and study the chemistry of the *cis* isomers of these cations.

3.   Acid catalyzed *cis/trans* photoisomerization. The photochemical *cis/trans* isomerizations of α,β-unsaturated carbonyl compounds can be catalyzed by the presence of Lewis or Bronsted acids in the reaction media. These reactions, which involve as the key step the photoisomerization of oxygen-substituted allyl cations, represent an important new and improved method of generating the *cis*-isomers of a wide variety of unsaturated carbonyl compounds.

Lewis and Oxman have shown that the position of the photostationary state reached on irradiation of dilute solutions of a α,β-unsaturated ester can be significantly modified by the addition of a catalytic amount of a strong Lewis acid to the reaction medium.[95] The data obtained for photoisomerization of ethyl cinnamate in the presence of various Lewis acids provides a typical example of the large changes in photostationary state composition which can be achieved, Table 3.

The changes in composition of the photostationary state reached between

Table 2. *Cis/trans* photoisomerizations of protonated carbonyl compounds

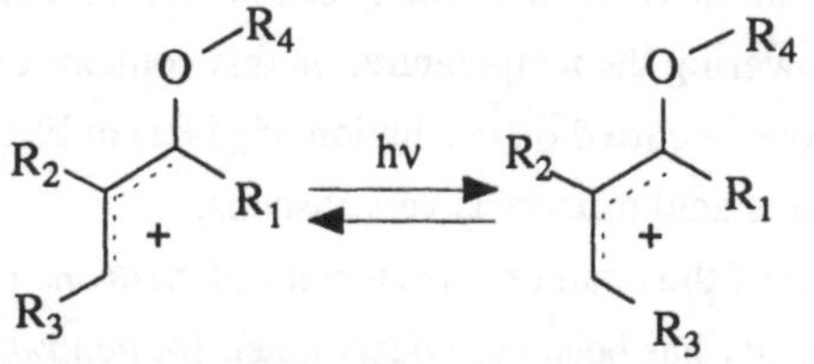

R_1	R_2	R_3	R_4	Refs.
H	H	Me	H	23,79
OH	H	Me	H	23,54,79
H	Me	Me	H	79
Me	H	Me	H	23,79
Me	Me	Me	H	79
Me	H	iPr	H	101,103
OH	H	iPr	H	101
H	H	iPr	H	102
tBu	H	iPr	H	102
OEt	H	Ph	H	82
H	H	Ph	Me	83
H	H	p-PhMe	Me	83
H	H	p-PhCl	Me	83
H	H	p-PhCF$_3$	Me	83
H	H	p-PhNO2	Me	85
OEt	H	Me	OAlEtCl$_2$⁻	82
H	H	Me	Me	23,80
OH	H	CO$_2$H	H	54

Table 3. Photostationary state compositions of ethyl cinnamate in the presence of Lewis acids

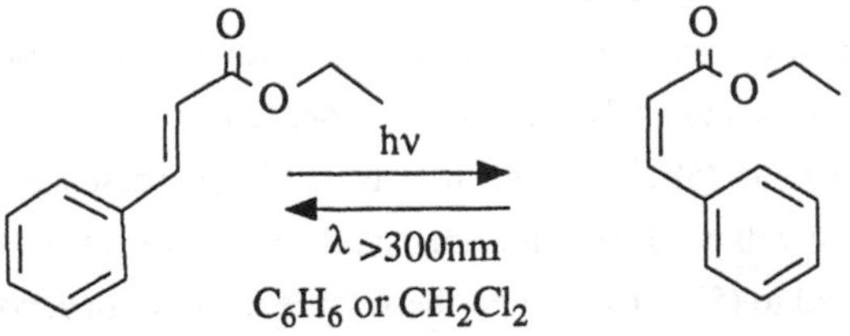

Lewis Acid	% (58) at PSS
none	42
AlCl$_3$	87
AlEtCl$_2$	85
BF$_3$.OEt$_2$	81
AlEtCl$_2$ (fully complexed)	58

(57) and (58) on addition of a Lewis acid are dramatic with close to 90% of the *cis* isomer being present in the optimum cases. The final composition reached is dependent not only on the Lewis acid used as a catalyst but also on the ratio of Lewis acid to ester. As can be seen from the data in Table 3, if an excess of Lewis acid is used such that all of the esters present in solution exist in the form of their Lewis acid complexes, then the photostationary state only contains 58% of (58)[82]; a value typical of the compositions reached with the protonated α,β-unsaturated carbonyl compounds.

In order to understand the origin of the catalytic effect of the Lewis acids on the photoisomerization of (57) and (58) it is important to bear in mind that there are four species present in solutions of these esters containing less than a stoichiometric amount of a strong Lewis acid. These species are (57) and (58) coupled with the complexes of these esters (57.LA) and (58.LA).[95-97] As a result of the red shift associated with the complexation of the carbonyl oxygen of an unsaturated ester, the complexes are the major light absorbing species present in solution when light of wavelength > 300nm is used.[96] A second factor is that the quantum efficiencies for *cis/trans* isomerization of the Lewis acid complexes are enhanced as compared to the neutral esters.[96] This change in efficiency of the photoisomerization parallels that found with the protonated materials discussed in a preceding section. The third and important factor is that the equilibrium constants for complex formation of (57) and (58) with a strong Lewis acid are different. The *trans* isomer (57) binds more strongly to a variety of Lewis acids than does the *cis* isomer (58).[82,96] This means that in the presence of a deficiency of Lewis acid there will be an equilibrium established in which the majority of the Lewis acid is bound to (57). Combined these three factors can account for the enhanced formation of the *cis* isomers in the photostationary state reached in the presence of a catalytic amount of a Lewis acid, Scheme 27.

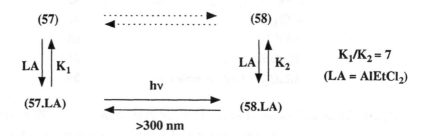

Scheme 27.

     Similar catalytic effects of Lewis acids have been observed with a variety of other α,β-unsaturated esters[95,96,82,98,99] and ketones.[82,100] These include several cinnamate esters[96], 4-phenyl-but-3-en-2-ones[82], dibenzylidene acetone[100], β-furylacrylic esters[98], and but-3-enoic esters.[99] In the latter aliphatic systems the typical deconjugation which occurs on the irradiation of neutral systems is absent in the presence of a Lewis acid. The extension of this catalytic effect to dienyl systems has met with mixed success.[99]

     The catalysts used in this type of reaction are not restricted to homogeneous Lewis acids. At McMaster it has been shown that alumino-silicates and Nafion, a perfluorinated acid resin containing sulfonic acid groups, can be used as heterogeneous catalysts for the photoisomerization of (57) and (58).[82]

4.   Photoinduced migration from C_4 to C_3. In addition to *cis/trans*
isomerizations, a further general type of photoisomerization observed with these
protonated α,β-unsaturated carbonyl compounds involves migration of an atom or
group of atoms from C_4 to C_3. For example, such a reaction has been shown to
occur with cations having an isopropyl group at C_3 (R_3= *i*-Pr in Table 2) as they
undergo an irreversible photoreaction to form protonated dihydrofurans, Scheme
28.[101]

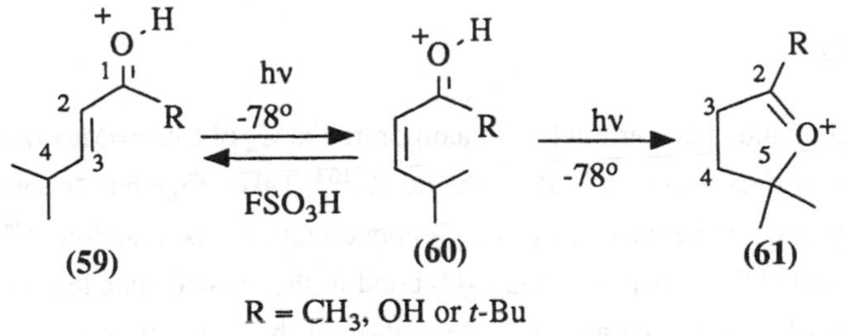

R = CH_3, OH or *t*-Bu

Scheme 28.

The photoisomerizations of cations (59) differ depending on the nature of
the substituent R. In the case where R=CH_3 the cyclic cation (61) is only produced
from the *cis* isomer (60), whereas when R=OH, or *t*-Bu both (59) and (60) yield
(61).[102] This effect would seem to be due to the requirement for an *s-cis*
conformation about the C_1C_2 bond of one of these cations in order for ring closure
to occur.[102] In the case of (59,R=OH) one of the OH groups is always in an *s-cis*
conformation. With (59,R=*t*Bu) and (60,R=*t*Bu), the size of the R substituent is
such that the preferred conformation in each case is *s-cis*. On the other hand,
cation (59,R=Me) was shown to preferentially exist in the *s-trans* conformation
while with (60,R=Me) the *s-cis* conformation is preferred.[102] In the case of
(59,R=H) and (60,R=H), which appear to exist exclusively in the *s-trans*
conformations, only photochemical *cis/trans* isomerizations were observed and no
ring closure reactions could be detected.

The mechanism of this photochemical ring closure is of interest as it
provides information on the charge distribution in the excited states of these
cations. Deuterium labelling experiments showed that deuterium in the solvent

pool was incorporated specifically at C_3 of the product during the photoisomerization of (59,R=Me)/(60,R=Me)) to (61,R=Me), Scheme 29.[101]

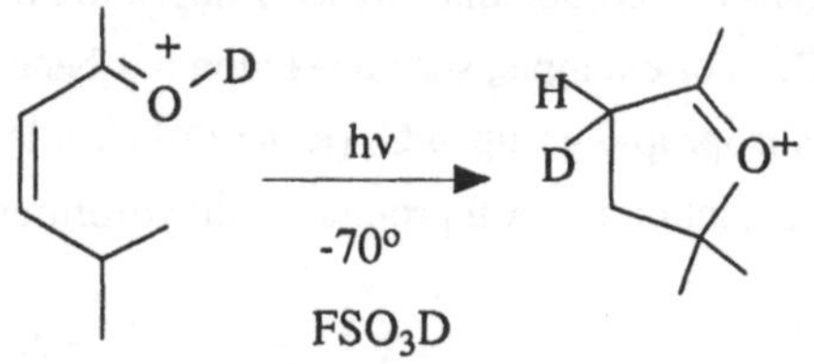

Scheme 29.

On the other hand, a deuterium label incorporated at C_4 of (59,R=Me) was shown to end up specifically at C_4 of (61), Scheme 30.[103] Taken together, these results completely rule out the possibility of a photoenolization type reaction.[89,90] Instead it was suggested that twisting of the C_2C_3 bond in the excited state leads to localization of positive charge at C_3. Migration of the isopropyl proton (or deuteron) to C_3 would yield a stable tertiary cation which could subsequently close to form (61). A similar charge polarization has been suggested to occur during the twisting of the singlet excited states of the unsaturated iminium cations involved in the visual chromophores.[104]

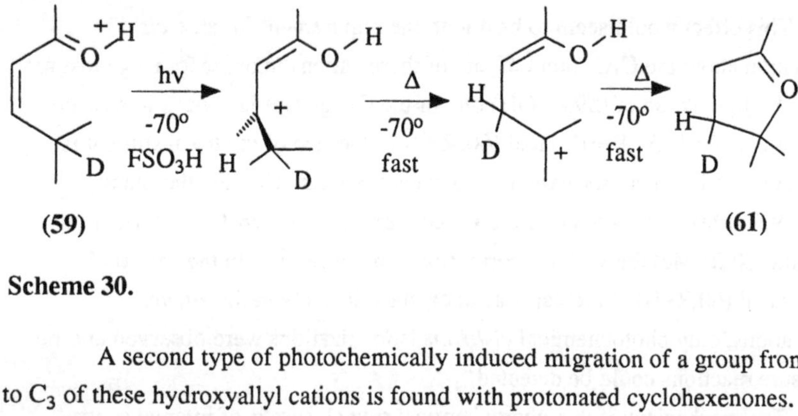

(59)   (61)

Scheme 30.

A second type of photochemically induced migration of a group from C_4 to C_3 of these hydroxyallyl cations is found with protonated cyclohexenones. The simplest example of this type of reaction is provided by the photoisomerization of protonated 4,4-dimethylcyclohexenone, (62), Scheme 31.[105] Irradiation of (62) at

254 nm leads to the formation of the two products (63) and (64).[105] The bicyclic
cation (63) was also shown to be photolabile and rearrange to (64). The
photochemical conversion of (62) to (63) and (64) can be thought of in terms of a
migration of the C_4,C_5 bond of (62) to yield (65) as an intermediate. Again partial
twisting about the C_2,C_3 bond of (62) in its π,π^* excited state may be occurring
with localization of positive charge at C_3. The intermediate (65) could either form
a bond between the tertiary cation center and C_2 to give (63) or undergo a hydride
shift to give (64).[105]

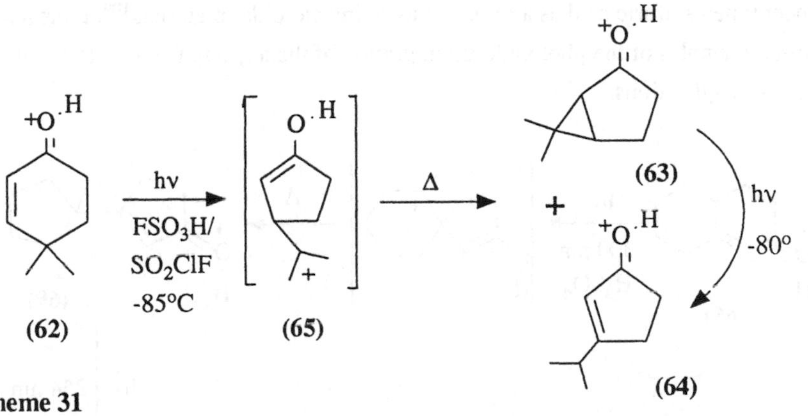

Scheme 31

Evidence to support this type of mechanism comes from the observation
that the protonated cyclohexenones (66) and (67) are photochemically inert under
the conditions where (62) photoisomerizes.[105] Migration of the C_4C_5 bonds in (66)
and (67) would give a primary or secondary carbenium ion, respectively, and this
would appear to be energetically inaccessible.

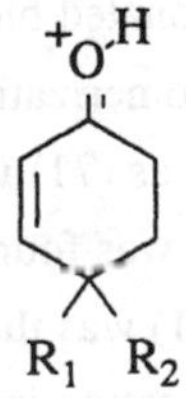

(66,R_1=R_2=H)
(67,R_1=H,R_2=Me)

The photoisomerization of (62) described above parallels that reported for

4,4-dimethylcyclohexenone in conventional media.[106] In the latter case it has been suggested that the excited state involved in the photoreaction is a π,π^* triplet state.[107] No information on the multiplicity of the excited state involved in the photoisomerization of (62) is available.

Cornell and Filipescu have reported that the protonated bicyclic ketone (68) undergoes isomerization to give (69) on irradiation at 300 nm in H_2SO_4 at room temperature. This product subsequently undergoes a further photoisomerization on irradiation at 254 nm to give (70), Scheme 32.[108] These reactions, which are of interest in terms of the use of fluorescence measurements in concentrated sulfuric acid as a means of assaying steroidal materials[109], represent further examples of the photoinduced migration of the a group from C_4 to C_3 of hydroxy allyl cations.

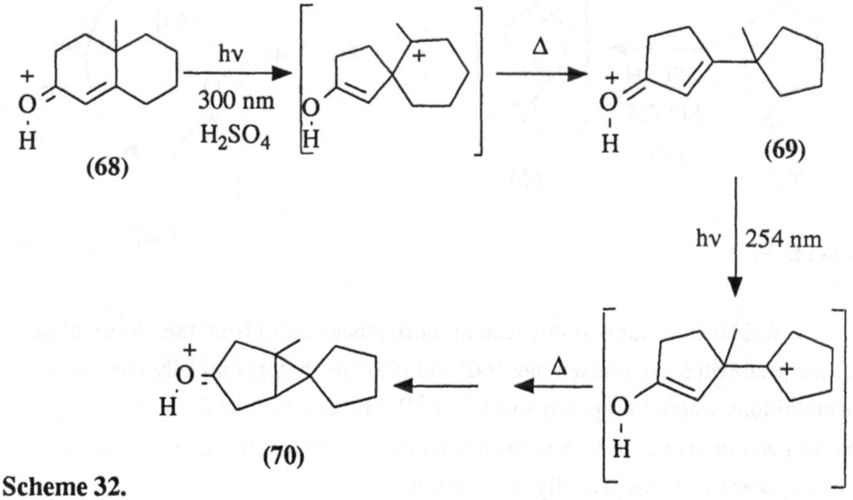

Scheme 32.

In view of the formation of the protonated bicyclic ketone (63) from (62), Childs and co-workers examined the photoisomerization of (68) in FSO_3H at low temperatures to see if bicyclic materials such as (71) were being produced as a photoproduct. No evidence for its formation was found on irradiation of (68) at -70°C, however, as it was also shown that (71) was thermally unstable at this temperature the involvement of this tricyclic cation in the photoisomerization of (68) remains an open question.[105] It should be noted that the irradiation of the ketone corresponding to (68) in *t*-butanol gives rise to the tricyclic ketone corresponding to (71) as the major product.[110] This latter isomerization would appear to involve a π,π^* triplet state of the bicyclic ketone.

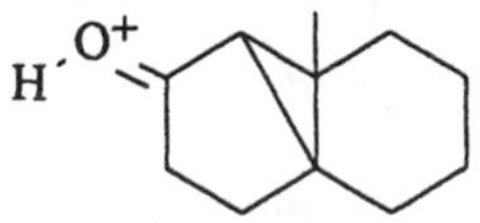

(71)

More recently Bonet and co-workers have described the photoisomerizations of a series of protonated steroidal enones.[101] A typical example is given in Scheme 33. These latter workers again showed that this type of photo-initiated migration of C_4C_5 bond in a protonated cyclohexenone only occurs when a tertiary carbocation can be produced as an intermediate.

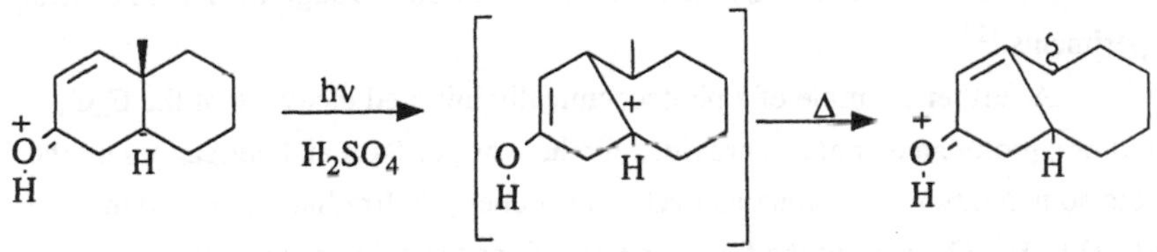

Scheme 33.

5.   Cleavage of the C_4C_5 bond. The photoisomerization of the Lewis acid complex (72) has been examined.[112] Irradiation of (72) at -50°C in $CHCl_3$ leads to the formation of (73). The reverse photoisomerization also occurs with (73) being converted to (72). A photostationary state was established between (72) (64%) and (73) (36%). While the interconversion of (72) and (73) can be regarded as a suprafacial 1,3-sigmatropic shift, in view of the other reactions of protonated enones presented above, an equally plausible pathway is the cleavage of the C-C bond to give the stable allyl cation as shown in Scheme 34.

The photochemical isomerization of (72) is also of interest on two additional accounts. In the first place the presence of the Lewis acid has substantially modified the course of the photoisomerization of the neutral ketone. Irradiation of the parent ketone leads to a 2 + 2 cycloaddition between the two

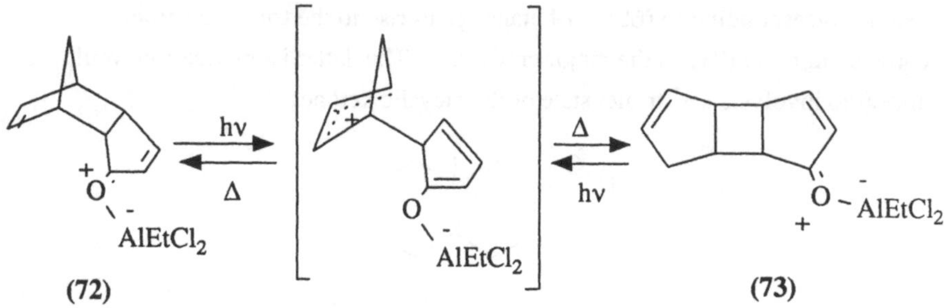

Scheme 34.

double bonds and the formation of a cage compound.[113] Second, the other possible type of photoreaction of (72) would be a 2 + 2 addition involving the double bond of the bicyclo[2.2.1] fragment with the 3 carbon fragment of the allyl cation. Both of these possible cycloaddition reactions were ruled out through the use of control experiments.[112]

A further example of a photochemically initiated cleavage of the C_4C_5 bond of a protonated enone is provided by the work of Sato and colleagues on the photoisomerization of protonated cyclic 1,3-diones.[114] Irradiation of (74) in FSO_3H or H_2SO_4 leads to the formation of (75) as the sole product, albeit in relatively low overall yield, Scheme 35. It would seem that the presence of a *gem* dimethyl group on C_5 of the cyclohexa-1,3-dione is essential in order for this reaction to occur and protonated diones lacking this feature are photochemically inert. The most likely course for this photoisomerization involves cleavage of the C_4C_5 bond of (74) to give the ring opened tertiary cation (76). Subsequent ring closure by bonding between C_5 and oxygen would eventually lead to (75), Scheme 35. A similar type of photoisomerization was also found to occur on irradiation of protonated thiolane-2,4-diones.[114]

6.   Acid catalyzed cycloaddition reactions. The photochemical dimerization of α,β-unsaturated carbonyl compounds, while generally an inefficient reaction, is of importance in terms of the synthesis of substituted four-membered ring compounds.[115] Typically these reactions are the most effective with cyclic systems where relaxation by twisting of the C=C bond is restricted. Recent work by Lewis and his group have shown that these photo-dimerizations can be

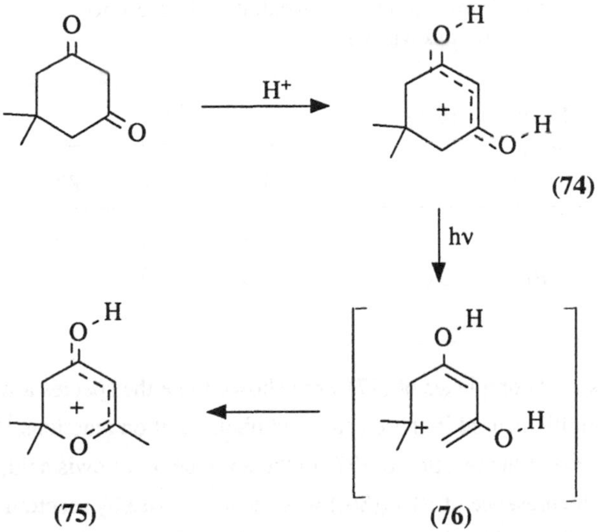

(74)

(75)   (76)

Scheme 35.

effectively catalyzed by the addition of a Lewis acid to the solutions.[95,116-118] For example, irradiation of coumarin (77) in CH_2Cl_2 solution leads to the formation of three dimers, (78), (79) and (80) with low quantum efficiency, Scheme 36.[116,118] The principle product, (79), would seem to arise from the reaction of a triplet coumarin with a ground state coumarin.[119] Addition of BF_3 to the solution substantially modifies the ratio of the products obtained as well as enhancing the quantum efficiency of the reaction by a factor of more than 200, Table 4.[116]

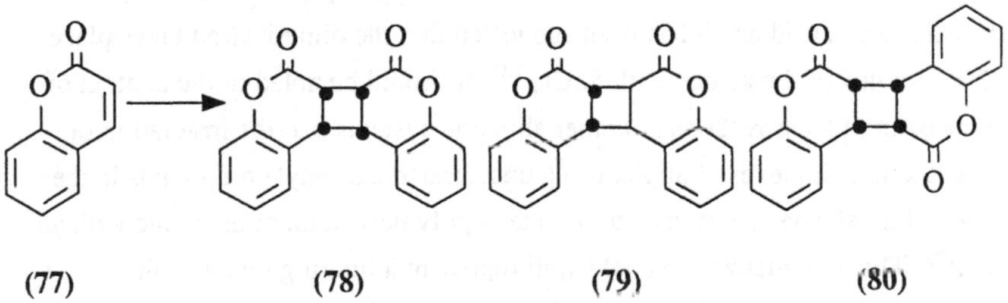

(77)   (78)   (79)   (80)

Scheme 36.

Table 4. Products formed from coumarin on irradiation in presence
of a Lewis acid.

[(77)]	Lewis Acid	[LA]	Yield %		
M	(LA)	M	78	79	80
0.25	none		20	71	10
0.20	BF$_3$	0.02	0	10	90

The Lewis acid complexes of (77) were shown to be the species absorbing light under the conditions used in these catalyzed photoaddition reactions.[113] In contrast to the course of the reaction of (77) in the absence of a Lewis acid, the major product in the presence of BF$_3$ ((80)) was shown to arise by reaction of a singlet state of (77.BF$_3$) with a ground state coumarin molecule. The minor product derives from a triplet excited state of the complex. As was noted before, protonation, or in this case complexation, of a carbonyl oxygen by moving the n,π^* states to much higher energy enhances the lifetime of the singlet excited state.[24] It has been suggested that the greatly enhanced efficiency of the reactions and the substantial proportion of the products which originate from the singlet state of the complex derive from this effect.[85,120]

The Lewis acid catalysis of the photochemical dimerization of esters of cinnamic acid[85,117,118], cyclopentenone[121], and the cross additions of alkenes to coumarin[90] have also been examined. In the case of cyclopentenone, SnCl$_4$ was used as the Lewis acid and it has been suggested that the dimerization takes place in a 2:1 complex of the ketone with SnCl$_4$.[121] It should be noted in the context of the expansion of these reactions to other aliphatic systems that the irradiation of the Lewis acid complexes of aliphatic α,β-unsaturated carbonyl compounds in the presence of an alkene can be used to initiate copolymerization of an enone with an alkene.[122] This polymerization could well represent a limiting factor in the use of these acid catalyzed cycloaddition reactions.

A further class of Lewis acid catalyzed photoreactions of α,β-unsaturated ketones which can lead to cyclic as well as other products have been described by

Sato and colleagues.[123] As can be seen from the reactions outlined in Scheme 37, irradiation of methanol solutions of a Lewis acid such as $TiCl_4$ and an unsaturated ketone lead to the formation of a variety of products which contain extra carbon atoms derived from the solvent.

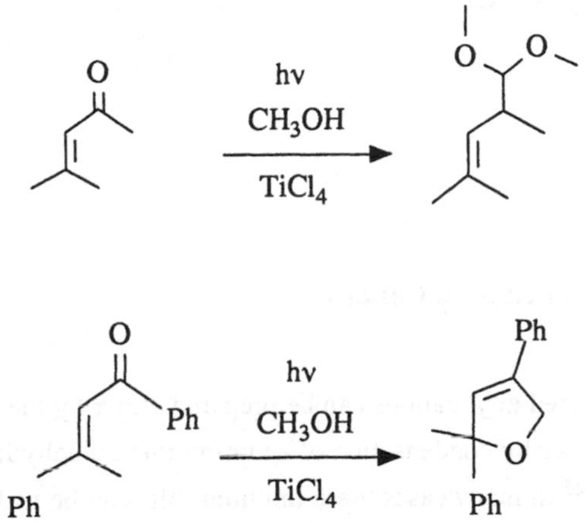

Scheme 37.

The chemistry involved in these interesting transformations is complex and many aspects still have to be sorted out. In simple terms the photoreactions can be thought of as involving a complex of the enone/$TiCl_4$/methanol.[123] Excitation of this complex could lead to a "metal assisted" electron transfer with the formation of the radical cation/radical anion complex shown in Scheme 38. Proton transfer to give a diradical species followed by coupling and dissociation of the titanium would form a 1,2-diol. Under the acidic conditions used in these reactions the intermediary 1,2-diols can then undergo a variety of acid catalyzed reactions to give the products observed. This mechanistic scheme is simplistic in that the exact nature of the initial complex is not known with any degree of certainty. For example, some of the chlorine atoms of the $TiCl_4$ will certainly have been replaced by methoxy groups. In addition, photochemically initiated valence changes of the titanium atom are possible. Overall, the reaction bears considerable similarity to the coupling of two carbonyl compounds in a pinacol reaction.[124]

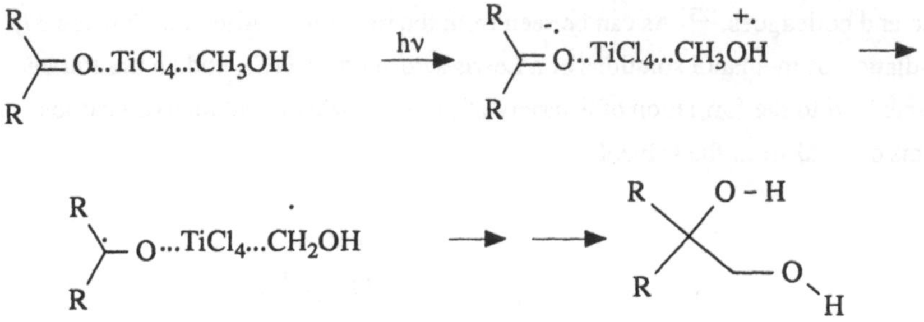

$$R \atop R \Big\rangle = O..TiCl_4...CH_3OH \quad \xrightarrow{h\nu} \quad R \atop R \Big\rangle \overset{-.}{=} O..TiCl_4...\overset{+.}{CH_3OH} \quad \longrightarrow$$

$$R \atop R \Big\rangle \overset{.}{-} O...TiCl_4...CH_2OH \quad \longrightarrow \quad \longrightarrow \quad R \atop R \Big\rangle\!\!\!\overset{O-H}{\underset{O\diagdown H}{\times}}$$

Scheme 38.

C.   Nitrogen Substituted Allyl Cations

Nitrogen substituted allyl cations can be prepared either by the protonation of α,β-unsaturated imines[35a] or by condensation of an unsaturated aldehyde or ketone with a dialkyl ammonium salt.[125] In many cases these iminium salts can be isolated as crystalline solids which are relatively air stable compared to their oxygen substituted analogs. The ground state properties of these cations are well established[126] and the stuctures of several ions have been determined using x-ray crystallography.[127]

Just as with the corresponding oxygen substituted allyl cations, α,β-unsaturated iminium salts such as (81) exhibit restricted rotation about the C_1,N, C_1,C_2 and C_2,C_3 partial double bonds. Typically the barriers to uncatalysed rotations about the C_1,N and C_2,C_3 bonds are higher than those for the oxygen substituted systems.[127a,128,129] However, stereomutations about these bonds can be catalysed by protonation on nitrogen[129] or the addition of a nucleophile to either C_1 or C_3.[130] As a result of both acid and base catalysed isomerizations about these bonds great care must be taken in photochemical studies to work in media where the rates of these thermal processes are slow and do not interfere with photochemically induced reactions.

The photochemical reactions observed with iminium salts such as (81) include isomerization about the C_1,N and C_2,C_3 bonds as well as photoadditon reactions.

1.   Photoisomerization about the C_1,N and C_2,C_3 bonds. Dickie working at McMaster investigated the photoisomerizations of a series of α,β-unsaturated iminium salts.[131] Irradiation of the iminium salt (82) was reported to lead to *cis/trans*

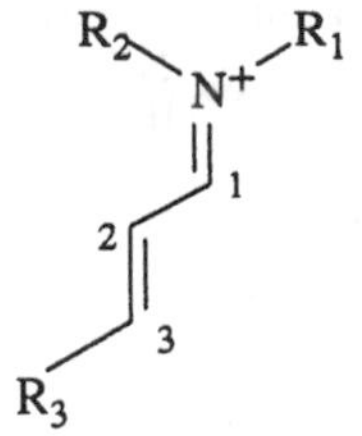

(81)

isomerization about the C_1,N bond to give a photostationary state consisting of (82) (65%) and (83) (35%), Scheme 39. This photostationary state represents a non-thermodynamic equilibrium mixture of the two isomers suggesting that the reaction proceeds by an excited state twisting about the C,N bond. The quantum efficiency of the conversion of (82) to (83) was found to be 0.12 in H_2SO_4 and 0.16 in CF_3CO_2H.

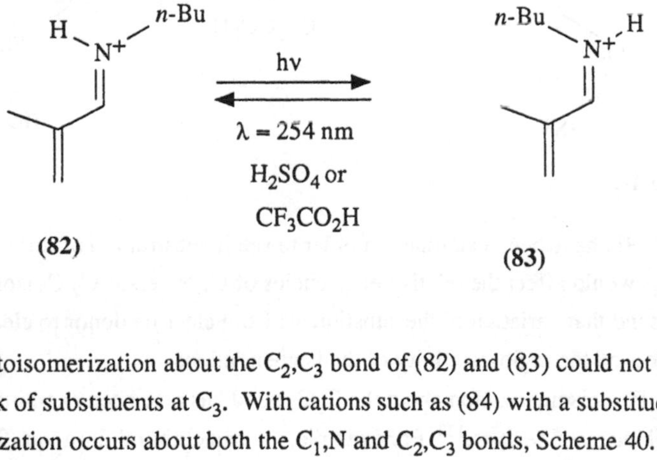

Scheme 39.

Photoisomerization about the C_2,C_3 bond of (82) and (83) could not be detected given the lack of substituents at C_3. With cations such as (84) with a substituent at C_3 photoisomerization occurs about both the C_1,N and C_2,C_3 bonds, Scheme 40.[131] In this case a photostationary state consisting of 50% (84), 12% (85), 28% (86), and 10% (87) was eventually reached.

Substitution at C_3 of these ions can have a large effect on the relative quantum efficiencies for the isomerization about the C_1,N and C_2,C_3 bonds.[131] Thus (89) is the only product obtained on irradlation of the iminium salt (88), Scheme 41. The quantum efficiency of isomerization about the C_1,N bond was estimated to be <0.05; a value substantially smaller than that observed for the conversion of (82) to (83).

The photoisomerizations of a series of aryl substituted iminium salts based on

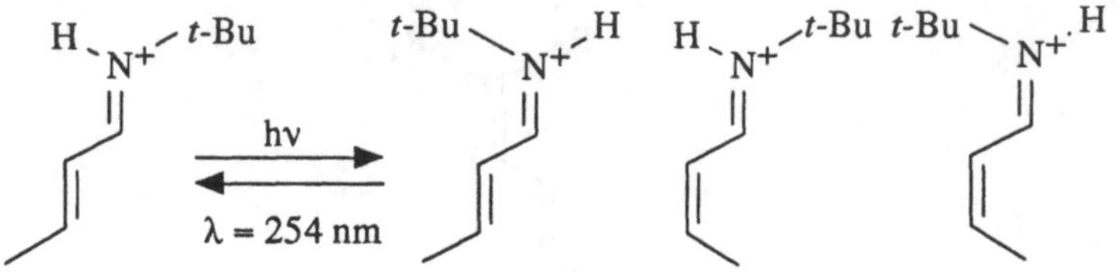

(84)   **(85)**   **(86)**   **(87)**

Scheme 40.

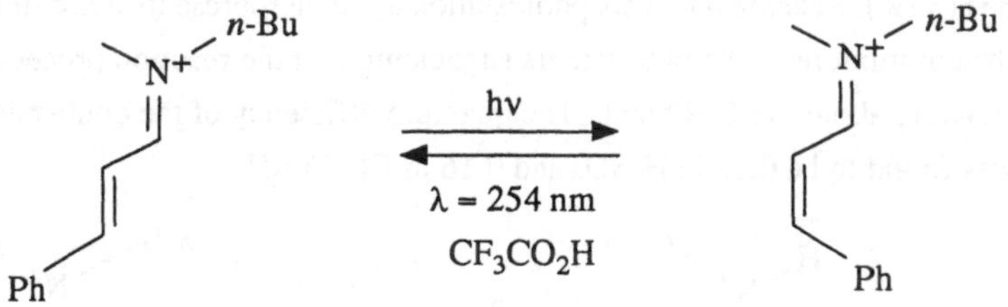

(88)   **(89)**

Scheme 41.

(90) and (91) have been examined in order to see if substitution of different groups on the aryl rings would affect the relative efficiencies of C_1,N versus C_2,C_3 isomerization.[131,132] It was found that variation of the substituents from electron donor to electron acceptor had little effect on the regioselectivity of the isomerizations.

Photoisomerization about the C_2C_3 bond has also been examined in a variety of polymethine cyanine dyes.[133] For example, the trimethiniminium ion (92) undergoes efficient *trans/cis* photoisomerization ($\phi=0.5$) over a wide range of temperatures, Scheme 42. The cis isomer (93) produced from this reaction undergoes facile thermal isomerizations to give back (92). The photoisomerizations of other cyanine dyes have been examined.[134]

The photoisomerizations about C_1N and C_2C_3 of unsaturated iminium salts described above would generally seem to occur from π,π^* singlet excited states. Excitation to this state can be followed by fluorescence, rotation of the double bond to a

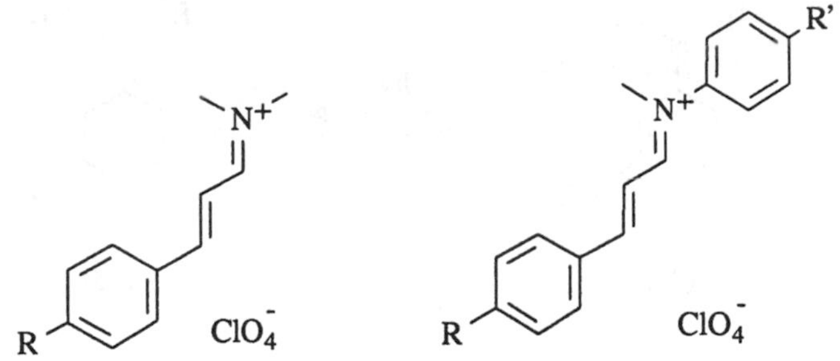

R= H, OMe, Me, Cl, NO_2

(90)

(91)

R'= H, OMe, Me, Cl, NO_2

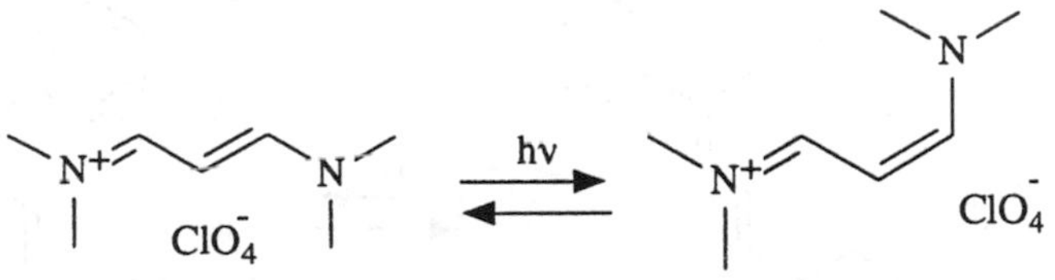

(92)

(93)

Scheme 42.

perpendicular geometry or intersystem crossing to the triplet state.[135] In most allyl iminium ions studied, no fluorescence has been observed, even in glass matrices.[131,132] This suggests that any barriers to rotation about the C_1N and C_2C_3 bonds in the excited states of these ions are very small.[136]

One exception to this general involvement of singlet excited states in these photoisomerizations is provided by the interconversion of (94) and (95), Scheme 43. This interconversion can be sensitized by xanthone suggesting the involvement of a triplet state.[66]

A further mechanism for the photoisomerization of unsaturated iminium salts involves an electron transfer process. For example, the major photoreaction of (96) involves an efficient isomerization about the C_1N bond, $\phi = 0.31$, Scheme 44.[132] It was suggested that this reaction involves a photoinduced electron transfer process to the iminium ion followed a facile isomerization about C_1N bond in the resulting radical.

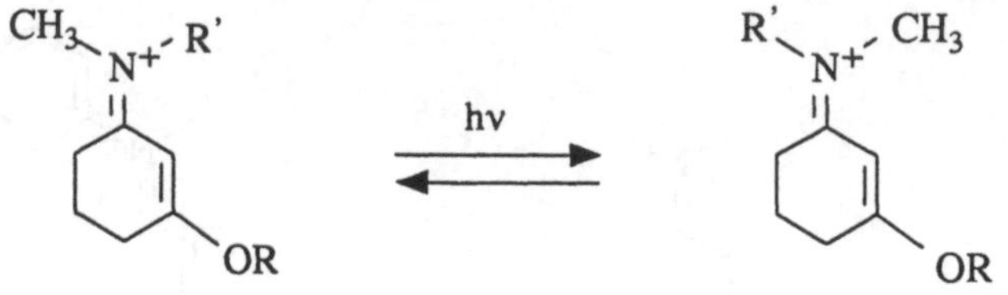

$$\text{(94)} \qquad \xrightarrow{\text{hv}} \qquad \text{(95)}$$

Scheme 43.

Irradiation of (96) in the presence of the electron donor, tris (2,2'-bipyridine) ruthenium (II) dichloride under conditions where the ruthenium complex was the light absorbing species led to the formation of a thermodynamic mixture of (96) and (97).

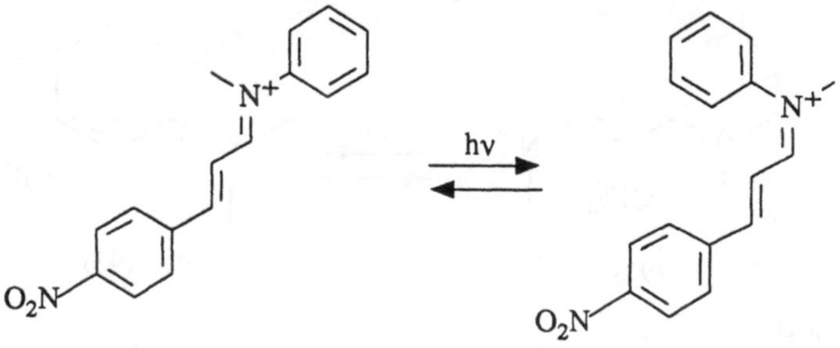

$$\text{(96)} \qquad \xrightarrow{\text{hv}} \qquad \text{(97)}$$

Scheme 44.

A detailed examination of the photoisomerization for several styrylpyridinium and styrylquinolinium ions has been reported by Görner and co-workers.[137-139] While these systems are not nitrogen-substituted allyl cations in a formal sense they are worth mentioning here in that they well illustrate the profound effect solvent and anion can have on the photoisomerizations of unsaturated nitrogen containing cations.

In general, cations such as (98) undergo photoisomerization about the central CC double bond to form the cis isomers (99), Scheme 45. It has been found that this isomerization can occur from either an excited singlet or triplet state and that the anion has an influence on the isomerization pathway.

The photostationary state for (98)/(99) where R=CN, H, CH_3 and OCH_3 and

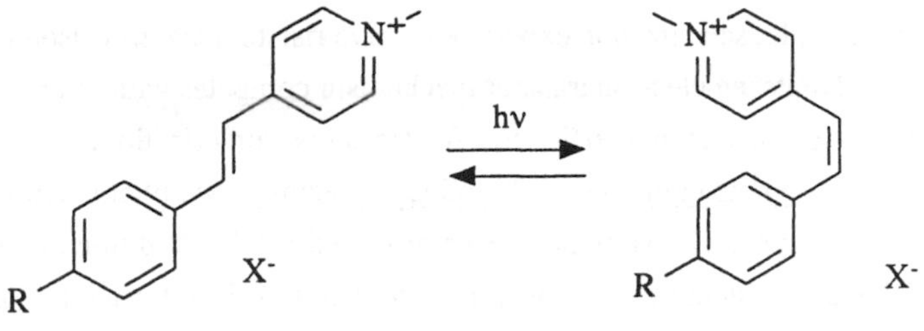

$$hv$$

(98)   (99)

Scheme 45.

$X^-=ClO_4^-$ is not affected by solvent polarity. Thus in each case irradiation of these cations in either dichloromethane or methanol yields a photostationary state which is composed of 80-85% (99).[138] The quantum yields for these photoisomerizations all lie between 0.40 and 0.55 for either solvent. The mechanism suggested for this isomerization involves a similar bond rotation in an excited singlet state to that suggested above for the allyl substituted iminium ions.

In the case where the anion is iodide, the composition of the photostationary state and the quantum yield for isomerization are both solvent dependent.[137,139] For example, irradiation of the cyanostyrylpyridinium ion (98,R=CN) in methanol yields a photostationary state comprised of 80% cis (99,R=CN) and 20% (98). In dichloromethane the photostationary state contains less than 30% of (99,R=CN). The quantum yield for this reaction is 0.4 (MeOH) and 0.02 (CH_2Cl_2). The difference in quantum yield in the two solvents has led to the suggestion that an electron transfer mechanism operates in non-polar solvents. It has been shown that the cation and anion exist as an ion pair in non-polar solvents such as dichloromethane. In the excited singlet state, electron transfer occurs from the iodide anion to the cation. This is followed by a rapid thermal decay of the intermediate to the *trans* ground state. This process competes with bond rotation of the excited singlet intermediate giving rise to a decreased quantum efficiency for *trans/cis* isomerization.

While most styrylpyridinium ions undergo a *trans/cis* photoisomerization from excited singlet states, the photoisomerizations of the nitro substituted ions occur through triplet excited states.[137] These compounds exhibit Stern-Volmer quenching with low energy triplet quenchers such as ferrocene, azulene and oxygen in polar solvents.

Furthermore triplet sensitization experiments give rise to a *trans/cis* isomerization. In non-polar solvents, an electron transfer mechanism competes with intersystem crossing which decreases the quantum efficiency for *trans/cis* isomerization.

        2.   Photoaddition to C_1N and C_2C_3 bonds. Two photoadditions to an isolated nitrogen substituted allyl cation have been observed.[140,141] Addition of methanol occurs to C_3 of the allyl system in a similar fashion to that described for methylium ions (Scheme 10). In some instances nucleophilic addition to the iminium bond can subsequently occur to give cyclic products, Scheme 46.

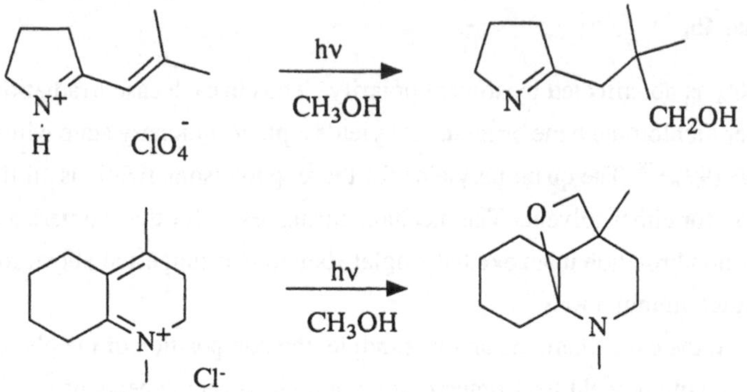

Scheme 46.

IV   PHOTOREACTIONS OF DIENYL AND POLYENYL CATIONS

A.   Alkyl/Aryl-substituted Dienyl Cations

      The photochemistry of acyclic alkyl and aryl substituted dienyl cations has not been examined. One reason for this omission is technical difficulties arising from the ease with which they undergo thermally induced ring closure to form cyclopentenyl cations.[142] On the other hand, the cyclohexadienyl, or benzenium cations are readily prepared and there have been several investigations of their photochemistry.[143,144]

      Benzenium cations are generated by the protonation or alkylation of benzenoid compounds. Alkyl substituted, monocyclic ions typically display strong, long wavelength absorption bands at ~400 nm and weak fluorescence

emission between 440-500 nm.[145,146] The singlet energies of the polymethylbenzenium cations are in the range 65-72 kcal/m.

   The polymethylbenzenium ions undergo an efficient photoisomerization
to give the corresponding bicyclo[3.1.0]hexenyl cations. For example, the
pentamethylbenzenium ion (100) is cleanly converted to (101) on irradiation at
400nm in FSO$_3$H at -70°C.[16,147] A photostationary state is established between
(100) (~20%) and (101) (~80%). Results obtained with the tetramethyl substituted
ions show that the photoisomerization involves primarily an electrocyclic ring
closure reaction rather than a more complex processes involving scrambling of the
ring carbons.[147,19]

$$\text{hv} \quad \text{FSO}_3\text{H} \quad -78°\text{C}$$

(100)   **(101)**

   This photochemical conversion of a benzenium cation to a bicyclohexenyl
ion has been extended to a range of alkyl substituted systems, Table 5, as well as
the hydroxybenzenium ions discussed in the next section. The reactions all proceed
in high yield. Indeed, as benzenium cations are colored materials and their
photoisomerizations proceed very cleanly without any apparent side reactions,
these cations have been considered as potential photochemical/latent heat solar
energy storage systems.[146] In this regard it is interesting that the cycle involving
the photoisomerization of (104) to (105) and reverse thermal conversion of (105) to
(104) was carried out some 22 times in an aerated FSO$_3$H solution with less than a
2% loss of material per cycle.[146] These photoreactions are not restricted to strong
acid media. The conversion of (104) to (105) in CH$_2$Cl$_2$ has also been reported.[146]

   The quantum efficiencies for the conversion of (100) to (101) (0.23 ∓ .03),
(102) to (103) (0.30 ∓ .05), and (104) to (105) (0.40 ∓ .04) have been measured in
FSO$_3$H at -70°C.[146] The quantum yields were shown to be temperature
independent over the range of temperatures studied and not affected by the
presence of oxygen in the solution. It would seem that the photoisomerizations of

Table 5. Photoisomerizations of benzenium ions

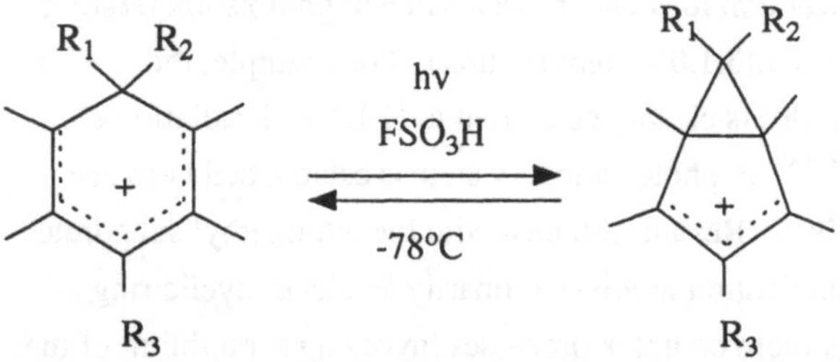

	R_1	R_2	R_3		Refs.
(100)	H	H	Me	**(101)**	16,146,147,149,151
(102)	H	Me	Me	**(103)**	16,146-151
(104)	Me	Me	Me	**(105)**	16,146,148,149,151
(106)	Me	Et	Me	**(107)**	149
(108)	Me	CHCl$_2$	Me	**(109)**	149
(110)	H	H	H	**(111)**	147

these benzenium ions involve singlet excited states of the cations. It is not clear why substitution at C_6 of these cations should increase the efficiency of the formation of the bicyclic cations.

Benzenium ions with different substituents at C_6 can in principle give two different stereoisomers on photochemical ring closure. In general the larger substituent is found preferentially in the C_6 endo position of the bicyclic ion.[16,148,149] For example, irradiation of the hexamethylbenzenium cation, (102), leads to the formation of the endo isomer as the sole product despite the corresponding exo compound being stable under the reaction conditions, Scheme 47.

The stereoselective ring closure of (102), parallels the regio-selectivities observed in the photoisomerizations of the cross conjugated cyclohexadienones[152] and could have a similar origin. The conversion of (102) to (103) requires one of the C_6 substituents to move through the plane of the C_1C_5 methyl groups and this process will be favored for the H atom, the smaller of the two C_6 substituents.

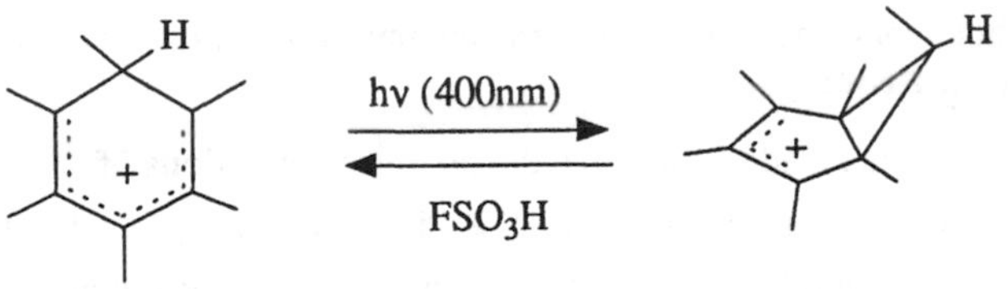

$$hv \ (400nm)$$

$$FSO_3H$$

(102)   (103-endo)

Scheme 47.

However, it should also be borne in mind that unsymmetrically substituted
benzenium ions such as (102) are not planar but adopt a shallow boat conformation
with the C_6 methyl group in a pseudo axial position as a result of steric interactions
between the C_6C_1, and C_5 methyl groups on the six-membered ring.[153]

The bicyclo[3.1.0]hexenyl cations are important non or
"anti"-homoaromatic carbenium ions[154] that show highly stereoselective
degenerate circumambulatory rearrangments.[16,148,155] They have been considered
as intermediates in the industrially important acid catalyzed isomerization of
xylenes, although it is likely that this reaction involves a series of proton and
methyl shifts of the benzenium ions rather than the bicyclic cations.[156] The
photochemical synthesis of bicyclo[3.1.0]hexenyl cations represents a very
convenient method of generating these cations from benzenium ions. The major
limiting factor in the extension of these isomerizations to the preparation of less
highly substituted bicyclo[3.1.0]hexenyl cations would seem to be their thermal
instability and the ease with which they revert to the isomeric benzenium
ions.[147-149,157,158]

One way to circumvent the limitation imposed by the thermal instability
of the bicyclohexenyl cations would be to use less strongly acidic media for the
photoisomerizations and capture the products with a nucleophile. Indeed, several
years ago Bryce-Smith and Gilbert raised the suggestion that the
photoisomerizations of benzene in weakly acidic media might involve benzenium
ions.[159] Such a pathway is not inconceivable given the large enhancement in
basicity of aromatic compounds on excitation to their first excited singlet states[145]
and the demonstrated ease of photorearrangement of benzenium ions. However,

in the case of the photoisomerizations of benzene the involvement of benzenium ions has been ruled out.[160-162]

Attempts have been made to catalyze the photoisomerizations of polymethyl substituted benzenes with a variety of acid catalysts using conditions where benzenium ions, in equilibrium with the aromatic hydrocarbons, would be the light absorbing species.[146] While these attempts have been unsuccessful with the aromatic hydrocarbons, it has been shown that the photoisomerization of the basic hydrocarbon (112) (pK_a 1.38)[163], the precursor of cation (104)[164], can be catalyzed in the presence of acid.[165,166] The use of a CH_2Cl_2 swollen Nafion resin in its as acid form, either as a membrane or as beads suspended in CH_2Cl_2 was explored by Mika-Gibala at McMaster University.[165] Irradiation of a CH_2Cl_2 solution of (112) at 400 nm in the presence of the Nafion led to the formation of one major (95%) and two minor products, Scheme 48. The hydrocarbon (113) is known to be formed from (105) by loss of a proton.[167] No photoreaction took place in the absence of the Nafion. It was shown that the photoreactive species was the benzenium ion (104), formed as (112) was absorbed by the Nafion. The quantum efficiency of the conversion of (104) to (105) absorbed on Nafion was found to be the same as that in solution. The Nafion functioned as a catalyst for the reaction with a turn over of material between the contacting solution and the active sites in the polymeric phase.

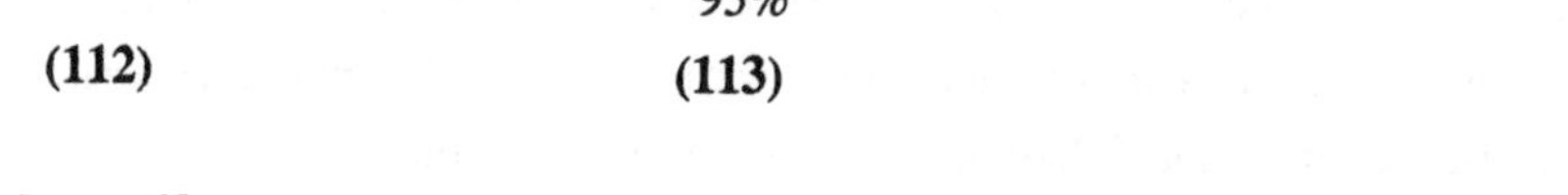

(112)

hv
400 nm
$\xrightarrow{\hspace{2cm}}$
Nafion/H⁺
CH_2Cl_2

95%
(113)

+

H

+

Scheme 48.

The use of aqueous acids in the photoisomerizations of (112) was examined by Zeya and Dain.[166] The photoisomerizations were again shown to involve (104). The bicyclohexenyl cation (105) formed on irradiation of these systems is unstable in the aqueous acids and reacts with the medium to give pentamethylcyclopentadiene as the major product, Scheme 49. It has been suggested that this photochemical reaction could be used as the second step of a convenient two stage synthesis of pentamethylcyclopentadiene from hexamethylbenzene.

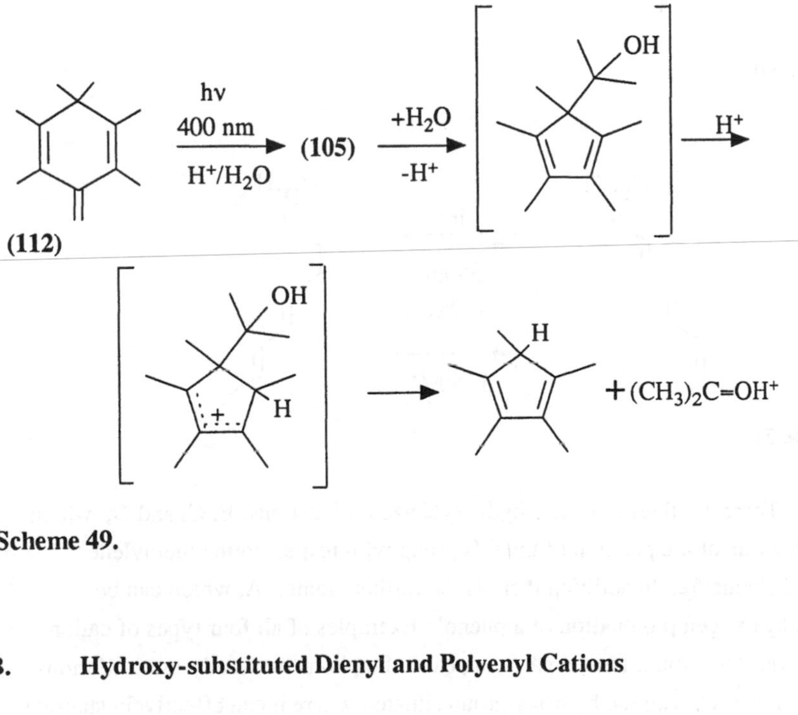

(112)

Scheme 49.

B.   Hydroxy-substituted Dienyl and Polyenyl Cations

1.   Acyclic systems. Preliminary reports of the photoisomerizations of the protonated unsaturated aldehydes shown in Schemes 50 and 51 have appeared.[168] In both instances cis/trans isomerization about the C_2C_3 bonds of the cations would appear to be the preferred photoreaction. The occurrence of facile thermal isomerizations of these acyclic cations limit the utility of these reactions.

2.   Hydroxybenzenium ions. The photoisomerizations of hydroxy-substituted benzenium ions have been extensively investigated. These reactions, which have considerable synthetic utility, bear a strong relationship to the photoisomerizations of the 2,4 and 2,5-cyclohexadienones.[169,170]

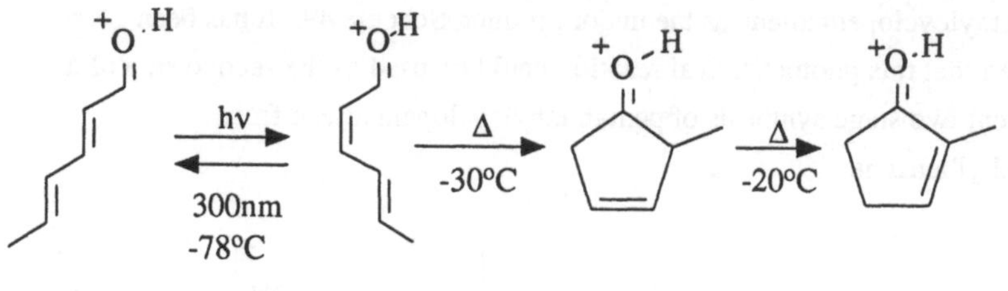

Scheme 50.

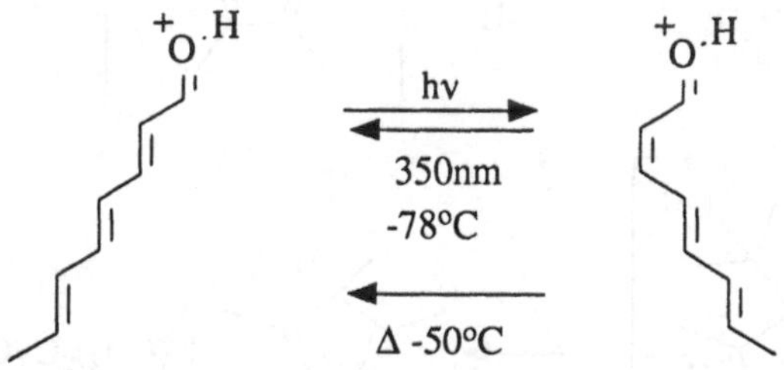

Scheme 51.

There are three isomeric hydroxybenzenium cations, B, C, and D, which differ in terms of the position of the OH group with respect to the methylene carbon, Scheme 52. In addition there is the further isomer, A, which can be derived by oxygen protonation of a phenol. Examples of all four types of cation are known. The onium ions, type A, appear to be photochemically inert. Cations of types B and C, with the hydroxy group situated where it can effectively stabilize the positive charge on the benzenium ion, can be formed either by O-protonation of the corresponding 2,4- or 2,5-cyclohexadienones[171] or by C-protonation of phenols in super acid media.[172,173] In the latter case the site of protonation depends on the substituent pattern on the benzene ring and in some instances more than one isomer is present in solution. Cations of the type D are less frequently encountered as they are generally less stable than the other three isomers.[172] Only one photo-reaction of a cation of this type has been described. However, type D cations are important as they are formed as short lived intermediates in the photo-isomerizations of these systems.

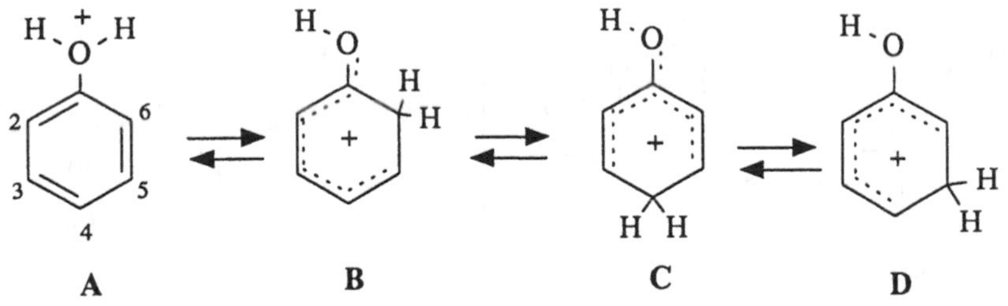

Scheme 52.

The photochemical transformations of cations of types B and C involve the same type of electrocyclic ring closure observed with the benzenium ions. However, frequently the photoisomerizations are more complex than those of the corresponding benzenium ions. In the first place the initially formed photoproducts do not have an optimum placement of the OH group to stabilize the positive charge. As a result there is generally a thermal isomerization step which follows the photoreaction. Second, under the conditions used in these isomerizations, the products of the photoreactions also absorb light and undergo photoisomerization reactions. As a result the distribution of products obtained in these photoisomerizations is frequently wavelength dependent.

The simplest photoisomerizations to unravel involve cations of type B. These cations absorb at longer wavelengths than the other hydroxybenzenium ions or the products of the reactions and as such they can be excited reasonably selectively. For example, Parrington in this group showed that irradiation of (114) in FSO_3H at low temperatures leads to the formation of (115).[19] The bicyclic product (115) undergoes a further, slower photoisomerization to give (116) and (117), Scheme 53. Comparable isomerizations have been reported for protonated hexamethyl-2,4-cyclohexadienone[19] and 6-dichloromethyl-6-methylcyclohexadienone.[174,175] In the latter case work up of the solutions following reaction leads to the formation of methylhydroxybenzaldehydes.

The photoisomerization of (114) can be understood in terms of an initial ring closure reaction to form (118). The intermediate (118) can undergo two competing thermal isomerizations; a symmetry restricted thermal ring opening to regenerate (114) or a circumambulation of C_6 around the five-membered ring to

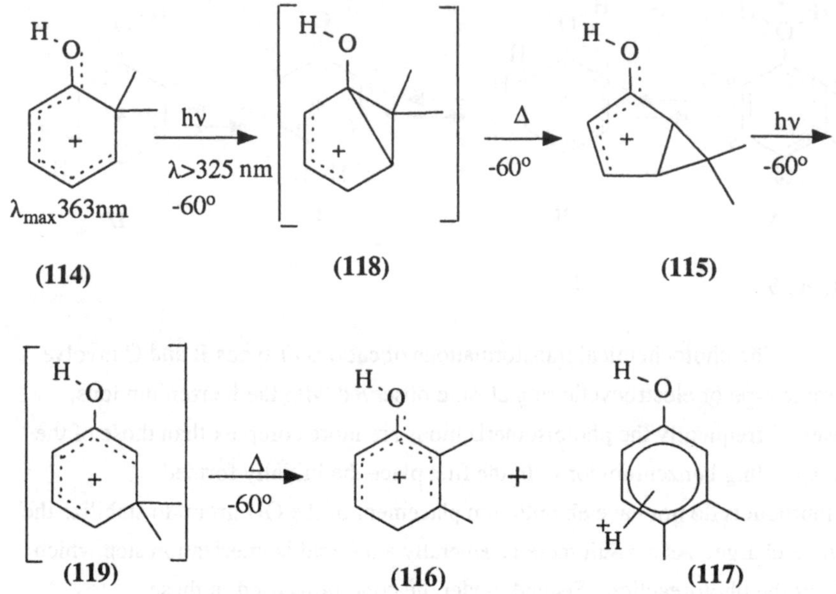

Scheme 53.

give (115). Ample precedent exists for both of these reactions.[148,155,157,158]

A marked parallel exists between the photoisomerizations of these protonated 2,4-cyclohexadienones and their neutral precursors.[169,178] In both instances bicyclo[3.1.0]hexenones, or their protonated counterparts, are formed. Two different types of excited states, n,π^* and π,π^*, can be involved in the formation of bicyclohexenones from neutral cyclohexadienones and the course of the reactions varies depending on which state is populated. The lowest excited states of the protonated cyclohexadienones are of π,π^* character.

Protonated bicyclo[3.1.0]hexenones, which typically exhibit absorption maxima at ca. 330 nm, are also photolabile. Thus (115) is converted to a 1:3 mixture of (116) and (117) on irradiation. It has been suggested that this transformation involves a photochemical electrocyclic ring opening of (115) to (119), a type D cation, followed by a series of methyl and hydrogen shifts to give the more stable ions (116) and (117), Scheme 53. The quantum efficiencies of the photoinduced ring opening reactions of a series of protonated bicyclo[3.1.0]hexenones have been measured.[157]

The photoisomerizations of cations of type C follow a similar sequence of steps to those outlined for protonated 2,4-cyclohexadienones. A much wider range of systems have been examined including not only protonated cyclohexadienones but also cations derived from the ring protonation of phenols.[19,179,157] The photoisomerization of para-protonated phenols represents a very considerably expansion of the scope of the well studied 2,5-cyclohexadienone photorearrangement.

The photoisomerizations of the tetramethylphenols provide an interesting example of the interconversions which can occur in these systems, Scheme 54.[157,179,180] George at McMaster has shown that the irradiation of either (120) or (122) in CF_3SO_3H at room temperature leads to the establishment of a photostationary state consisting of the various isomers (120)-(124).[157] This series of photoisomerizations is remarkable in that there is no apparent loss of material even when high concentrations of the substrates are used. Phenols themselves are relatively inert photochemically and few useful photoisomerizations have been reported.[181] The photochemistry of simple benzenoid materials generally leads to a considerable loss of material as a result of the formation uncharacterized polymeric material.[159]

When (122) was irradiated at low temperatures in FSO_3H the bicyclic cation (125) was produced in addition to (123) and (124). This product was not detected at room temperature as (125) is thermally unstable at this temperature and isomerizes to give (122).[157] The regio- and stereoselectivities of the photoisomerizations of cations of Type C have been examined in detail by Pavlik and Pasteris.[174]

The photoisomerizations of the tetramethylphenols shown in Scheme 54 represent a situation where all the cations are photochemically reactive and undergo isomerization on irradiation. With many other protonated phenols, one product is frequently photochemically inert and this then ends up as the major product on continued irradiation. For example, protonated 2,3-, 2,4-, 2,5-, 2,6- and 3,4-dimethyl phenols all readily undergo photoisomerization on irradiation. However, protonated 3,5-dimethylphenol is essentially photochemically inactive and builds up as the eventual end product on continued irradiation of many of these protonated dimethylphenols.[157,179]

The quantum efficiencies of the photoisomerizations of a variety of

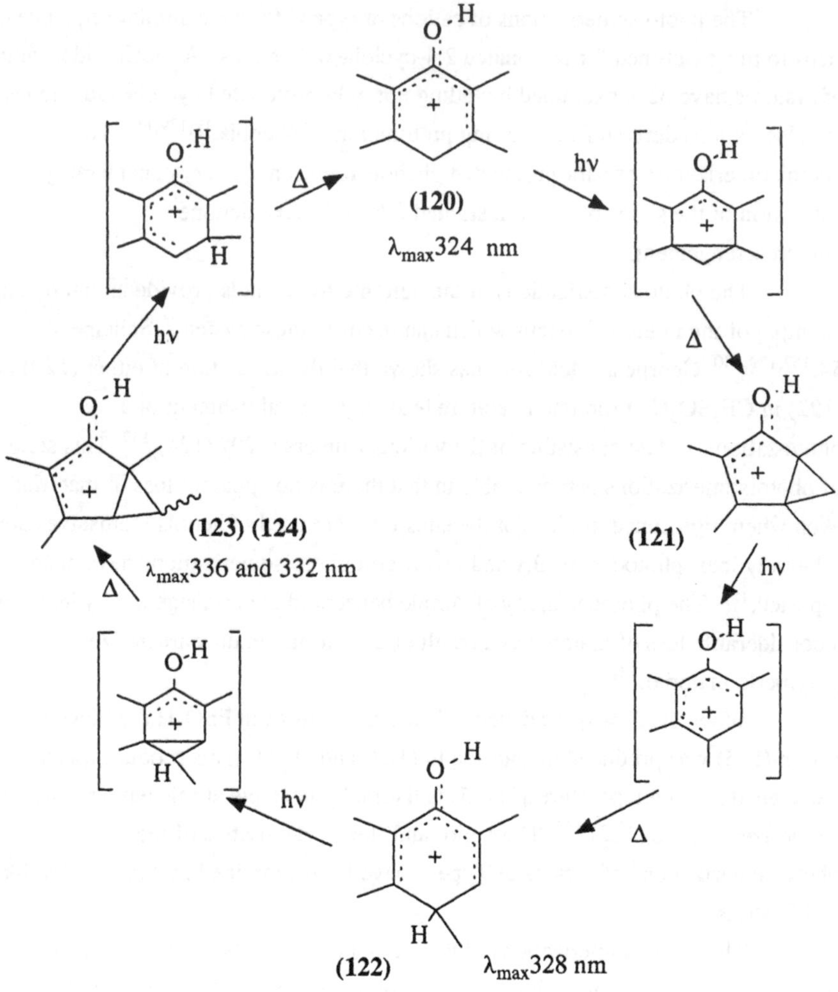

Scheme 54.

para-protonated phenols (126) to the corresponding protonated bicyclic-ketones (128) have been measured by George, Table 6.[157] Considerable variation in the efficiencies of these photo-reactions was found. Methyl groups at positions C_2 and C_6 of the starting protonated phenol increase the efficiency of the reactions while methyls at C_3 and C_5 have the opposite effect. Surprisingly, methyl groups at C_4 were found to have little effect on the efficiencies of the reactions despite the

(125)

importance of methyl substitution at the migrating cyclopropyl carbon in the circumambulatory rearrangements of bicyclo[3.1.0]hexenyl cations.[148] It was suggested that this methyl group dependence can be understood in terms of a competition between the two thermal isomerizations of the initial photoproduct (127). Methyl groups at the bridgehead positions in cations such as (127) will facilitate the ring opening reaction whereas methyls at C_2 and C_6 will stabilize the intermediate (127).[157]

The ratio of k_1/k_2 for the competing isomerizations of (127,R_2=R_3=R_5=R_6=Me) has been shown from independent work to be 0.125.[157] This means that even if (127,R_2=R_3=R_5=R_6=Me) is formed with a quantum efficiency of 1 on irradiation of (120) the observed quantum yield for the formation of (121) can only be 0.2. Indeed the observed quantum yield for the formation of (121) (0.24) is of this magnitude suggesting that energy loss in the conversion of (120) to (121) stems almost entirely from the thermal reversion of (127,R_2=R_3=R_5=R_6=Me) to the starting material.[176]

The parallel between the photoisomerizations of these type C hydroxybenzenium ions and the corresponding cyclohexadienones is most marked. The primary photochemical step of these latter systems is a photochemical ring closure to give a zwitterion intermediate corresponding to (127). This is followed by either a circumambulatory rearrangement or ring opening to regenerate the starting material.[170,182,183] As such it is perhaps not surprising that similar stereoselectivities and regioselectivities are observed in the neutral and protonated systems.

The photoisomerizations of the type C hydroxybenzenium cations represent a convenient synthetic entry into the bicyclo[3.1.0]hexenones. For example, Baeckström and co-workers have used the photoisomerization of protonated thymol as the basis of a one step synthesis of umbellone.[179] The

Table 6. Quantum efficiencies for isomerization of protonated phenols.

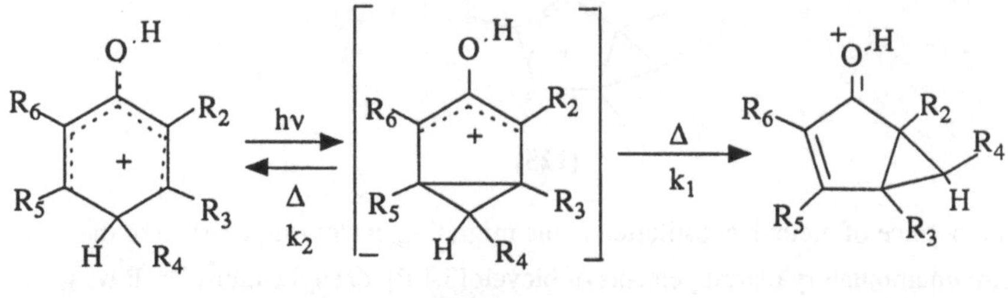

	R_2	R_3	R_4	R_5	R_6	$\phi_{126\rightarrow128}$
	Me				Me	0.65
	Me	Me				0.38
		Me		Me		0.018
	Me	Me			Me	0.50
	Me	Me	Me		Me	0.49
	Me	Me		Me	Me	0.24
	Me	Me	Me	Me	Me	0.20

McMaster group has shown that in terms of preparative scale reactions it is possible to carry out the irradiations of many of these protonated phenols, in triflic acid with concentrations of the phenols up to 20% (w/v) without significant loss of material.[184] Photochemical reactions are not normally carried out at such high concentrations, however, the cationic nature of the protonated phenols reduces their propensity to undergo intermolecular reactions. The reaction can be used with protonated phenol itself to give the parent bicyclic ketone, however, SbF$_5$ has to be added to the FSO$_3$H in order to generate the starting cation and under these conditions oxidation of the organic materials occurs.[184]

    One of the difficulties with the use of these reactions to prepare the bicyclo[3.1.0]hexenones is the formation of various isomers in these photoreactions and the ensuing difficulty of separating the products of the reactions. This problem has been partly overcome by the use of strong Lewis acids such as Al_2Br_6 to catalyze the photoreactions of methyl substituted phenols.[185]

One example of the photoisomerization of a type D cation has been reported, Scheme 55.[179] Irradiation of (129H) in FSO_3H at -70°C using light of wavelength greater than 360 nm leads to the formation of a photostationary state consisting of (130) and (129H) in approximately equal amounts. The reaction would seem to involve the selective excitation of the meta-protonated isomer (131) which is in equilibrium with other species in solutions of (129H) in FSO_3H.[179] Irradiation of (129H) using shorter wavelength light leads to the formation of products deriving from the para-protonated isomer.[178]

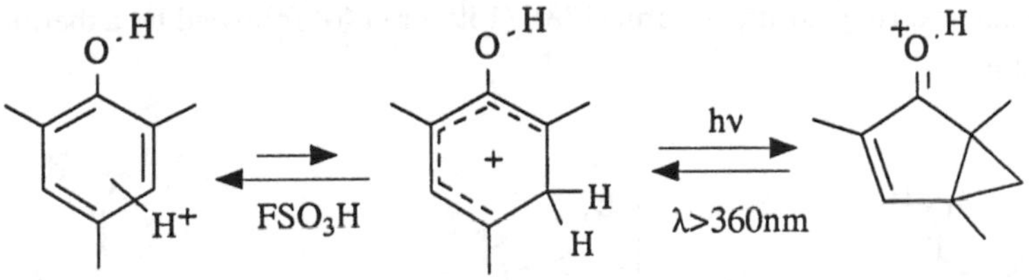

(129H)   (131)   (130)

Scheme 55.

3.   Protonated cycloheptadienones and larger ring dienones. The photochemistry of protonated 2,4- and 2,6-cycloheptadienone and a variety of their alkyl substituted derivatives has been investigated. Three different types of photoisomerizations of protonated cycloheptadienones have been reported -

◊   ring closure to form a bicyclo[3.2.0]heptenyl cation;

◊   ring opening of the seven membered ring to give a heptatrienyl cation;

◊   ring contraction to form a six-membered ring.

As with the hydroxybenzenium ions considered in the previous section, the initial photochemical step usually gives rise to a carbenium ion in which the OH group is not optimally placed to stabilize the positive charge. As a result this photochemical step is followed by a series of thermal rearrangements which result in the formation of more stable cations.

The ring closure of a protonated cycloheptadienone is the most widely encountered photoisomerization. This reaction, which would seem to take place with both the protonated 2,4- and 2,6-cycloheptadienones, is directly comparable to the ring closures observed with the benzenium and hydroxybenzenium ions. For example, Hine has shown that irradiation of protonated cyclohepta-2,4-dienone (132H) or its 2-methyl derivative (133H) lead to the formation of the protonated 7-norbornenones (134H) and (135H), respectively, in high chemical yield, Scheme 56.[186] A similar photoisomerization is observed with (136H) although in this case (137H) is only a minor product (4%) of the overall reaction.[187] (The major products formed for (136) are discussed in a later section of this review.) The formation of the protonated norbornenones can be understood in terms of an initial photochemical ring closure to form (138), (139) or (140) followed by a thermal 1,2 alkyl shift.

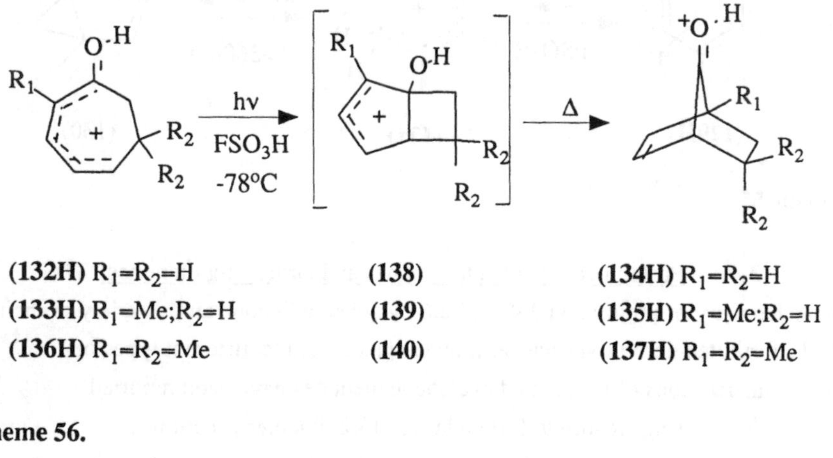

(132H) $R_1=R_2=H$	(138)	(134H) $R_1=R_2=H$
(133H) $R_1=Me;R_2=H$	(139)	(135H) $R_1=Me;R_2=H$
(136H) $R_1=R_2=Me$	(140)	(137H) $R_1=R_2=Me$

Scheme 56.

The photoisomerizations of the protonated cycloheptadienones (132H) and (133H) can be compared to those of the corresponding neutral ketones.[186] Irradiation of (132) or (133) in ether or methanol as solvent leads to the formation of (141) and (142), respectively, rather than 7-norbornenones, Scheme 57. These reactions of (132) and (133) involve bonding between C_2 and C_5 of the dienone rather than C_1 and C_5 as takes place with the protonated materials. However, when (133) is irradiated in aqueous acetic acid a mixture of (142) and (134) is obtained.[186] The neutral dienone (133) was shown to be the species absorbing light in this weakly acidic medium and it would appear that protonation of (133) occurs at some stage after photoexcitation. The exact timing of this protonation has not been determined. However, it is clear that the two types of ring closure are closely related processes.

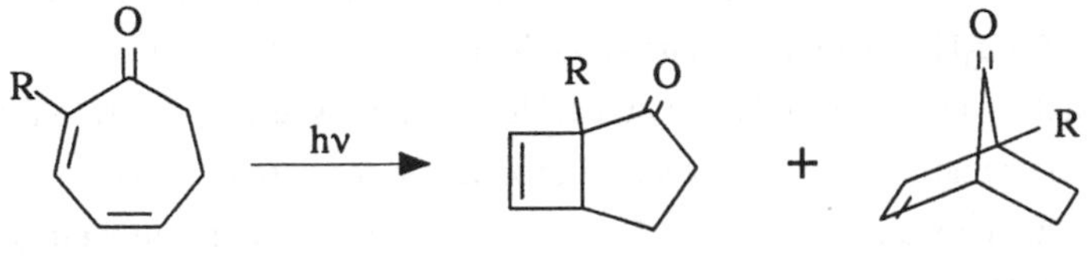

(132) R=H   (141) R=H   (134) R=Me
(133) R=Me   (142) R=Me

Scheme 57.

The photoisomerizations of (136) are complex and a variety of products are formed including the bicycloheptenone (143), the product of C_2,C_5 bonding.[188,189] However, once more a competition between the two types of ring closure (C_2C_5 versus C_1C_5 bonding) is observed in weakly acidic media. The onset of C_1C_5 bonding was shown to occur as the pH of the solvent becomes less than 2.[188] The irradiation of (136) in acidic methanol solutions was shown to give rise to the methanol addition product (144) as well as isomers of the starting ketone.[188] This product can be considered to result from nucleophilic trapping of either the carbenium ion intermediate (140), or the corresponding zwitterion. The isolation and characterization of (144) gives strong evidence for the correctness of the sequence of steps shown in Scheme 56.

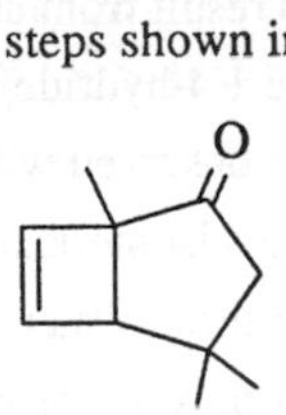

(143) 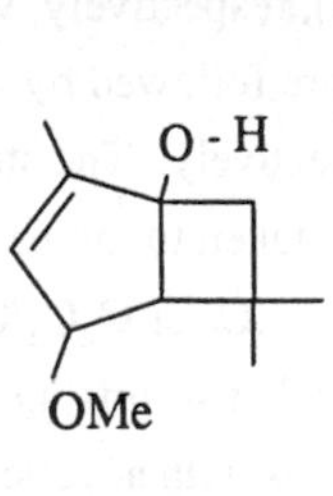   (144)

The photoisomerization of protonated 2,6-cycloheptadienone, (145), has been reported by Noyori and colleagues to lead to the formation of products arising from a ring opening reaction, vide infra.[190] However, it would again seem likely that a photochemical ring closure reaction is also occurring with this cross

conjugated system but that (146), the product of this 5-C ring closure, is unstable
and reverts to the starting material. Evidence for this suggestion comes from the
irradiation of (145) in acetic acid solutions where (147) is formed as one of the
products.[191] In FSO_3H, where there is no suitable nucleophile to trap an
intermediate such as (146), this relatively unstable cation must either revert to the
starting material or rearrange to a more stable product. In the case of (146)
1,2-alkyl shifts do not readily lead to the formation of a stable cation and thermal
reversion to (145H) would be the preferred pathway.

| (145) | (147) | (146) |

Further examples of 5-carbon ring closure reactions of protonated
cross-conjugated dienones have been reported by Noyori and co-workers for the
larger ring systems shown in Schemes 58 and 59.[192] Irradiation of protonated
2,8-cyclononadienone, (148), and 2,11-cyclododecadienone (149) lead in each case
to the formation of two types of product. In each of these reactions one of the
products, (150) and (151), respectively, would appear to result from an initial
photochemical ring closure followed by a stereoselective 1,4-hydride shift,
Schemes 60 and 61, respectively. The stereoselectivities observed with these
isomerizations have been taken to indicate that no cis/trans isomerization is
occurring about the C_2C_3/C_8C_9 or $C_2C_3/C_{11}C_{12}$ bonds in these systems on
irradiation. This is somewhat surprising in these larger ring systems given the
efficiency of these reactions with acyclic protonated enones. The steps involved in
the formation of (152) and (153) are not known.

The second type of photoisomerization which has been found with
protonated cycloheptadienones involves opening of the ring to give a hydroxy
substituted heptatrienyl cation. Heptatrienyl cations are well known to rearrange
to substituted cyclopentenyl cations at low temperatures[142] and comparable
rearrangements of these photochemically generated hydroxy-heptatrienyl cations

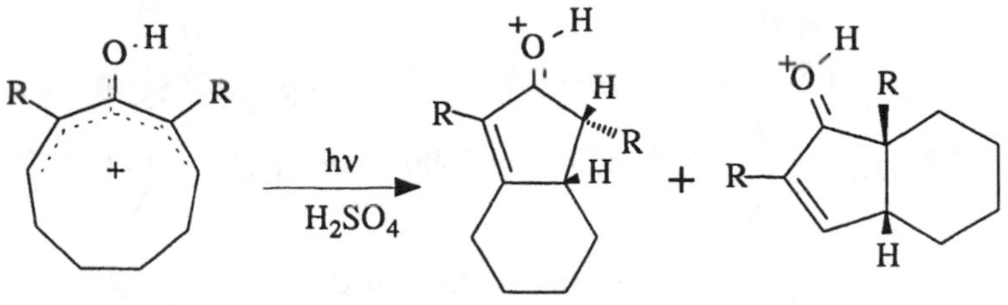

(148)   (150)   (152)

R=H or D

Scheme 58.

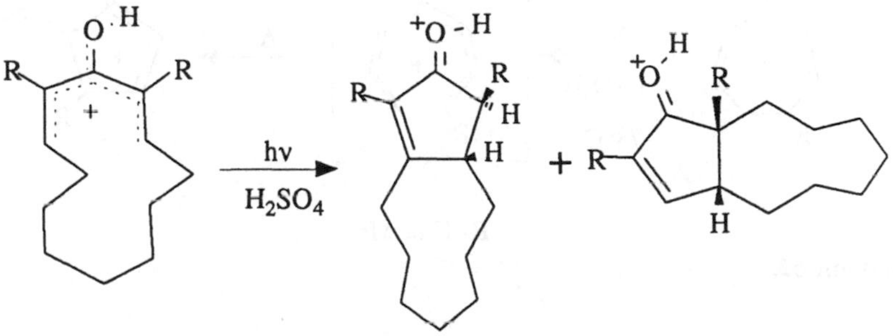

(149)   (151)   (153)

R=H or D

Scheme 59.

occur to give protonated cyclopentenones. Three examples of this type of photoreaction have been reported, Schemes 62[190] and 63.[193] Both protonated 2,4- and 2,6-cycloheptadienones undergo this type of reaction. These photochemical 6-electron electrocyclic ring openings correspond to those observed with the isoelectronic 1,3-cyclohexadienes.[194]

The third type of photoreaction of cycloheptadienones involves ring contraction to form a six-membered ring. Only a single example of this type of photoisomerization has been reported, Scheme 64.[187] As is shown, three primary

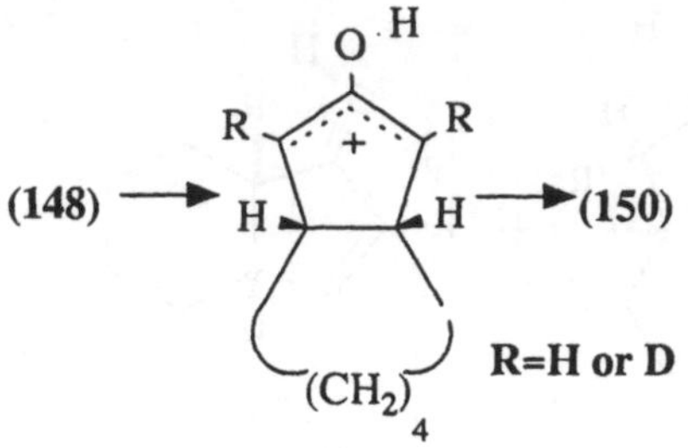

$(148) \longrightarrow$

$R=H$ or D

Scheme 60.

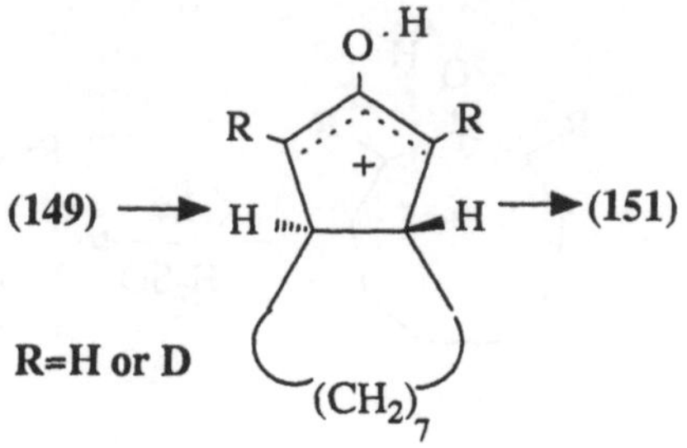

$(149) \longrightarrow$

$R=H$ or D

Scheme 61.

$R=H$ or Me

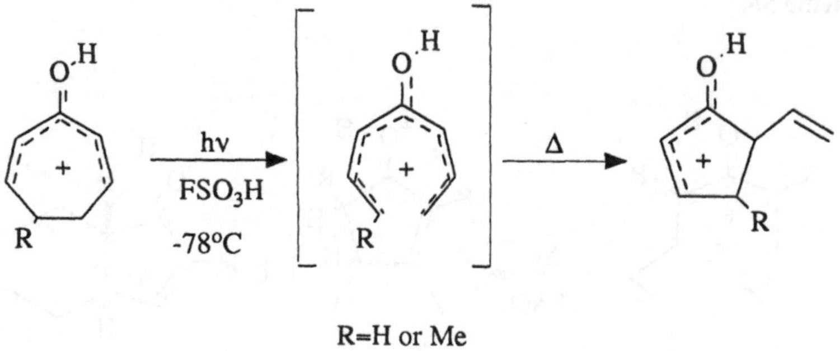

Scheme 62.

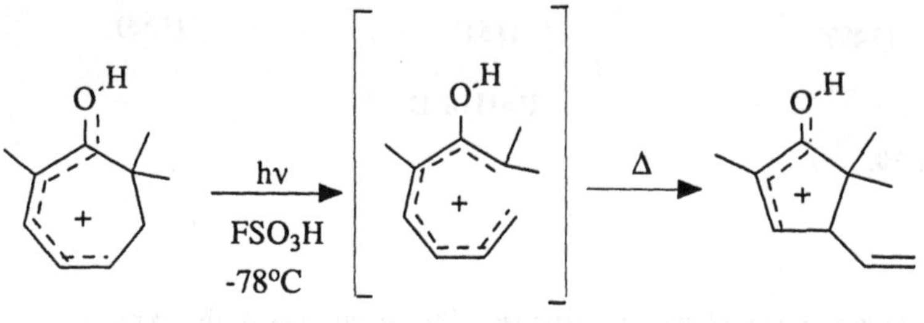

Scheme 63.

photoproducts are obtained on irradiation of (136H). The formation of the minor product, (137H) has already been discussed. The major products of this isomerization are (154) and (155). The bicyclic cation, (154) is both thermally and photochemically converted to (155).

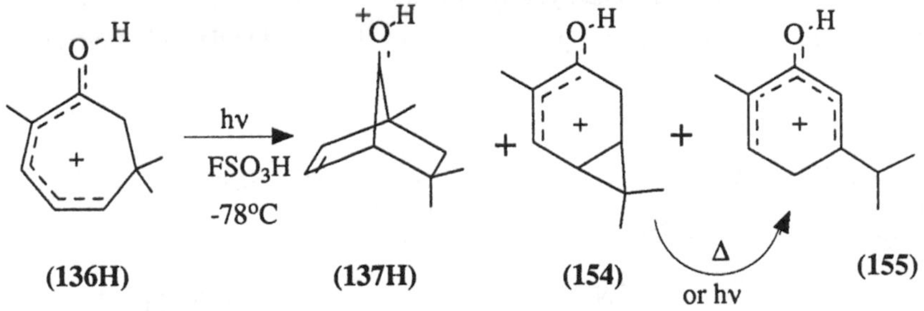

(136H)   **(137H)**   **(154)**   Δ or hv   **(155)**

Scheme 64.

The formation of (154) and (155) on irradiation of (136H) parallels the photoisomerizations observed with protonated 4,4-dimethylcyclohexenone (62) and the isopropyl-substituted enones discussed in section III,B,4. It would appear that the conversion of (136H) to (154) and (155) (and also the conversion of (154) to (155)) involves the tertiary cation (156) as an intermediate. The suggestion that the formation of (154) and (155) from (136H) involves an electrocyclic ring opening to form an open chain heptatrienyl cation would seem unlikely given the preference for this type of cation to thermally close to a protonated cyclopentenone.[187,193]

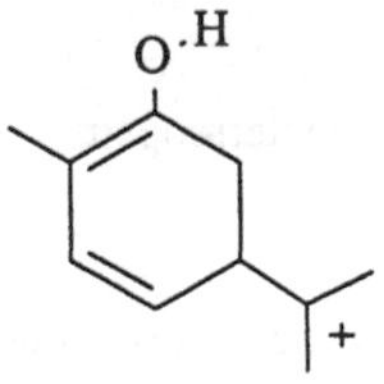

(156)

C.   Nitrogen Substituted Dienyl and Polyenyl Carbenium Ions

The major interest in these cations stems from their involvement in two important biological processes. The visual pigment rhodopsin contains an 11-cis retinal chromophore linked to lysine 296 via a protonated Schiff base.[5,195] During the visual process the retinal moiety undergoes a photoisomerization about the $C_{11}C_{12}$ bond followed by a thermal hydrolysis to form all-trans retinal and opsin, Scheme 65. The mechanism of this process has been discussed in detail in several reviews.[5]

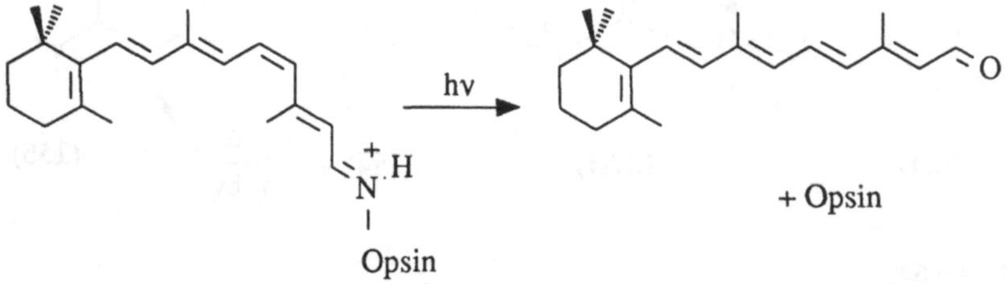

+ Opsin

Opsin

Scheme 65.

A photoisomerization is also the primary event for the photoprotein bacteriorhodopsin. In the light adapted state, bacteriorhodopsin contains an all-trans retinal chromophore which undergoes a photoisomerization about the $C_{13}C_{14}$ bond to form the 13-cis isomer, Scheme 66.[196-198] Unlike rhodopsin, the chromophore and protein do not dissociate in this process.[199]

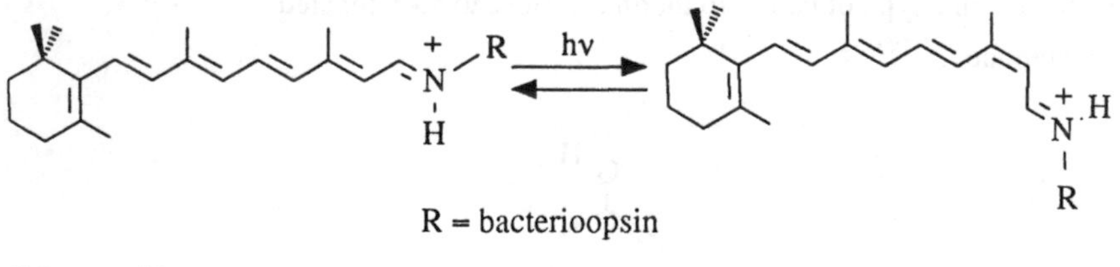

R = bacterioopsin

Scheme 66.

The photochemistry of nitrogen substituted polyenyl cations is similar to that of the allyl systems with the major decay pathway being *cis/trans* isomerization. However, the increased conjugation with the iminium function in these polyenyl systems can result in reduced barriers to thermal *cis/trans* isomerization.[130,200-202] Thus a careful choice of counterion, solvent and temperature must be made.

1.   <u>Dienyl and trienyl nitrogen substituted</u> cations. The photophysical properties of only a few dienyl and trienyl iminium ions have been studied. In general the properties of these ions are similar to the nitrogen substituted allyl

cations. The absorption spectra are characteristic of a π,π^* transition. Additional
conjugation to a cation such as (157) increases the λ_{max} by about 55 nm per
additional double bond.[203] Fluorescence emission has been examined in
compounds (158-160) and in each case has been found to be very weak ($\phi_f \leq 0.028$),
suggesting that only a small energy barrier to rotation exists in the excited singlet
state. Photoisomerization reactions of (158-160) have not been reported.[204]

(157)

(158) R = NH$^+$-nBu

(159) R = CH-C(CH$_3$)=NH$^+$$n$Bu

(160) R = CH-C(CH$_3$)=CH-CH=NH$^+$$n$Bu

Photoisomerization about the C$_2$C$_3$ and C$_4$C$_5$ bonds has been studied for
two dienyl iminium ions, Scheme 67.[203] Thus irradiation of (161) at 313 nm for an
extended period of time yielded primarily the 2-cis product (162). A similar result
was noted for (163). The quantum efficiencies for the formation of the 2-cis
isomers (162 and 164) were found to be 0.56 and 0.60 respectively. The selectivity
of these photoisomerizations agree well with theoretical work by Dormans and
co-workers[128b] which suggested that photoisomerization about the C$_2$C$_3$ bond in a
pentadienylidene iminium ion should be the favored pathway.

The photoisomerization of an octatrienyl iminium ion has also been
investigated qualitatively.[203] Irradiation of N-t-butyl octatrienylidene iminium
perchlorate (165) at 350 nm yielded the 2-cis isomer (166) as the major product,
Scheme 68.

2.   Photoisomerization of retinylidene iminium ions. The
photoisomerization of an all-trans retinylidene ion can in principle occur about the
C$_{15}$N, C$_{13}$C$_{14}$, C$_{11}$C$_{12}$, C$_9$C$_{10}$ or C$_7$C$_8$ bonds. Photoisomerization about two or
more double bonds has not been observed for these iminium salts as a result of a
single excitation. No photoisomerization about the C$_{15}$N bond of a retinylidene
iminium salt has been reported even when these reactions were carried out at low
temperatures.[205] Facile *cis/trans* thermal isomerization about C$_{15}$N occurs in

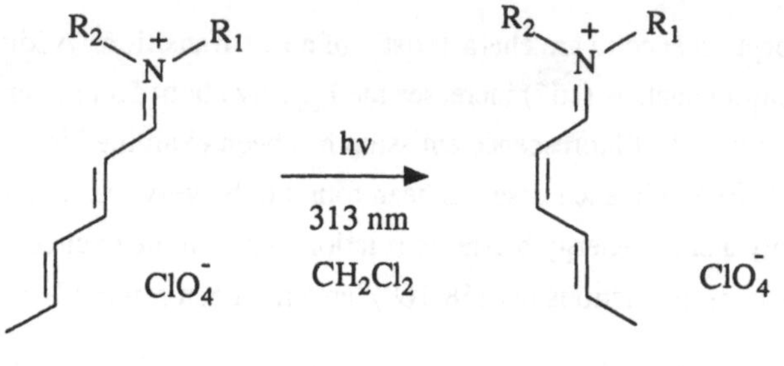

(161) R_1= *t*-Bu,R_2=H
(163) R_1=R_2=Me

(162) R_1= *t*-Bu,R_2=H
(164) R_1=R_2=Me

Scheme 67.

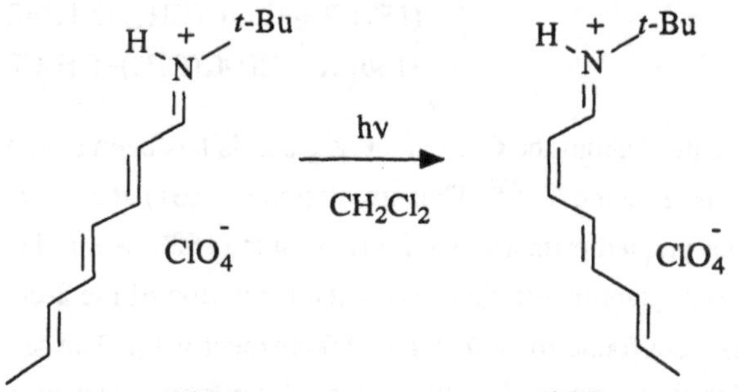

(165) (166)

Scheme 68.

retinylidene ions and this could be the reason for the failure to detect photoisomerization about the $C_{15}N$ bond.[128d,200,202]

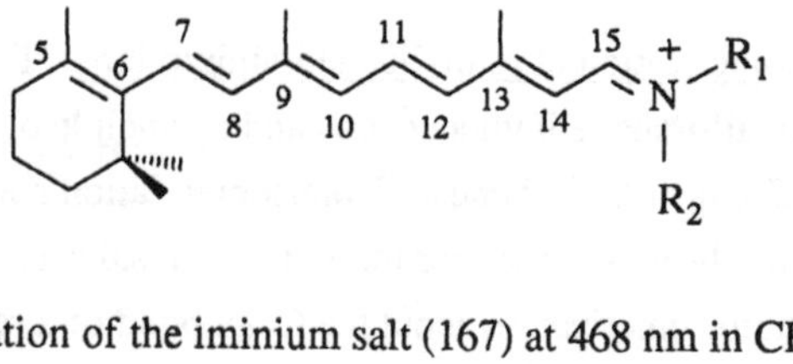

Irradiation of the iminium salt (167) at 468 nm in CH_2Cl_2 produces the

11-cis iminium ion (168) as the primary isomer, Scheme 69.[205] The quantum yield for this isomerization is 0.25. The 11-cis isomer is also formed regioselectively on irradiation of the n-butyl and N,N-dimethyl iminium derivatives of all trans retinal. It is worth noting that the regioselectivity observed in these photoisomerizations is different from that found with bacteriorhodopsin where the 13-*cis* isomer is formed. The position of the counterion and any other secondary negative "point" charges would seem to be important in determining the regioselectivity.

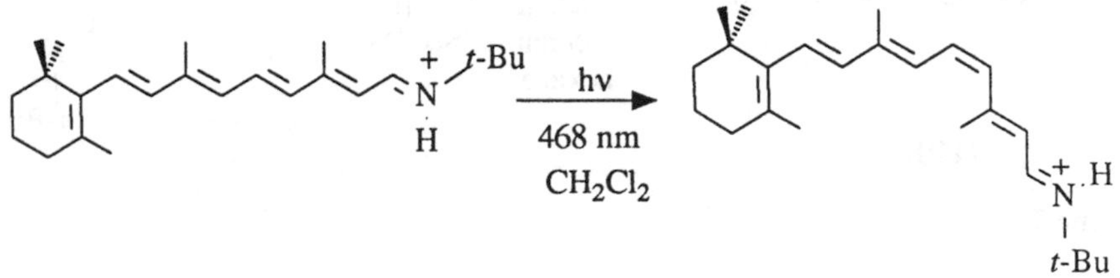

(167)   (168)

Scheme 69.

The photoisomerization of retinylidene iminium ions has been examined under a variety of different conditions.[206,207] For example, Freedman and Becker have reported the photoisomerization of (169) in two different solvents.[207] Irradiation of (169) in hexane yields four cis isomers, Scheme 70, with an overall quantum efficiency of 0.14. In methanol, the quantum efficiency remains the same as that in hexane ($\phi = 0.13$), however, the ratio of the cis isomers is markedly different.

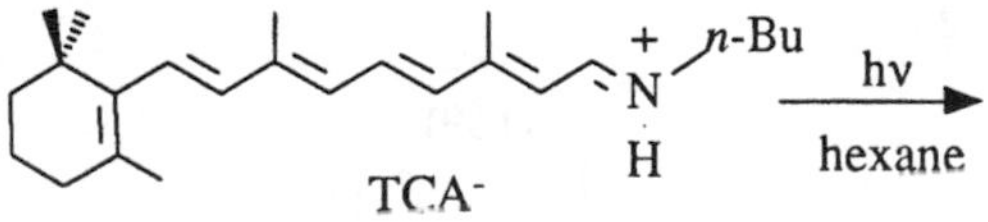

	13-*cis*,	11-*cis*,	9-*cis*,	7-*cis*
hexane	19%	70%	11%	trace

(169)

Scheme 70.

The *trans/cis* photoisomerization of retinylidene iminium salts such as (169) are very different from the corresponding neutral Schiff base.[207] Thus,

irradiation of (170) at 355 nm yields only the 13-cis isomer in an inefficient
photoisomerization process ($\phi \leq 0.01$), Scheme 71. In methanol the quantum
efficiency increases to a similar value as found for (169), Scheme 70. However,
even in methanol the product distribution obtained for (170) differs from that
derived from (169).

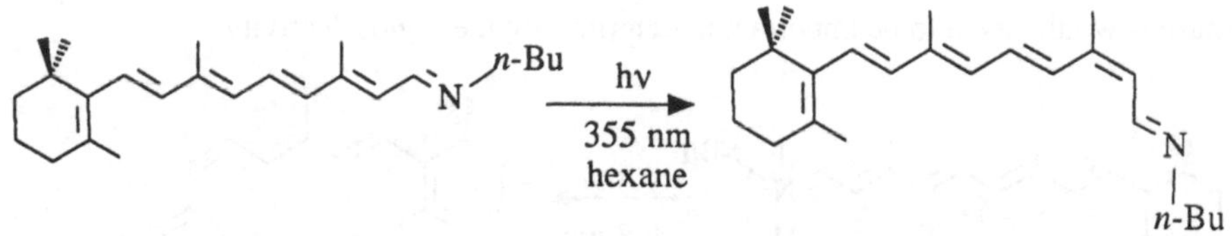

(170)

Scheme 71.

Photochemical *cis/trans* isomerization of various *cis* retinylidene iminium
salts has also been investigated. Becker and colleagues have reported a
comprehensive investigation of the photoisomerization about the $C_{11}C_{12}$ bond of
(171), Scheme 72.[207-210] Irradiation of (171) at 355 nm gives only the all-trans
isomer (169). The quantum yield for this process appears to be independent of the
wavelength of light used as well as the reaction medium.

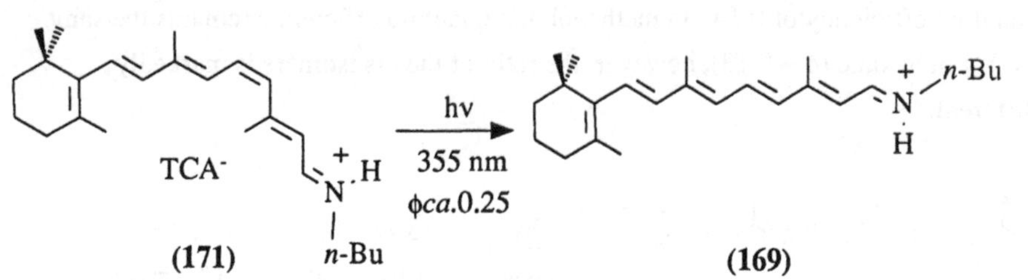

(171)   *n*-Bu   **(169)**

Scheme 72.

The photoisomerization of a series of 11-cis retinylidene ions having
oligopeptide substituents on nitrogen has been investigated, Table 7.[211] These
compounds, which are better models of the natural chromophore, exhibit higher
quantum efficiencies for the formation of the all-trans iminium ion than the n-butyl
derivatives (171). Recently calculations have been done in an attempt to

understand the quantum yield differences for *trans/cis* and *cis/trans* isomerizations in bacteriorhodopsin and rhodopsin as compared to the retinylidene ions, above.[128b]

Table 7. Photoisomerization of 11-*cis*-retinylidene ions.

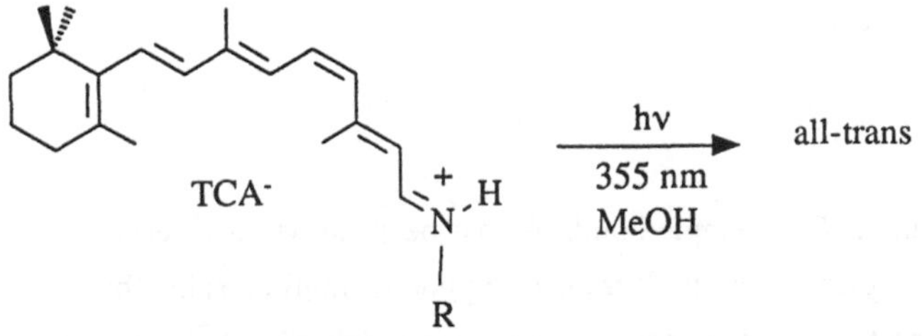

Cmpd.	R	ϕ
(172)	PEG-NH$_2$	0.22
(173)	PEG-Ala-Lys-Boc	0.28
(174)	PEG-Ala-Lys-Glu	0.27
(175)	PEG-Ala-Lys-Trp	0.26

(PEG = polyethylene glycol)

The photophysics of retinylidene iminium ions have recently been reviewed.[212] It would appear that these isomerizations occur from singlet excited states. In general these compounds exhibit little or no fluorescence ($\phi_f \leq 10^{-3}$) at room temperature.[213,214] Most triplet sensitization experiments have shown that triplet excited states are not involved.[207,214,215] However, one recent account has claimed that triplet excited states can be observed following laser excitation[216], although it is not clear whether isomerization can occur from this triplet state.

Cis/trans isomerization about the C$_9$C$_{10}$ and C$_{13}$C$_{14}$ bonds of retinylidene iminium salts also occurs on irradiation. In general these reactions have been found to be less efficient than isomerization about the C$_{11}$C$_{12}$ bond.[207] The 9-cis retinylidene ion (176) forms the all-trans isomer (169) with a quantum efficiency of ≤ 0.05, Scheme 73. The 13-cis isomer also forms (176) on irradiation with a similar quantum yield.

One further factor in the photoisomerization of retinylidene iminium ions

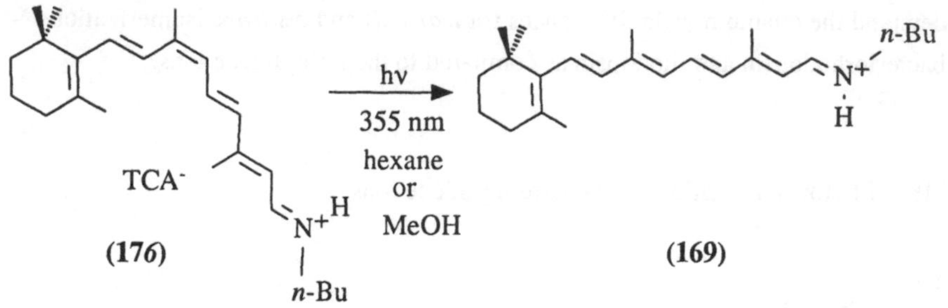

Scheme 73.

is the configuration about the C_6C_7 single bond. As has been shown in several studies with bacteriorhodopsin and with all-trans retinylidene iminium salts, this bond may exist in either a 6-s-cis or 6-s-trans configuration, Scheme 74.[217] These two conformers have different ground state and absorption properties.[128b] The influence of this conformational change on the regioselectivity or the quantum efficiency of the photoisomerization of retinylidene iminium salts is not understood at this time.

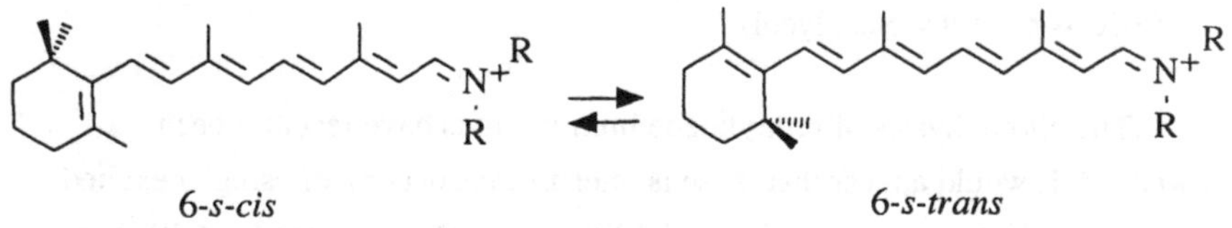

6-s-cis   6-s-trans

Scheme 74.

V   PHOTOREACTIONS OF AROMATIC AND HOMOAROMATIC CATIONS

A.   Cyclopropenium Cations

The photochemistry of the triphenylcyclopropenium cation (177) has been examined by van Tamelen and co-workers.[15,218] Irradiation of (177) in 10% aqueous sulfuric acid leads to the formation of hexaphenylbenzene in good overall yield. It was suggested that the photochemical step involved in this transformation is the reduction of an excited state of (177) to the corresponding radical by single electron transfer from an unspecified donor, Scheme 75. The dimerization of (178) and subsequent photochemical conversion of (179) to hexaphenylbenzene are known from previous work.[219]

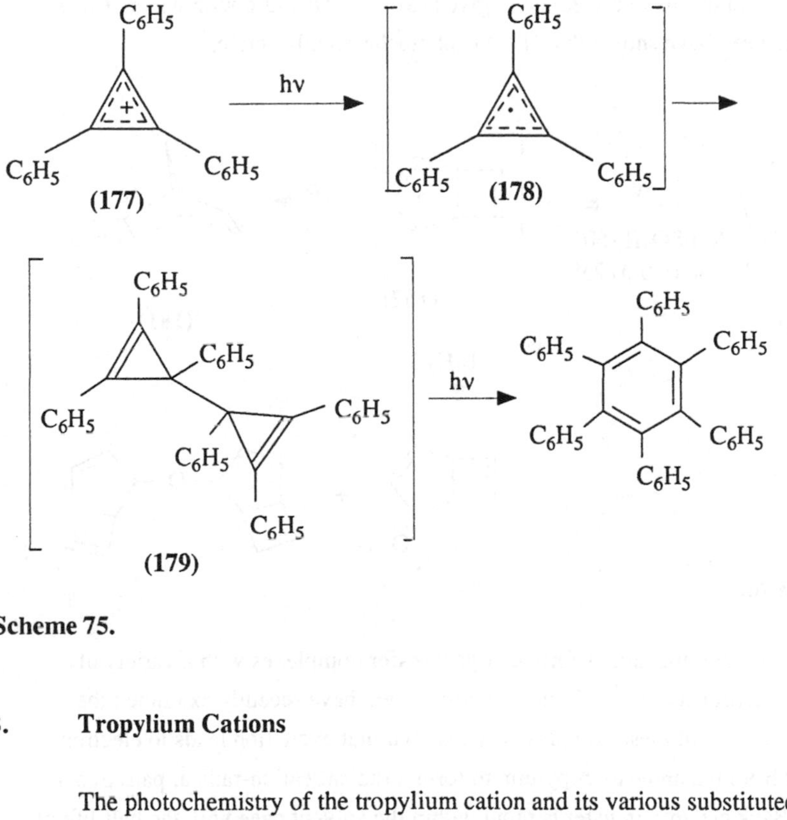

Scheme 75.

B. Tropylium Cations

The photochemistry of the tropylium cation and its various substituted derivatives has been studied under a variety of conditions. These range from super acid media where photoisomerizations are observed, to dilute aqueous acids where trapping of intermediate photoisomers takes place, to reactions in the presence of electron donors where single electron transfer reactions occur.

The photoisomerization of the tropylium ion in super acid media has been examined by Taguchi at McMaster[17] and Hogeveen and Gaasbeek.[18] Irradiation of

(180) in FSO_3H at -60°C leads to the formation of the 7-norbornadienyl cation, (181), as the sole product. Van Tamelen and colleagues have studied the photoisomerization of (180) in a variety of dilute acid media.[15,218] Under these conditions the alcohol and ether derivatives shown in Scheme 76 were formed. It is clear that all of these photoreactions of the tropylium ion proceed by an initial photochemical ring closure to form (182) followed either by a thermal isomerization of this intermediate to give (181)[220] or capture with water. Cabell and Hogeveen have shown that (181) is photochemically stable.[44]

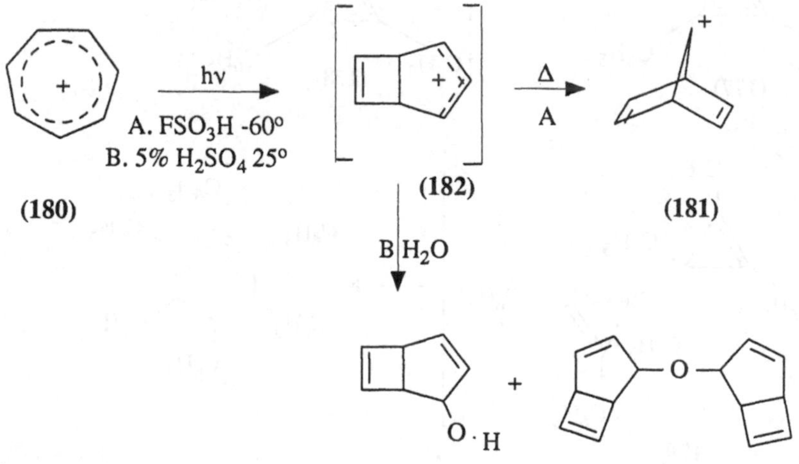

Scheme 76.

Tropylium cations form charge transfer complexes with a variety of aromatic hydrocarbons.[38] Kochi and co-workers have recently examined the photochemistry of these complexes and shown that excitation leads to electron transfer from the arene to tropylium to form a radical/cation-radical pair, Scheme 77.[221] Back electron transfer is rapid within the solvent cage with the half life of the cation radicals being typically of the order of 15 ps. This short lifetime is such that only very fast intramolecular reactions of arene cation radicals such as the intramolecular cycloreversions of dianthracene or hexamethyl Dewar benzene cation radicals, can compete with back electron transfer. Steady state irradiation of tropylium/arene charge transfer complexes for long periods of time did not lead to the formation of any detectable products.[221]

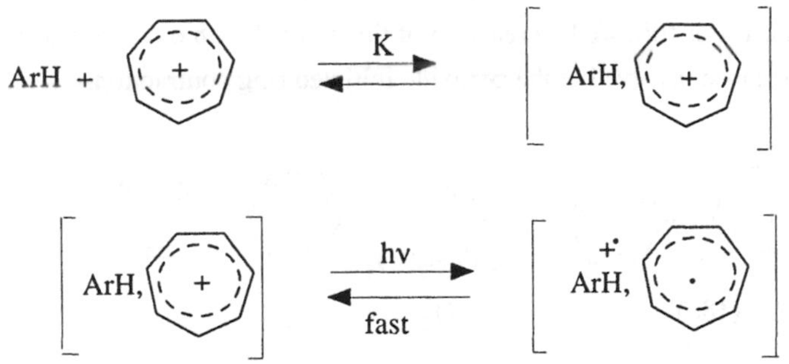

ArH + (tropylium cation +) $\xrightarrow{K}$ [ArH, (tropylium cation +)]

[ArH, (tropylium cation +)] $\underset{fast}{\overset{hv}{\longrightarrow}}$ [ArH, (tropylium radical cation)]

Scheme 77.

The photoisomerizations of several substituted derivatives of the tropylium cation have been examined. Methyltropylium cation is converted to 2-methyl-7-norbornadienyl cation on irradiation in FSO_3H.[17] This photoisomerization presumably goes by a similar route to that shown for the parent cation in Scheme 76.[17]

The phenyltropylium ion, (183) is photochemically stable when irradiated in FSO_3H or in aqueous sulfuric acid in the absence of oxygen.[17,218] The chemical stability of phenyl- and other aryltropylium cations towards light is perhaps disappointing in terms of the preparation of substituted norbornadienyl cations. However, the lack of reactivity of these colored ionic materials leads to several potential uses, for example, in photoconductive coatings and electrophotographic sensitive materials as well as photographic dye print out materials.[222]

The phenyltropylium cation does undergo photoreactions under some conditions. Van Tamelen et al showed that irradiation of aqueous sulfuric acid solutions of (183) in the presence of oxygen leads to the formation of biphenyl, Scheme 78.[218] The mechanism of this transformation is not known. One possibility which was suggested is that the reaction involves the photochemical production of hydrogen peroxide which then reacts in a dark reaction with (183) to give biphenyl. The formation of biphenyl on treatment of (183) with hydrogen peroxide has been previously described.[223] In view of the general occurrence of electron transfer processes in the photochemistry of carbenium ions, including the tropylium ion itself, a related possibility is formation of phenyltropylium radical by a single electron transfer followed by reaction of this radical with oxygen to give similar species to those involved in the peroxide initiated ring contraction.

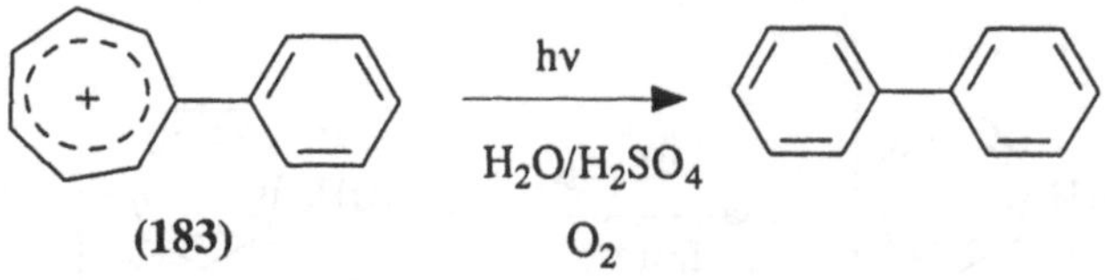

(183)   O$_2$

Scheme 78.

Irradiation of (183) in acetonitrile leads to the formation of *ortho*- and *para*-phenylbenzaldehyde together with *cis*- and *trans*-diphenylstilbene.[218] It is not clear whether this complex photochemical reaction involves the tropylium ion itself or an acetonitrile addition product present in solution.

The photoisomerization of the hydroxytropylium cation, (184), has been investigated.[17] As is shown in Scheme 79, two products, (185) and (186), are obtained on irradiation of (184) in FSO$_3$H at low temperatures. It would seem that the initial photochemical reaction is closure of the seven-membered ring to form either of two bicyclo[3.2.0]hexadienyl cations which differ in terms of the position of the hydroxy substituent. One of these ions, (185), is stable whereas the other, (187), undergoes a 1,2-alkyl shift and the addition of fluorosulfuric acid to give (186).

The vinyltropylium ion (188), formed on protonation of azulene, is converted to a mixture of (189) and (190) on irradiation in aqueous sulfuric acid, Scheme 80.[15,218] It is claimed that the isopropyl group in the products derives from the azulene moiety, however, no mechanism has been proposed to account for the products.

The photochemistry of the dibenzotropylium cation has been examined by Feldman and Thame.[27] Irradiation of an aqueous acid solution of (191) and the corresponding alcohol (192) under conditions where the cation is the light absorbing species leads to the formation of the three products shown in Scheme 81. While the same products can be formed thermally from (191) and (192), control experiments showed that the reaction occurred photochemically, probably from the singlet state of (191).[224] No mechanistic study was undertaken, however, electron transfer from either (192) or another species present in solution to excited (191)

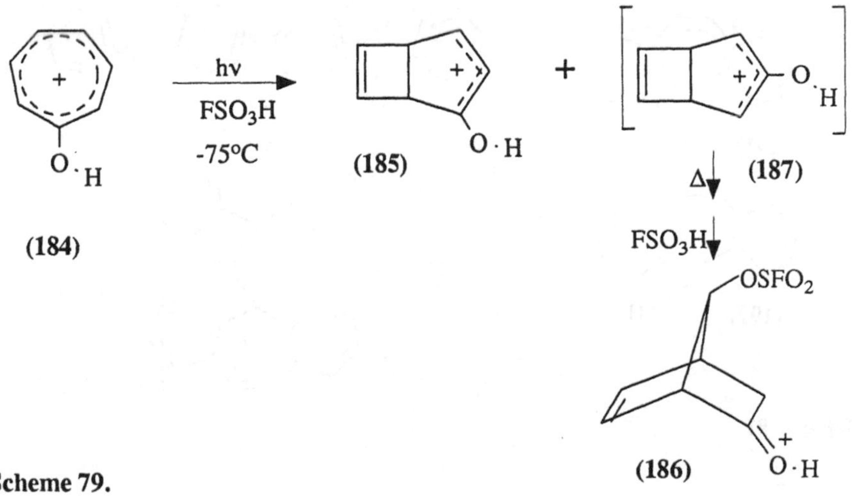

Scheme 79.

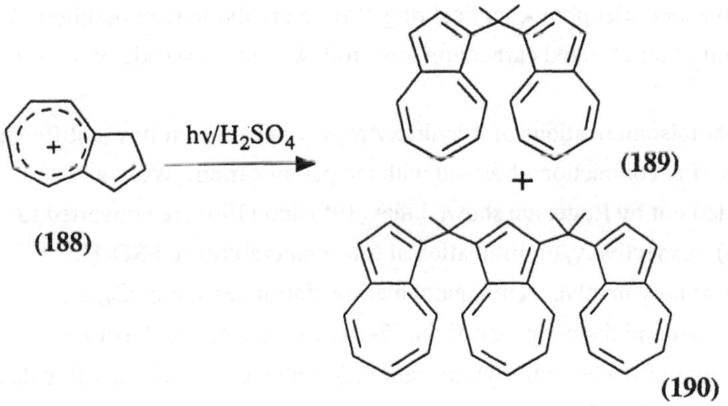

Scheme 80.

could account for the products.

C.   Homotropylium Cations

Hogeveen and Gassbeck reported that the homotropylium cation, (193) was photochemically labile and was converted to a single isomer on irradiation in FSO₃H at low temperature.[18] The identity of this product as (194) was established

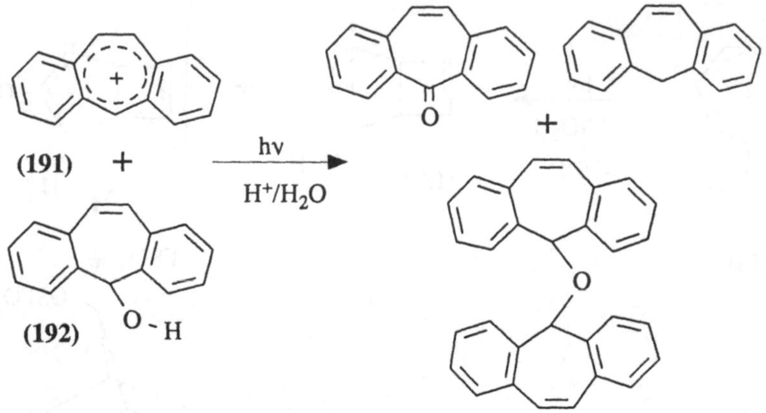

(191)   +

$\xrightarrow{\text{hv}}$
H^+/H_2O

(192)   O-H

Scheme 81.

by Sorensen and colleagues by preparing the ion independently by protonation of 2,4-dihydropentalene, Scheme 82.[225] The formation of (194) from (193) would appear to involve an initial photochemical ring closure involving the ubiquitous 1,5 bonding reaction of unsaturated carbenium ions, followed by a cascade of hydride shifts.[226]

The photoisomerizations of 2-hydroxytropylium cations follow a different course from the ring contraction observed with the parent cation. Work at McMaster carried out by Rogerson showed that (195) and (196) are converted to (197) and (198), respectively, on irradiation at low temperatures in FSO_3H.[227,228] The reactions formally involve a circumambulatory rearrangement of C_8, the bridging carbon, around the periphery of the "7-membered ring" of the closed resonance structure of the homotropylium cation, Scheme 83.[155] It is possible that such a circumambulation also takes place with the parent cation (193), however, in the absence of ring labels it would not have been detected.

This photochemical conversion of 2-hydroxyhomotropylium ions to their 1-hydroxy counterparts occurs with high stereoselectively. In accordance with orbital symmetry, an inversion of configuration at C_8, the migrating carbon occurs, leading to an overall retention of stereochemistry. For example, (199) is converted to (200), the thermodynamically least stable of the two isomeric

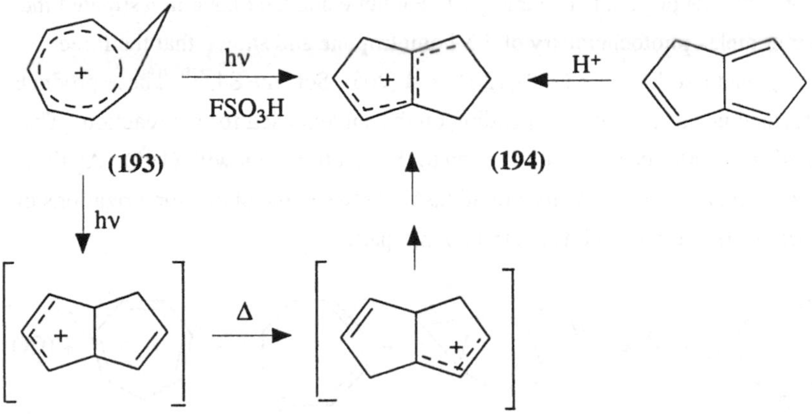

Scheme 82.

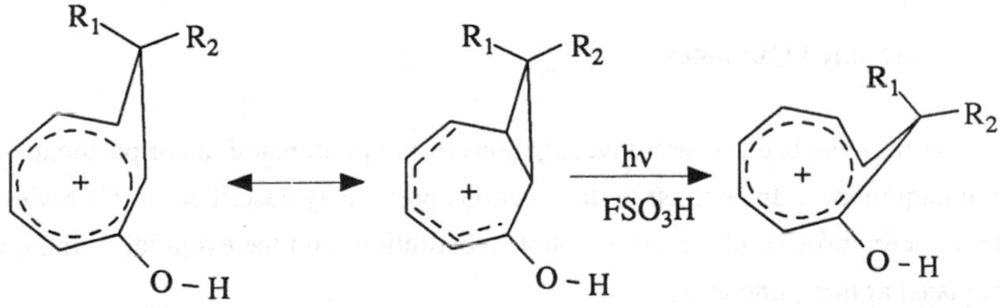

(195) $R_1=R_2=H$		**(197)** $R_1=R_2=H$
(196) $R_1=R_2=Me$		**(198)** $R_1=R_2=Me$
(199) $R_1=Me, R_2=H$		**(200)** $R_1=Me, R_2=H$
(201) $R_1H, R_2=Me$		**(202)** $R_1=H, R_2=Me$

Scheme 83.

8-methyl-1-hydroxyhomotropylium cations, with a 94% stereoselectivity.[227,228] Conversely, (201) is converted solely to (202).

It is interesting to compare the photochemistry of 2,3-homotropone with

that of (195), its protonated counterpart. Paquette and Cox have investigated the
rather complex photochemistry of 2,3-homotropone and shown that the three
primary photoproducts are (203), (204) and (205), Scheme 84.[229] These products
are formed in varying ratios depending on the solvent used for the reaction. The
formation of (204) corresponds directly to the reaction seen with (195). Again it
would seem that there is a close relationship between the photoisomerizations of
the protonated ketone and its neutral counterpart.

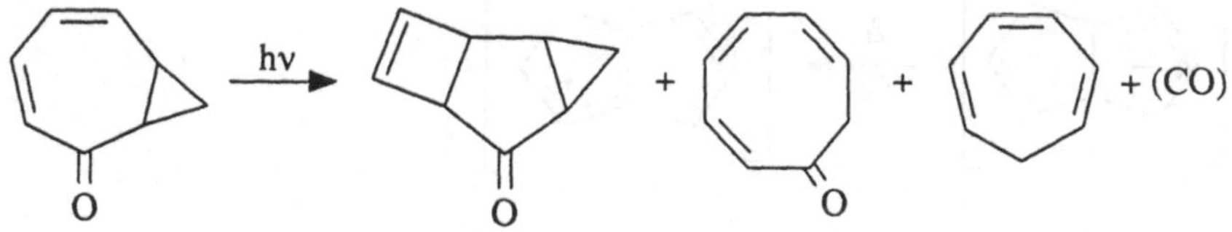

(203)   (204)   (205)

Scheme 84.

D.   Protonated Quinones

There have been several investigations of the photoreactions of protonated
9,10-anthraquinones. In contrast to the reactions previously described in this review
the predominant process observed is a photosubstitution with the eventual formation of
hydroxylated anthraquinones.

Filipescu and co-workers reported that irradiation of (206H) in concentrated
sulfuric acid leads, after an aqueous work up, to the formation of (207) as the major
product with minor amounts of (208) and (209) Scheme 85.[230,231] Broadbent and
Stewart re-examined this reaction in more detail and showed that when the irradiation
is carried out in the absence of oxygen the reduction products (210) and (211) are also
formed.[232] As oxygen is introduced into the reaction mixture the rate of formation of
all products is reduced.

As the strength of the acid medium is varied the site of hydroxylation is
changed. Studzinskii and El'tsov reported that irradiation of (206) in sulfuric acid
solutions with concentrations less than 80% leads primarily to the formation of the
1-hydroxy isomer (208).[232] Anthraquinone is only partly protonated in these less acidic
media[234] and it was suggested that the 1-hydroxylation reaction proceeds via excitation

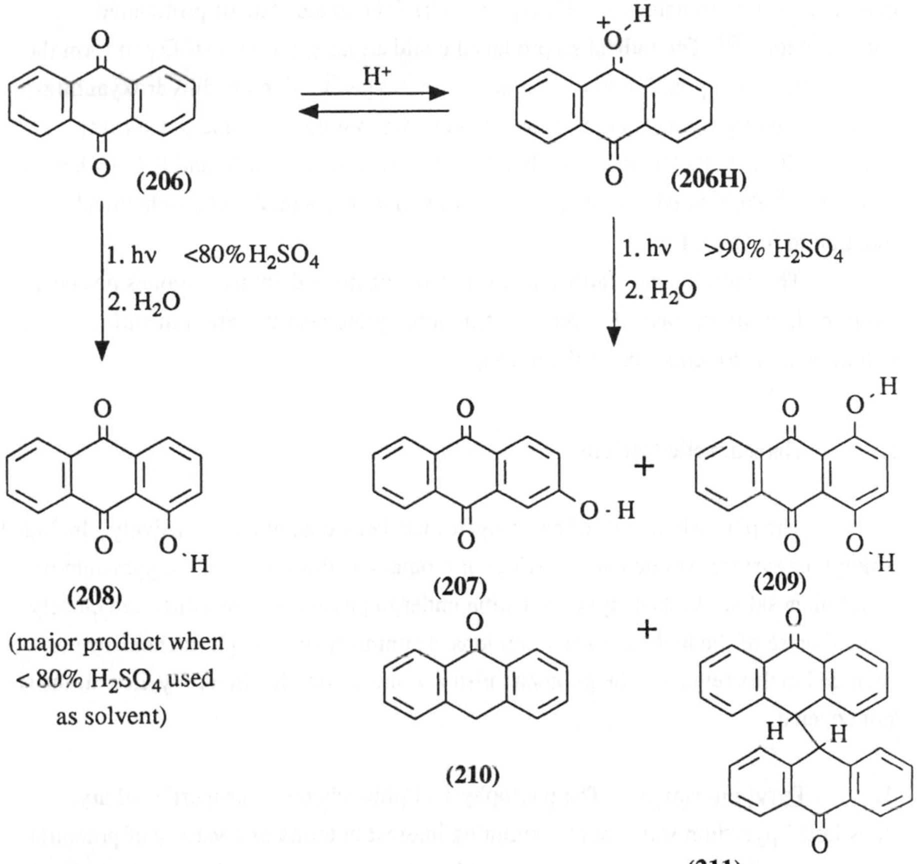

Scheme 85.

of the unprotonated anthraquinone, electron transfer from the solvent and coupling to form (208).

Anthraquinone is essentially completely protonated in the stronger acid media (>90% H_2SO_4) and under these conditions it would appear that the photoreactions involve excitation of the protonated quinone (206H). It was originally suggested that the reaction proceeded by way of nucleophilic attack by HSO_4^- on the excited triplet state of (206H)[230,235]; a type of photochemical reaction which is not observed with carbenium ions. However, based on the detection of the reduced products mentioned above and other evidence, Broadbent and Stewart have proposed that the reaction involves electron transfer from HSO_4^- to the triplet excited state of protonated anthraquinone.[232] The radical so produced could either react with HSO_4 to form the sulfate ester corresponding to (207), react with oxygen to give the dihydroxyanthra-quinones, or disproportionate to give 9,10-dihydroxyanthracene and the starting quinone (206). 9,10-Dihydroxyanthracene is known to give (206) and (210) when in solution.[236] Protonated anthrone was shown to undergo a facile photochemical reaction to form (211).[232]

The photohydroxylations of a series of substituted anthraquinones has been examined.[237] In the case of 1- and 2-nitro-anthraquinones irradiation in sulfuric acid solutions leads to reduction of the nitro groups.

E.   Heterocyclic Cations

The photochemistry of heterocycles has been examined extensively. Included among the systems studied are a variety of aromatic cations such as the pyrylium or pyridinium salts. As these types of cation undergo photoreactions which are closely related those of the hydroxycarbenium ions, a summary of their photochemistry is included in this review. The photochemistry of more complex heterocyclic cations is not covered.

1.   Pyrylium cations.   The photophysical/photochemical properties of aryl substituted pyrylium salts are of continuing interest in terms of a variety of potential applications.[238] Just as with the other stable cations such as the triphenylmethyl and tropylium ions, photoexcitation of a pyrylium salt in the presence of an oxidizable molecule can lead to single electron transfer and the formation of a radical/cation pair. This type of photo-induced electron transfer to stable aryl-substituted pyrylium cations has been used extensively with to initiate a variety of different reactions.[58,239]

The photoisomerizations of alkyl substituted pyrylium salts have been investigated by Barltrop and Day and colleagues. For example, it was shown that the salt (212) is quantitatively converted to (213) on irradiation in acetonitrile solution[240], Scheme 86. The isomerization was shown to involve a π,π^* singlet state of (212).

While this type of isomerization has been found to occur with a range of other alkyl substituted pyrylium cations, a major limitation is that there must be alkyl groups

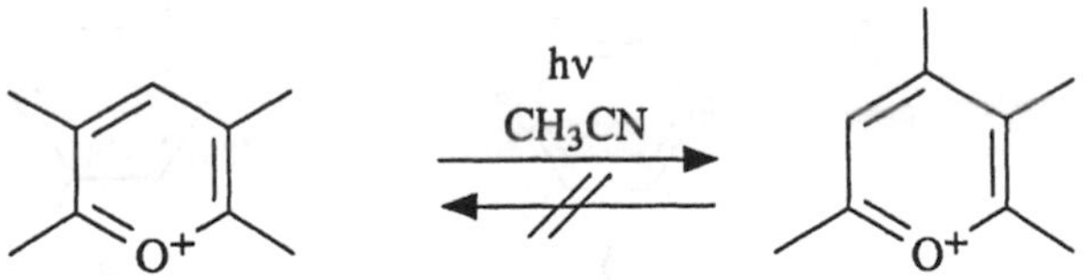

(212) **(213)**

Scheme 86.

present at both C_3 and C_5 of the cation in order for the reaction to proceed. Thus (213) and 2,4,6-trisubstituted cations such as (214) and (215) do not undergo photoisomerization when irradiated in CH_3CN.[240] In contrast to this photochemical stability in CH_3CN, earlier work of Barltrop and Day showed that cations such as (214) and (215) undergo photohydration/ring opening reactions when irradiated in aqueous solution, Scheme 87.[240,241]

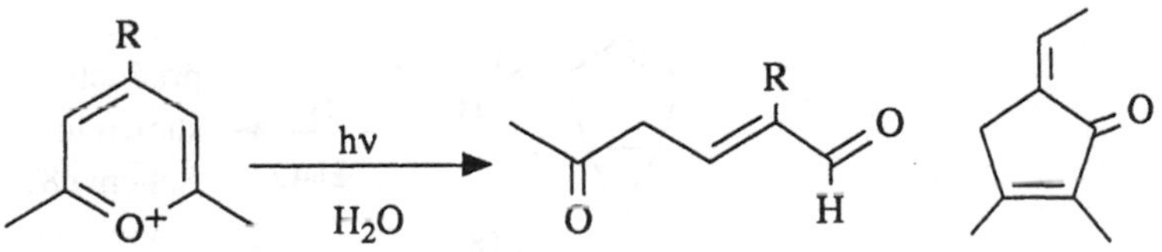

(214) R=Me major minor
(215) R=Et (when R=Et)

Scheme 87.

In order to account for the photoisomerizations of (212) and related cations Barltrop and Day suggested that the key photochemical step is the formation of (216) or (217), Scheme 88.[240,241] Both of these intermediates could be formed from the pyrylium ion and indeed, good precedent exists for both types of photoreaction either in the photochemistry of benzenoid materials[162] or dienyl cations. The oxoniabenzvalene (216) could open directly to give a pyrylium cation with the required transposition of ring carbon atoms. Alternatively, circumambulation of the oxygen atom in (217) followed by a reverse electrocyclic ring opening could accomplish the same transformation. Capture of either intermediate with water would give (218) which could undergo acid catalyzed rearrangements to give the products shown in Scheme 88.

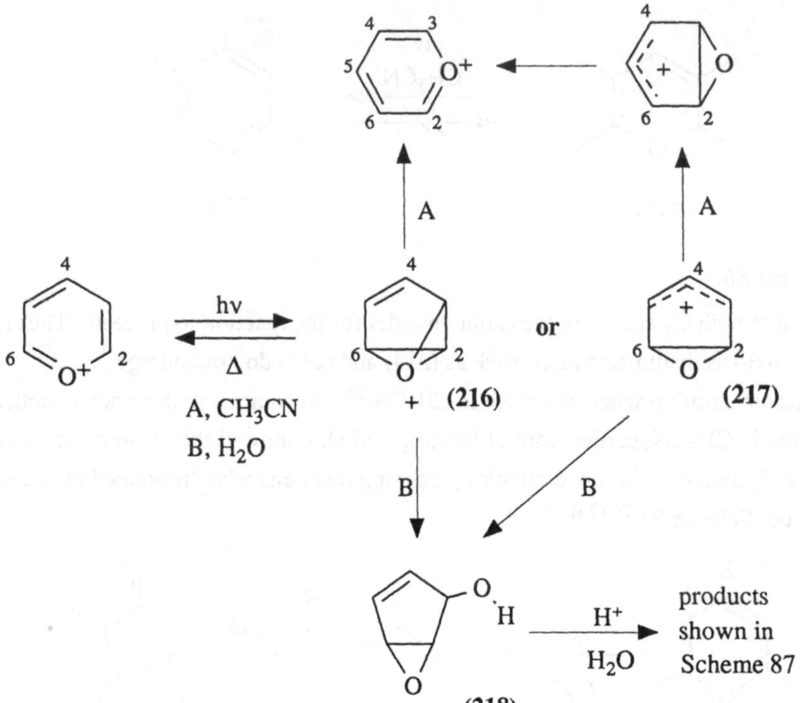

Scheme 88.

As was recognized by Barltrop and Day, the difficulty with the mechanism outlined in Scheme 88 is in accounting for the large solvent effect on the isomerization of cations such as (214) and (215). It is interesting to note, however, that the ring closure of a pyrylium cation to form (217) and the subsequent circumambulation of the oxygen atom closely parallels the reactions observed with protonated phenols and 2,5-cyclohexadienones. These systems exhibit a similar large substituent dependence on the quantum efficiency of formation of the bicyclo[3.1.0]hexenyl cations that was accounted for by the effects of the substituents on the relative rates of cyclopropyl circumambulation versus ring opening to give back the starting material.[157,158] The intermediate (217) in Scheme 88 could similarly undergo competitive thermal reversion to the starting pyrylium ion or oxygen migration. Based on the results with the protonated phenols, groups at C_3 and C_5 of (217) should be particularly effective in promoting circumambulation while groups at C_2 and C_6 should enhance ring opening and the return to starting material. In the presence of water, collapse of the initial cation (219) could well be faster than the reverse isomerization.

2.   2- and 4-Hydroxypyrylium cations. The photoisomerizations of the 2- and 4-hydroxypyrylium cations have been investigated by the groups of Barltrop/Day and Pavlik. For example, Pavlik and Clennan reported in 1973 that the 4-hydroxypyrylium cation (219) was converted to (220) on irradiation.[242] The product (220) was shown in turn to be in photoequilibrium with (221). Subsequent work showed that (222) is also formed as a minor product during the course of the conversion of (219) to (220), Scheme 89.[243]

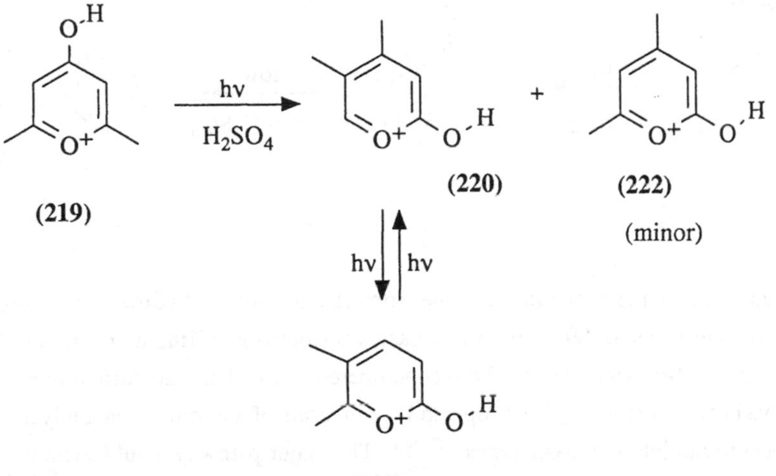

Scheme 89.

Comparable photoisomerizations to those shown in Scheme 89 also occur with a wide range of other alkyl-substituted pyrylium salts.[244-248] In the case of the *t*-butyl substituted cation (223) the only product observed is (224), Scheme 90; the product corresponding to the minor pathway in the photoisomerization of (219), Scheme 89.[246] In some examples trapping of photochemically produced intermediates takes place with the formation of covalent materials, Schemes 91[245] and 92.[248] In the case of the photoisomerization of (225), a mixture of the three sulfate esters shown in Scheme 92 was shown to be present in the acid medium.[248] The cyclic sulfate ester (226) could be obtained by careful neutralization of the H_2SO_4 solution.

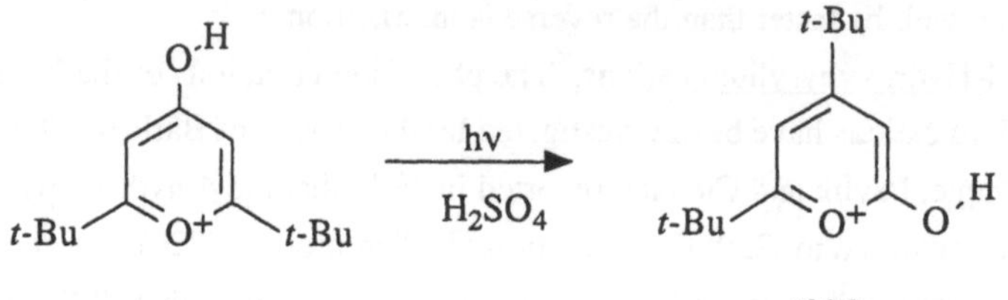

(223)   (224)

Scheme 90.

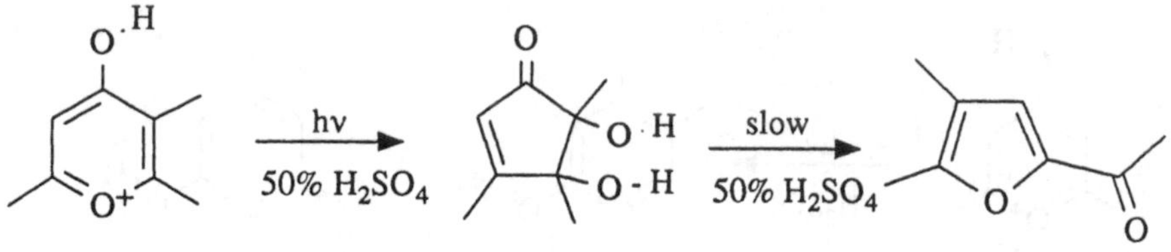

Scheme 91.

Just as with the pyrylium salts, the isomerizations of the hydroxy substituted cations shown in Schemes 89-92 involve a basic transposition of ring atoms rather than a movement of substituents around the six-membered rings. The transformations, which have been examined by Barltrop and Day in terms of a permutation analysis, would seem to fall into two basic types.[247-248] The major pathway would seem to involve the photochemical formation of an oxabicyclo[3.1.0]hexenyl cation which can either undergo oxygen migration, capture by a nucleophile in solution, or presumably the hidden process of ring opening of the initially formed intermediate to give back the starting hydroxypyrylium salt, Scheme 93. The alternative but normally minor pathway except in the case (223) likely involves C_2C_5 bonding and formation of an oxa-Dewar benzene. Rearrangement of the Dewar isomer and ring opening would give the rearranged product as shown in Scheme 93.

It is interesting to compare the photo-reactions of the hydroxypyrylium cations with those of the corresponding pyrones.[249] Irradiation of dilute solutions of 4-pyrones leads largely to the formation products which are similar to those outlined above for the cationic systems. It has been proposed that zwitterions corresponding to the bicyclic

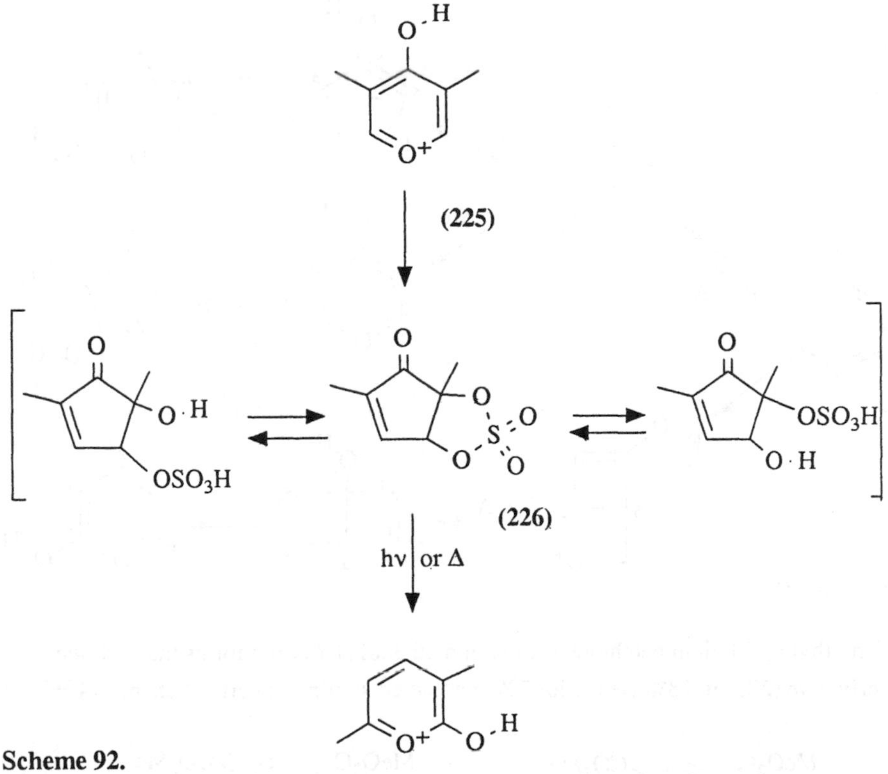

Scheme 92.

cations formed from the protonated materials are intermediates in the photoreactions of the neutral materials. Dimerization is the other major type of photoreaction observed with pyrones. Such bimolecular reactions are not encountered with the protonated materials, presumably as a result of the charge repulsion which would result from the dimerization of two cationic species.

      3.   Pyridinium Cations. Pyridinium ions can undergo a large variety of photochemical reactions. In general, these are alcohol addition, alkylation, rearrangement including cyclization, and dimerization. As many of these processes have been presented in extensive reviews[61,250], only a sample of the reactions which encompass the photochemistry of pyridinium ions will be presented.

      One of the simplest and earliest photoreactions of pyridinium cations to be described involves alcohol addition to the heterocyclic ring. Thus, irradiation of

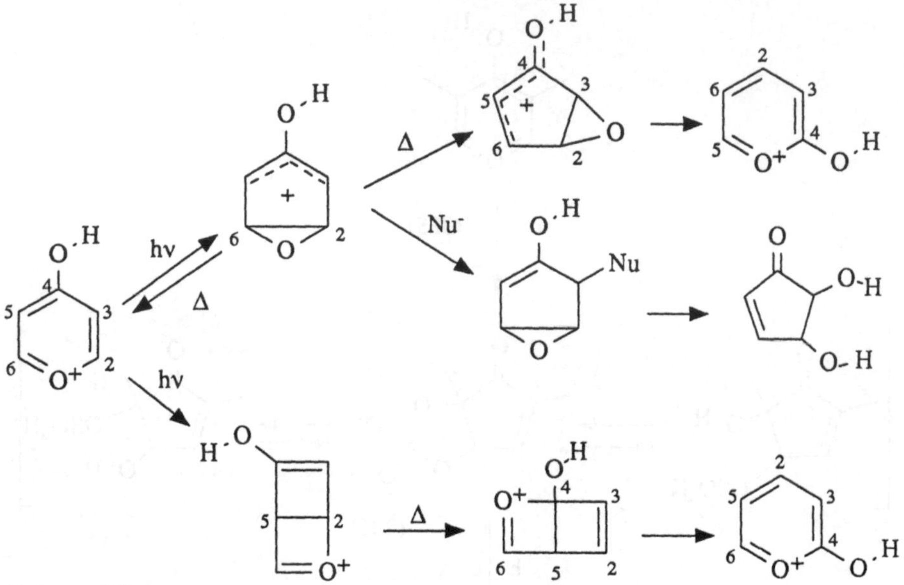

Scheme 93.

N-methyl pyridinium perchlorate (227) in methanol at 254 nm forms the pyridine derivative (228) in 18% yield with 50% unreacted starting material, Scheme 94.[251]

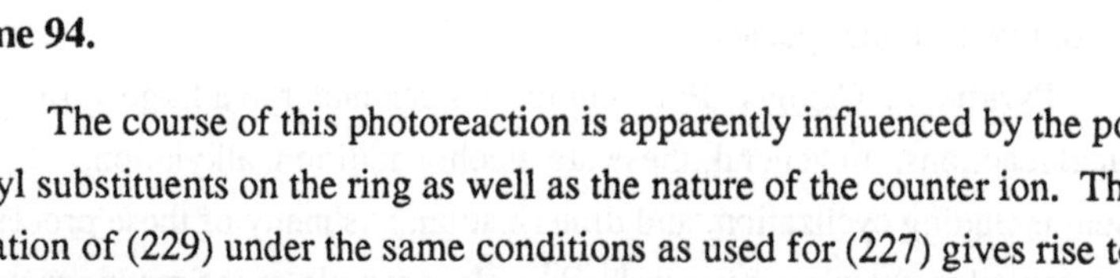

Scheme 94.

The course of this photoreaction is apparently influenced by the position of alkyl substituents on the ring as well as the nature of the counter ion. Thus irradiation of (229) under the same conditions as used for (227) gives rise to the photoreduced product (230) in 11% yield, Scheme 95, while the methyl sulfate salt (231) is unreactive. Van Bergen and Kellogg have suggested that photoinduced enolizations compete with isomerization with (229) and (231), Scheme 96.[251] In the case of (231), this was tested by irradiating in a solution of

CH$_3$OD when deuterium incorporation was found at the C$_2$, C$_6$ and C$_4$ methyl groups.

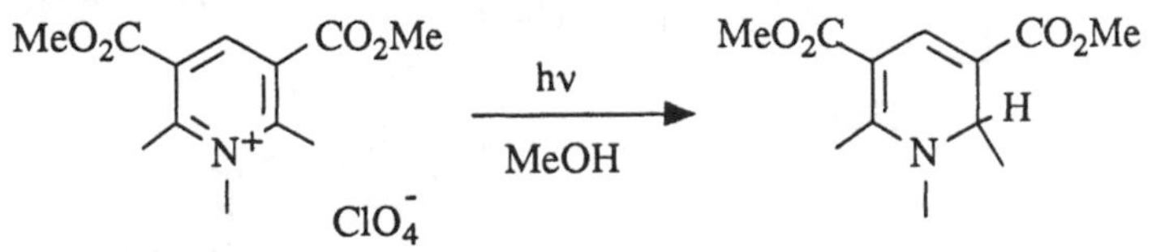

Scheme 95.

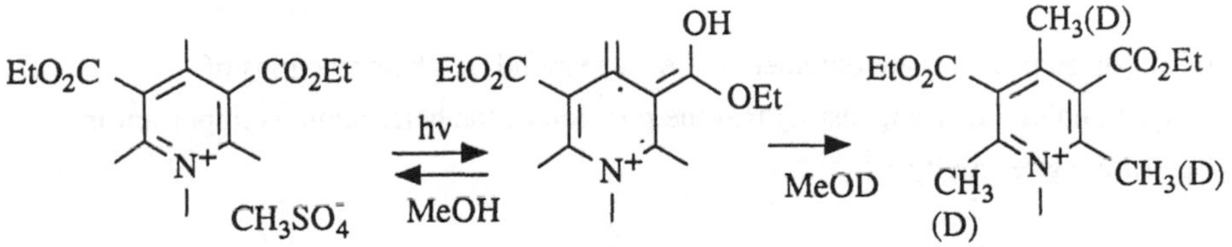

Scheme 96.

The photoalkylation of pyridine has been reported by several groups.[252-255] Travecedo and Stenberg[252] observed that irradiation of pyridine in methanolic-HCl led to the formation of several addition products, Scheme 97. It would appear that a photo-induced alcohol addition occurs, followed by elimination of H$_2$O. Thus pyridinium chloride (232) undergoes alcohol addition to yield the 2- and 4-methylpyridines, (233) and (234) respectively, as the major products. The intermediate for this reaction has been suggested to be the hydroxymethyl pyridine derivative (235), Scheme 97. Furihata and Sugimori[253] have found that a similar alcohol addition occurs on irradiation of 2-cyanopyridine in alcoholic H$_2$SO$_4$. In this case however, elimination of HCN occurs. Similar mechanisms have been suggested for the photoalkylation of quinolinium, isoquinolinium, and phenanthridinium ions.[254,255]

Photoalkylation also may occur through an electron transfer mechanism from a donor carboxylic acid. For example, the quinolinium ion (236) undergoes photoalkylation in the presence of carboxylic acids, Table 8.[256] Much lower yields

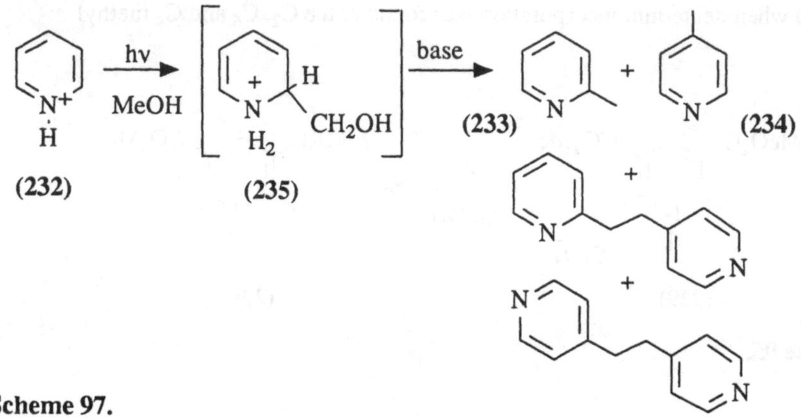

Scheme 97.

of alkylated products are obtained in the corresponding photoreactions of isoquinolinium ions suggesting that the position of the heteroatom is important in determining reactivity.[257]

Table 8. Photoalkylation of quinolinium salts

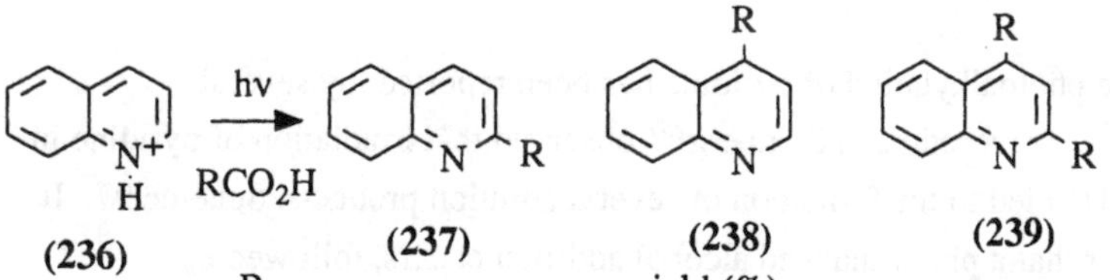

R	(237)	(238) yield (%)	(239)
Me	20	10	5
Et	23	-	6
i-Pr	27	20	6
i-Am	37	-	-

Kaplan and co-workers found that irradiation of 1-methylpyridinium chloride (240) and several other derivatives at 254 nm in either basic aqueous or methanol solutions yielded the bicyclo [3.1.0] derivatives (241), Scheme 98.[258] The formation of these compounds was studied using a variety of deuterated isomers and on this basis it was suggested that the reaction proceeds via an azoniabenzvalene intermediate, (242), followed by nucleophilic attack and subsequent ring opening. A similar bicyclic photoproduct has also been obtained from the irradiation 1-phenyl-3-oxidopyridinium ion.[259]

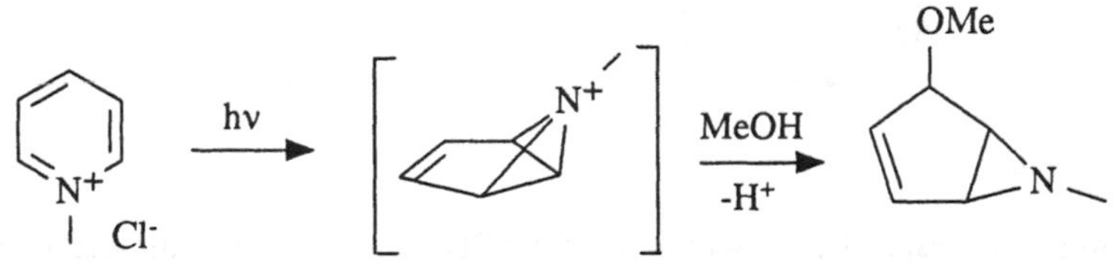

(240)   (242)   (241)

Scheme 98.

The photocyclization of several 1-styrylpyridinium ions has been observed and offers a practical synthetic route for substituted phenanthridizinium ions.[260] In each case the starting pyridinium ion (243) undergoes a *trans/cis* photoisomerization to yield the cis isomer (244) which subsequently undergoes photocyclization/oxidation to yield (245), Scheme 99.

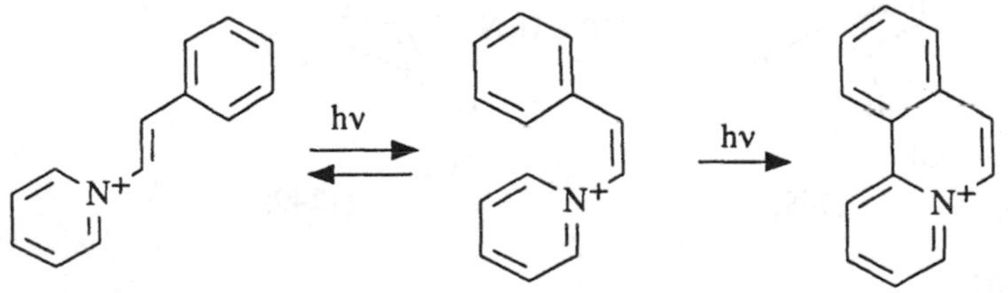

(243)   (244)   (245)

Scheme 99.

Photocyclization of 2-halo substituted arylmethyl pyridinium salts (246) has been found to occur, Scheme 100.[261-263] It has been suggested that this reaction involves electron transfer from the benzene to the pyridinium ring, followed by radical coupling and re aromatization of the aryl ring through elimination of either H• or X•. The cyclizations of several other heterocyclic compounds having more extensive π systems have also been studied.[61,264,265]

One type of the photoproduct frequently encountered upon irradiation of a

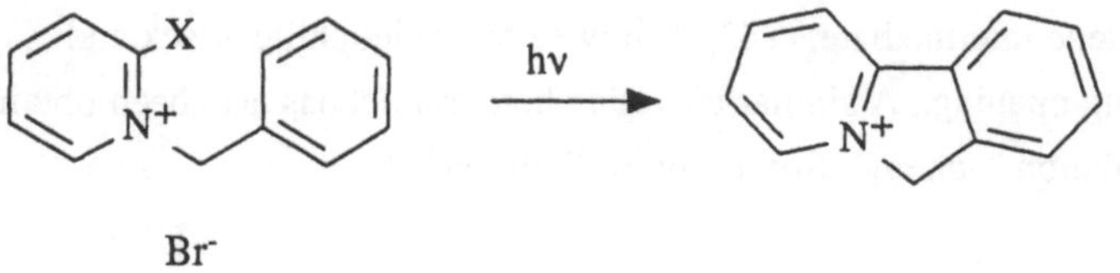

Br⁻

(246)   (247)

Scheme 100.

pyridinium salt results from the dimerization of two cations. This dimerization
would seem to occur through either a cycloaddition reaction or through radical
coupling after electron transfer. For example, irradiation of the 2-amino
pyridinium salt (248) was found to yield the dimer (249) as the primary
photoproduct. It was suggested that this reaction proceeds by way of an
intermolecular $4\pi + 4\pi$ cycloaddition[266,267] Scheme 101. Comparable
photodimerization reactions are well known with several naphthalene and
anthracene derivatives.[268]

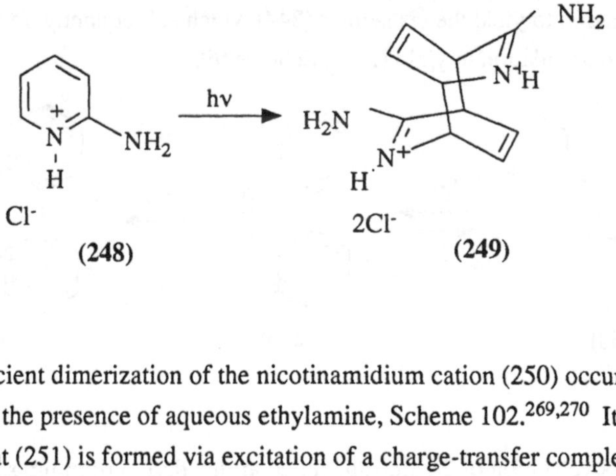

Cl⁻   2Cl⁻

(248)   (249)

Scheme 101.

Efficient dimerization of the nicotinamidium cation (250) occurs on
irradiation in the presence of aqueous ethylamine, Scheme 102.[269,270] It was
suggested that (251) is formed via excitation of a charge-transfer complex of (250)
with ethylamine.

A number of photoinduced nucleophilic substitutions are known to occur
with pyridinium ions.[271-273] Thus irradiation of the 3-substituted pyridinium
chloride ions (252) yields the 2-hydroxy pyridines (253) after neutralization,
Scheme 103.[271]   Oxygen inhibits the formation of (253)

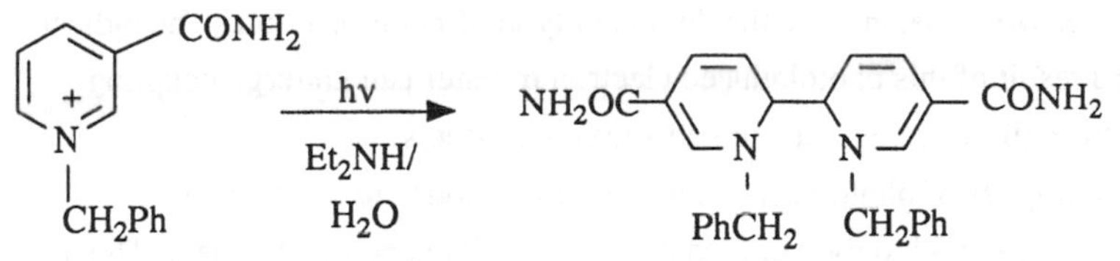

(250) (251)

Scheme 102.

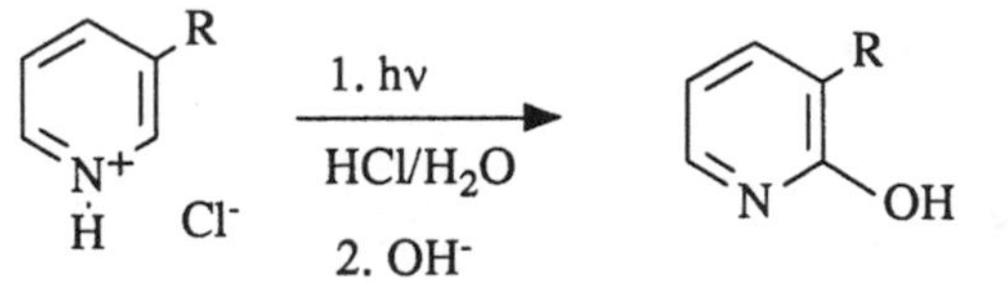

(252) (253)

R=COOH, CN, CONH$_2$

Scheme 103.

VI CONCLUSION

It will be clear from the material presented in this review that while a considerable number of photoreactions of carbenium ions have been described, for the most part no detailed mechanistic studies have been carried out. Indeed there is considerable scope for more detailed work in this area that not only seeks to elucidate the course of the photoreactions but also studies the reactions of carbenium ions produced as transient species in many of these reactions.

Despite the lack of mechanistic information on the photoreactions of carbenium ions it is clear that there are two broad classes of reaction observed. Thus in media such as FSO$_3$H the typical type of reaction which occurs appears to be photoisomerization. Few bimolecular reactions are encountered in this superacid medium. In less strongly acidic media, and particularly with relatively rigid polycyclic aromatic ions or with salts which contain a readily oxidized anion, electron transfer would seem to be the dominant type of photoreaction. The radical produced as a result of this photoinduced electron transfer can undergo coupling reactions to form dimeric materials or substitution products.

In both types of photoreactions of carbenium ions the reactions are frequently very clean with only a limited range of products being obtained. This is particularly the case with the photoisomerizations in strong acid media where the photoreactions can be carried out at unusually high concentrations as compared to the comparable transformations of neutral molecules. In many instances the photoreactions of carbenium ions offer considerable promise as the basis of useful synthetic transformations. This represents a further area in which continuing work would be of considerable interest.

VII   REFERENCES

1.   G. A. Olah and P. von R. Schleyer, Editors, *Carbonium Ions, I-V* (1968-1975). D. Bethel, *Comprehensive Organic Chemistry, 1,* 411 (1975). G. A. Olah, G. K. S. Prakash, R. E. Williams, L. D. Field, and K. Wade, *Hypercarbon Chemistry,* J. Wiley and Sons, New York (1987).

2.   P. Vogel, *Carbocation Chemistry,* Elsevier, Amsterdam 1985.

3.   R. J. Gillespie, *Accounts Chem. Res., 1,* 202 (1968). R. J. Gillespie and T. E. Peel, *Adv. Phys. Org. Chem., 9,* 1 (1971). G. A. Olah, G. K. S. Prakash and J. Sommer, *Superacids,* Wiley Interscience, New York (1985).

4.   a) M. Sundaralingam and A. K. Chwang, *Carbonium Ions, 5,* 2428 (1975). b) R. F. Childs, M. Mahendran, S. D. Zweep, G. S. Shaw, S. K. Chadda, N. A. D. Burke, B. E. George, R. Faggiani, C. J. L. Lock, *Pure and Appl. Chem. 58,* 111 (1986). c) T. Laube, *J. Amer Chem. Soc, 111,* 9224 (1989).

5.   V. Balogh-Nair, K. Nakanishi, *New Comprehensive Biochem., 3,* 283 (1982). D. S. Kliger and E. L. Menger, *Accounts Chem. Res. 8,* 81 (1975). C. Longstaff and R. Rando, *Biochemistry, 24, 8137* (1985).

6.   For previous reviews of the photochemistry of carbocations see a) P. W. Cabell-Whiting and H. Hogeveen, *Adv. Phys. Org. Chem. 10,* 129 (1973). R. F. Childs, *Rev. Chem. Intermed. 3,* 285 (1980).

7.   J. L. Faria and S. Steenken, *J. Amer. Chem. Soc. 112,* 1277 (1990). S.

Steenken, J. Buschek and R. A. McClelland, *J. Amer. Chem. Soc. 108*, 2808 (1986).

8. S. Cristol and T. H. Bindel, *Organic Photochemistry, 6*, 327 (1983). E. O. Alonso, L. J. Johnston, T. C. Scaiano and V. G. Toscano, *J. Amer. Chem. Soc. 112*, 1270 (1990). S. Kobayashi, Q. Q. Zhu and W. Z. Schnabel, *Z. Naturforsch. 43b*, 825 (1988). R. A. McClelland, N. Banait and S. Steenken, *J. Amer. Chem. Soc., 108*, 7023 (1986).

9. P. Wan and K. Yates, *Rev. Chem. Intermed. 5*, 157 (1984). R. A. McClelland, V. M. Kanagasabapathy and S. Steenken, *J. Amer Chem. Soc. 110*, 6913 (1988). P. Wan and P. Wu, *J. Chem. Soc. Chem. Commun., 822* (1990).

10. S. Steenken and R. A. McClelland, *J. Amer. Chem. Soc. 111*, 4967 (1989).

11. For a review of the chemistry of heteroatom substituted carbenium ions see G. A. Olah, A. M. White and D. H. O'Brien, *Carbonium Ions, 4*, 1697 (1973).

12. For a review of amino-substituted carbenium ions see F. L. Scott and R. N. Butler, *Carbonium Ions, 4*, 1644 (1973)

13. See for example, J. L. Dektar and N. P. Hacker, *J. Org. Chem. 55*, 639 (1990).

14. H. Dauben, Jr., L. R. Honnen and K. M. Harmon, *J. Org. Chem., 25*, 1442 (1960)

15. E. E. van Tamelen, T. M. Cole, Jr., R. Greeley and H. Schumacher, *J. Amer. Chem. Soc., 90*, 1372 (1968).

16. R. F. Childs and S. Winstein, *J. Amer. Chem. Soc. 90*, 7146 (1968). R. F. Childs, M. Sakai and S. Winstein, *J. Amer. Chem. Soc, 90*, 7144 (1968).

17. R. F. Childs and V. Taguchi, *J. Chem. Soc. Chem. Commun., 695* (1970).

18. H. Hogeveen and C. J. Gaasbeek, *Rec. Trav. Chim. Pays-Bas, 89*, 1079 (1970).

19. R. F. Childs and B. Parrington, *J. Chem. Soc. Chem. Commun.,* 1540 (1970).

20. For early reports on the photoreactions of carbonyl compounds in strong acid media or as Lewis acid complexes see A. Eckert, *Chem. Ber., 58*, 322 (1925). S. G. Korenman, H. Wilson and M. B. Lipsett, *Steroids, 3*, 203 (1964). M. Finkelstein, *Acta Endocrin., 10*, 149 (1952). H. A. Jones,

Can. J. Spectrosc., 16, 10 (1971). A. R. Butler and H. A. Jones, *J. Chem. Soc. Perkin II,* 963 (1976). H. Stobbe and E. Färber, *Chem. Ber. 58,* 1548 (1925). P. Praetorius and P. Korn, *Chem. Ber. 43,* 2744 (1910).

21.   G. Wald, *Science, 162,* 230 (1968). R. Hubbard and G. Wald, *J. Gen. Physiol., 36,* 673 (1968). For a review see E. W. Abrahamson, *Accounts Chem. Res., 8,* 101 (1975).

22.   H. H. Freedman, *Carbonium Ions, 4,* 1501 (1973).

23.   R. F. Childs and A. W. Cochrane, *J. Org. Chem. 46,* 1086 (1981).

24.   N. Filipescu, S. K. Chakrabarti and P. G. Tarassoff, *J. Phys. Chem., 77,* 2276 (1973).

25.   J. F. Ireland and P. A. H. Wyatt, *J. Chem. Soc. Farad. Disc. I, 68,* 1053 (1972) and *69,* 161 (1973).

26.   P. G. Tarassoff and N. Filipescu, *J. Chem. Soc. Chem. Comm.,* 208 (1975).

27.   M. R. Feldman and N. G. Thame, *J. Org. Chem., 44,* 1863 (1979).

28.   E. Vander Donckt, *Progr. React. Kinetics, 5,* 273 (1970). S. G. Schulman in *Modern Fluorescence Spectroscopy,* E. L. Wehry, Ed., Plenum Press, New York, *2,* 239 (1976). J. F. Ireland and P. A. H. Wyatt, *Adv. Phys. Org. Chem., 12,* 131 (1976). H. Shizuka and E. Kimura, *Can. J. Chem., 62,* 2041 (1984).

29.   G. A. Olah, C. U. Pittman, Jr. and M. C. R. Symons, *Carbonium Ions, 1,* 153 (1968).

30.   N. C. Deno, J. Bollinger, N. Friedman, K. Hafer, J. D. Hodge and J. J. Houser, *J. Amer. Chem. Soc., 85,* 2998 (1963).

31.   D. H. Williams and I. Flemming, *Spectroscopic Methods in Organic Chemistry,* McGraw Hill, London, p. 1 (1980).

32.   M. Arnaboldi, M. G. Motto, K. Tsujimoto, V. Balogh-Nair, and K. Nakanishi, *J. Amer. Chem. Soc., 101,* 7082 (1979). H. Kakitani, T. Kakitani, H. Rodman, and B. Honig, *Photochem. Photobiol., 41,* 471 (1985). M. Sheves, K. Nakanishi, and B. Honig, *J. Amer. Chem. Soc., 101,* 7086 (1985). M. G. Motto, M. Sheves, K. Tsujimoto, V. Balogh-Nair, and K. Nakanishi, *J. Amer. Chem. Soc., 102,* 7947 (1985). M. Sheves and K. Nakanishi, *J. Amer. Chem. Soc., 105,* 4033 (1983).

33.   R. Rusakowicz, G. W. Byers and P. A. Leermakers, *J. Amer. Chem. Soc.,*

93, 3263 (1971). N. S. Bayliss and E. G. McRae, *J. Phys. Chem. 58*, 1006 (1954). H. H. Jaffe and M. Orchin in *Theory and Applications of Ultraviolet Spectroscopy*, Wiley, New York, p. 186 (1962).

34. J. Griffiths and H. Hart, *J. Amer. Chem. Soc., 90*, 5296 (1968). T. Takino and H. Hart, *J. Chem. Soc. Chem. Commun.*, 450 (1970). H. Hart and T. Takino, *J. Amer. Chem. Soc., 93*, 720 (1971). A. A. Lamola, *J. Chem. Phys., 47*, 4810 (1967). J. B. Gallivan, *Can. J. Chem., 50*, 3601 (1972). R. D. Rauh and P. A. Leermakers, *J. Amer. Chem. Soc., 90*, 2246 (1968).

35. a) P. E. Blatz, J. H. Mohler, and H. V. Navangul, *Biochemistry, 11*, 848 (1972) b) P. E. Blatz and J. H. Mohler, *Biochemistry, 11*, 3240 (1972) c) P. E. Blatz and J. H. Mohler, *Biochemistry, 14*, 2304 (1975) d) J. Favrot, D. Vocelle, and C. Sandorfy, *Photochem. Photobiol., 30*, 417 (1979) f) F. I. Harosi, J. Favrot, Leclercq, D. Vocelle, and C. Sandorfy, *Rev. Can. Biol., 37*, 257 (1978).

36. J. P. Honovich and R. C. Dunbar, *J. Phys. Chem., 85*, 1558 (1981). B. S. Freiser and J. L. Beauchamp, *J. Amer. Chem. Soc., 99*, 3214 (1977).

37. E. M. Arnett and G. Scorrano, *Adv. Phys. Org. Chem., 13*, 83 (1976).

38. K. M. Harmon, *Carbonium Ions, 4*, 1579 (1973).

39. E. M. Kosower, *An Introduction to Physical Organic Chemistry*, Wiley, New York, p. 293 (1968). Cf. K. Dimroth, C. Reichardt, T. Siepmann and F. Bohlmann, *Justus Liebig's Ann. Chem., 661*, 1 (1963). C. Reichardt in *Solvent Effects in Organic Chemistry* Verlag Chemie, New York, p. 189 (1979).

40. N. J. Turro, Modern Molecular Photochemistry Benjamin/Cummings, Menlo Park, CA., 1978.

41. H. Kuhlewind, H. J. Neusser and E. W. Schlag, *J. Phys. Chem., 89*, 5593 and 5600 (1985).

42. L. Andrews and B. W. Keelan, *J. Amer. Chem. Soc., 103*, 99 (1981).

43. F. W. McLafferty and J. Winkler, *J. Amer. Chem. Soc., 96*, 5182 (1974).

44. P. W. Cabell and H. Hogeveen, unpublished results referred to in reference 6a.

45. H. El-Akabi, H. Kawata and P. de Mayo, *J. Org. Chem., 53*, 1471 (1988).

46. E. E. van Tamelen and T. M. Cole Jr., *J. Amer. Chem. Soc., 93*, 6158 (1971).

47.   E. D. Owen and D. M. Allen, *J. Chem. Soc. Perkin II*, 95 (1973). D. M.
        Allen and E. D. Owen, *J. Chem. Soc. Chem. Commun.*, 848, (1971).

48.   T. M. Cole, Jr., *J. Amer. Chem. Soc., 92*, 4124 (1970).

49.   E. E. van Tamelen and T. M. Cole, Jr., *J. Amer. Chem. Soc., 92*, 4123
        (1970).

50.   D. Bethel and P. N. Clare, *J. Chem. Soc. Perkin II*, 1464 (1972).

51.   K. Okada, K. Hisamitsu and T. Mukai, *Tetrahedron Lett.*, 1251 (1981).

52.   R. S. Velichkova, V. D. Toncheva and I. M. Panayotov, *Makromol.
        Chemie., 181*, 671 (1980). V. D. Toncheva and R. S. Velichkova,
        Makromol. Chimie., 186, 1739 (1985) and *184*, 2231 (1983).

53.   D. H. R. Barton, G. Leclerc, P. D. Magnus and I. D. Menzies, *J. Chem.
        Soc. Chem. Commun.*, 447 (1972). D. H. R. Barton, R. K. Haynes, G.
        Leclerc, P. D. Magnus and I. D. Menzies, *J. Chem. Soc., Perkin 1*, 2055
        (1975).

54.   A. M. Amat, G. Asensio, M. J. Castello, M. A. Miranda and A.
        Simon-Fuentes, *Tetrahedron, 43*, 905 (1987).

55.   P. J. Kropp, *Accounts Chem. Res., 17*, 131 (1984).

56.   Y. Ogata, T. Itoh and Y. Izawa, *Bull Chem. Soc. Jpn., 42*, 794 (1969).

57.   G. Opitz, H. Hellmann and H. W. Schubert, *Liebigs Ann. Chem., 623*, 117
        (1959). H. D. Bartfeld and W. Flitsch, *Chem. Ber., 106*, 1423 (1973).

58.   G. M. Badger, C. P. Joshua, and G. E. Lewis, *Tett. Lett., 49*, 3711 (1964).

59.   H. H. Perkampus and B. Behjati, *J. Heterocyclic Chem., 11*, 511 (1974).

60.   J. Saltiel and J. L. Charlton, *Rearrangements in the Ground and Excited
        States*, Ed. P. DeMayo, Academic Press, N.Y. p. 25 (1980).

61.   P. S. Mariano, *Tetrahedron, 39*, 3845 (1983). M. I. Knyazhanskii, Y. R.
        Tymyanskii, V. M. Feigelman, and A. R. Katritzky, *Heterocycles, 26*,
        2963 (1987).

62.   W. von Dörscheln, H. Tiefenthaler, H. Goth, P. Cerutti, and H. Schmid,
        Helv. Chim. Acta, 50, 1759 (1967).

63.   J. L. Stavinoha and P. S. Mariano, *J. Amer. Chem. Soc., 103*, 3136 (1981).

64.   A. J. Y. Lan, R. O. Heuckeroth and P. S. Mariano, *J. Amer. Chem. Soc.,
        109*, 2738 (1987).

65.   C. L. Tu and P. S. Mariano, *J. Amer. Chem. Soc., 109*, 5287 (1987).

66.   I. S. Cho, C. L. Tu and P. S. Mariano, *J. Amer. Chem. Soc., 112*, 3594

(1990).

67. a) I. S. Cho and P. S. Mariano, *J. Org. Chem.*, *53*, 1591 (1988). b) G. Dai-Ho and P. S. Mariano, *J. Org. Chem.*, *53*, 5113 (1988).

68. G. Dai-Ho and P. S. Mariano, *J. Org. Chem.*, *52*, 704 (1987).

69. R. Ahmed-Schofield and P. S. Mariano, *J. Org. Chem.*, *52*, 1478 (1987).

70. W. Adam and M. A. Miranda, *J. Org. Chem.*, *52*, 5498 (1987).

71. S. Farid and S. E. Shealer, *J. Chem. Soc. Chem. Commun.*, 677 (1973). R. Searle, J. L. R. Williams, D. E. Demeyer and J. C. Doty, *J. Chem. Soc. Chem. Commun.*, 1165 (1967).

72. Q. B. Broxterman, H. Hogeveen and D. M. Kok, *Tetrahedron Lett.*, 173 (1981).

73. H. Bock, B. Roth and G. Maier, *Angew. Chem.*, *19*, 209 (1980).

74. Q. B. Broxterman, H. Hogeveen and R. F. Kingma, *Tetrahedron Lett.*, 2043 (1984).

75. P. B. J. Driessen and H. Hogeveen, *J. Amer. Chem. Soc.*, *100*, 1193 (1978).

76. J. L. Courtneidge and A. G. Davies, *Accounts Chem. Res.*, *20*, 90 (1987).

77. J. L. Courtneidge, A. G. Davies, P. S. Gregory and S. N. Yazdi, *Farad. Disc. Chem. Soc.*, *78*, 49 (1984).

78. D. M. Brouwer, *Tetrahedron Lett.*, 453 (1968). G. A. Olah, Y. Halpern, Y. K. Mo and G. Liang, *J. Amer. Chem. Soc.*, *94*, 3554 (1972). G. A. Olah and M. Calin, *J. Amer. Chem. Soc.*, *90*, 405 (1968). N. C. Deno, C. U. Pittman Jr. and M. J. Wisotsky, *J. Amer. Chem. Soc.*, *86*, 4370 (1964).

79. R. F. Childs, E. F. Lund, A. G. Marshall, W. J. Morrisey and C. V. Rogerson, *J. Amer. Chem. Soc.*, *98*, 5924 (1976).

80. R. F. Childs and M. E. Hagar, *J. Amer. Chem. Soc.*, *101*, 1052 (1979) and *Can. J. Chem.*, *58*, 1788 (1980).

81. R. F. Childs, D. L. Mulholland and A. Nixon, *Can. J. Chem.*, *60*, 801 and 809 (1982).

82. R. F. Childs, B. Duffey and A. Mika-Gibala, *J. Org. Chem.*, *49*, 4352 (1984).

83. C. Blackburn and R. F. Childs, *J. Chem. Soc. Chem. Commun.*, 812 (1984).

84. D. Cremer, J. Gauss, R. F. Childs and C. Blackburn, *J. Amer. Chem. Soc.*,

107, 2435 (1985). D. Farcasiu, J. O'Donnell, K. Wiberg and M. J. Matturro, *J. Chem. Soc. Chem. Commun.*, 1124 (1979). D. M. Brouwer, E. L. Mackor and C. Maclean, *Recl. Trav. Chim. Pays-Bas, 85*, 114, (1965). B. E. Smart and G. S. Ready, *J. Amer. Chem. Soc., 98*, 5593 (1976). R. K. Lustgarten, M. Brookhart and S. Winstein, *Tetrahedron Lett.*, 141 (1971). B. G. Ramsey and R. W. Taft, *J. Amer. Chem. Soc., 88*, 3058 (1966).

85.   C. Blackburn, R. F. Childs, D. Cremer and J. Gauss, *J. Amer. Chem. Soc., 107*, 2442 (1985).

86.   K. Mullen, E. Kotzamani, H. Schmickler and B. Frei, *Tetrahedron Lett.*, 5623 (1984).

87.   For a review on photochemically induced *cis/trans* isomerizations see J. Saltiel, J. D'Agostino, E. D. Megarity, L. Metts, K. R. Neuberger, M. Wrighton and O. C, Zafiriou, *Org. Photochem., 3*, 1 (1973).

88.   L. E. Friedrich and G. B. Schuster, *J. Amer. Chem. Soc., 91*, 7204 (1969) and *94*, 1193 (1972).

89.   A. Deflandre, A. Lheureux, A. Rioual and L. Lemaire, *Can. J. Chem., 54*, 2127 (1976).

90.   J. F. Graf and C. P. Lillya, *Mol. Photochem., 9*, 227 (1979).

91.   H. Morrison and O Rodriguez, *J. Photochem., 3*, 471 (1974).

92.   R. M. Weiss and A. Warshel, *J. Amer. Chem. Soc., 101*, 6131 (1979).

93.   D. Gegiou, K. A. Muszkat and E. Fischer, *J. Amer. Chem. Soc., 90*, 12 (1968).

94.   H. Görner and D. Schulte-Frohlinde, *J. Amer. Chem. Soc., 101*, 4388 (1979).

95.   F. D. Lewis and J. D. Oxman, *J. Amer. Chem. Soc., 103*, 7345 (1981)

96.   F. D. Lewis, J. D. Oxman, L. L. Gibson, H. L. Hampsch and S. L. Quillen, *J. Amer. Chem. Soc., 108*, 3005 (1986).

97.   F. D. Lewis, J. D. Oxman and J. C. Huffman, *J. Amer. Chem. Soc., 106*, 466 (1984)

98.   F. D. Lewis, D. K. Howard, J. D. Oxman, A. L. Upthagrove and S. L. Quillen, *J. Amer. Chem. Soc., 108*, 5964 (1986).

99.   F. D. Lewis, D. K. Howard, S. V. Barancyk and J. D. Oxman, *J. Amer. Chem. Soc., 108*, 3016 (1986).

100.   N. W. Alcock, P. de Meester and T. J. Kemp, *J. Chem. Soc. Perkin II*, 921
        (1979).

101.   R. F. Childs, T. DiClemente, E. F. Lund-Lucas, T. J. Richardson and C.
        V. Rogerson, *Can. J. Chem., 61*, 856 (1983).

102.   R. F. Childs, M. Mahendran, C. Blackburn and G. Antoniadis, *Can. J.
        Chem., 66*, 1355 (1988).

103.   R. F. Childs and G. S. Shaw, *J. Chem. Soc. Chem. Commun.*, 261 (1983).

104.   L. Salem and P. Brückmann, *Nature (London), 258*, 526 (1975)

105.   R. F. Childs, K. E. Hine and F. A. Hung, *Can. J. Chem., 57*, 1442 (1979).

106.   For reviews see K. Schaffner, *Tetrahedron, 30*, 1891 (1974) and O. L.
        Chapman and D. S. Weiss, *Organic Photochemistry, 3*, 197 (1973).

107.   W. G. Dauben, W. A. Spitzer and M. S. Kellogg, *J. Amer. Chem. Soc., 93*,
        3674 (1971).

108.   D. G. Cornell and N. Filipescu, *J. Org. Chem., 42*, 3331 (1977).

109.   H. A. Jones, *Steroids, 13*, 693 (1969).

110.   H. E. Zimmerman, R. G. Lewis, J. J. McCullough, A. Padwa, S. W. Staley
        and M. Semmelhack, *J. Amer. Chem. Soc., 88*, 1965 (1966). P.
        Margaretha and K. Schaffner, *Helv. Chim. Acta, 56*, 2884 (1973).

111.   P. Lupón, F. Canals, A. Iglesias, J. C. Ferrer, A. Palomer, J. J. Bonet, J. L.
        Brainsó, J. F. Piniella, G. Germain and G. S. D. King, *J. Org. Chem., 53*,
        2193 (1988). P. Lupón, J. C. Ferrer, J. F. Piniella and J. J. Bonet, *J.
        Chem. Soc. Chem., Commun.* 718 (1983).

112.   R. F. Childs, B. M. Duffey and M. Mahendran, *Can. J. Chem., 64*, 1220
        (1986).

113.   G. Jones and B. R. Ramachandran, *J. Org. Chem., 41*, 798 (1976).

114.   K. Saito, K. Nagumo, R. Nakane, T. Ohyama, T. Sato and H. Yuki, *Can.
        J. Chem., 59*, 1717 (1981) and T. Sato, K. Nagumo, K. Saito, T. Ohyama
        and R. Nakane, *Chem. Lett.*, 1203 (1979).

115.   For reviews of enone/alkene photoaddition reactions see S. W. Baldwin,
        Org. Photochem., 5, 123 (1981), W. Oppolzer, *Accounts Chem. Res., 15*,
        135 (1982), A. C. Weedon in *Synthetic Organic Photochemistry*, Edit by
        W. M. Horspool, Plenum Press, N. Y. 1984.

116.   F. D. Lewis and S. V. Barancyk, *J. Amer. Chem. Soc., 111*, 8653 (1989).

117.   F. D. Lewis, D. K. Howard and J. D. Oxman, *J. Amer. Chem. Soc., 105*,

3344 (1983).

118.   F. D. Lewis, S. L. Quillen, P. D. Hale and J. D. Oxman, *J. Amer. Chem. Soc., 110*, 1261 (1988).

119.   For earlier work on the dimerization of coumarin see G. O. Schenck, I. von Wilucki and C. H. Krauch, *Chem. Ber., 95*, 1409 (1962). C. H. Krauch, S. Farid and G. O. Schenck, *Chem. Ber., 99*, 625 (1966). G. S. Hammond, C. A. Stout and A. A. Lamola, *J. Amer. Chem. Soc., 86*, 3103 (1964). H. Morrison, H. Curtis and T. McDowell, *J. Amer. Chem. Soc., 88*, 5415 (1966). R. Hoffman, P. Wells and H. Morrison, *J. Org. Chem., 36*, 102 (1971). H. Morrison and R. Hoffman, *J. Chem. Soc. Chem. Commun.*, 1453 (1968). K. Muthuramu and V. Ramamurthy, *J. Org. Chem., 47*, 3976 (1982).

120.   S. C. Shim, E. I. Kim and K. T. Lee, *Bull. Korean Chem. Soc., 8*, 140 (1987).

121.   T. Ogawa, Y. Maui, S. Ojima and H. Suzuki, *Bull. Chem. Soc. Jpn., 60*, 423 (1987).

122.   H. Hirai, *J. Polym. Sci. Macromol. Rev., 11*, 47 (1976). J. Furukawa, K. Omura and K. Ishikawa, *Tetrahedron Lett.*, 3119 (1979).

123.   T. Sato, S. Yoshiie, T. Imamura, K. Hasegawa, M. Miyahara, S. Yamamura and O. Ito, *Bull Chem. Soc. Jpn., 50*, 2714 (1977), K. Saito, H. Yuki, T. Shimada and T. Sato, *Can. J. Chem., 59*, 1722 (1981), T. Sato, H. Kaneko and S. Yamaguchi, *J. Org. Chem., 45*, 3778 (1980) and *Tetrahedron Lett.*, 1863 (1979), and E. Murayama, A. Kohda, T. Sato, *J. Chem. Soc. Perkin I*, 947 (1980).

124.   R. C. Fuson, *Record Chem. Progress, 12*, 1 (1951).

125.   N.J. Leonard and J. V. Paukstelis, *J. Org. Chem., 28*, 3021 (1963).

126.   R. Merényi, *Adv. Org. Chem.*, (H. Böhme and H. G. Viehe, eds.), Wiley, New York, p. 23 (1976).

127.   a) R. F. Childs, G. S. Shaw, and C. J. L. Lock, *J. Amer. Chem. Soc., 111*, 5424 (1989) b) R. F. Childs, B. D. Dickie, R. Faggiani, C. A. Fyfe, C. J. L. Lock, and R. E. Wasylishen, *J. Cryst. Spectr. Res., 15*, 73 (1985) c) J. O. Selzer and B. W. Matthews, *J. Phys. Chem., 80*, 631 (1976). B. W. Matthews, R. E. Stenkamp, and P. M. Colman, *Acta Cryst., B29*, 449 (1973).

128. a) P. A. Kollman, *Adv. Org. Chem.*, (H. Böhme and H. G. Viehe, eds.), Wiley, New York, p1 (1976). S. Seltzer, *J. Amer. Chem. Soc., 109*, 1627 (1987). J. E. Johnson, N. M. Silk, E. A. Nalley, and M. Arfan, *J. Org. Chem., 46*, 546 (1981). b) G. J. M. Dormans, G. C. Groenenboom, W. A. van Dorst, and H. M. Buck, *J. Amer. Chem. Soc., 110*, 1406 (1988). c) P. Tavan, K. Schulten, and D. Oesterhelt, *Biophys. J., 47*, 415 (1985).

129. G. Scheibe, C. Jutz, W. Seiffert, and D. Grosse, *Angew. Chem. Int. Ed. Engl., 3*, 306 (1964). M. Pankratz and R. F. Childs, *J. Org. Chem., 50*, 4553 (1985).

130. D. Lukton and R. R. Rando, *J. Amer. Chem. Soc., 106*, 4525 (1984).

131. a) R. F. Childs and B. D. Dickie, *J. Chem. Soc. Chem. Commun.*, 1268 (1981). b) R. F. Childs and B. D. Dickie, *J. Amer. Chem. Soc., 105*, 5041 (1983).

132. M. Pankratz and R. F. Childs, *J. Org. Chem., 53*, 3278 (1988).

133. G. Scheibe, J. Heiss, and K. Feldmann, *Angew, Chem., 77*, 545 (1965) J. Heiss and K. Feldman, *Angew. Chem., 77*, 545 (1965). J. Heiss and K. Feldman, *Angew. Chem., 77*, 546 (1965).

134. V. A. Kuzmin and A. P. Darmanyan, *Chem. Phys. Lett., 54*, 159 (1978). F. Dietz and S. K. Rentsch, *Chem. Phys., 96*, 145 (1985).

135. J. Michl, *Mol. Photochem., 4*, 243 (1972).

136. M. C. Bruni, J. P. Daudey, J. Langlet, J. P. Malrieu, and F. Momicchioli, *J. Amer. Chem. Soc., 99*, 3587 (1977). P. Bruckmann and L. Salem, *J. Amer. Chem. Soc., 98*, 5037 (1976). R. R. Birge and L. Hubbard, *J. Amer. Chem. Soc., 102*, 2195 (1980).

137. H. Görner and D. Schulte-Frohlinde, *Chem. Phys. Lett., 101*, 79 (1983). H. Görner and D. Schulte-Frohlinde, *J. Phys. Chem., 89*, 4105 (1985). H. Görner, *J. Phys. Chem., 89*, 4112 (1985).

138. H. Görner, A. Fojtik, J. Wroblewski, and L. J. Currell, *Z. Naturforsch, 40a*, 525 (1985).

139. H. Görner and H. Gruen, *J. Photochem., 28*, 329 (1985).

140. J.L. Stavinoha, E. Bay, A. Leone and P. S. Mariano, *Tett. Lett.*, 3455 (1980).

141. R. Gault and A. I. Meyers, *J. Chem. Soc., Chem. Comm.*, 778 (1971).

142. T. S. Sorensen and A. Rauk, *Pericyclic Reactions,* Ed. A. P. Marchand

and R. E. Lehr, Academic Press, New York, Vol 2, p. 1 (1977).

143.   V. A. Koptyug, *Topics in Current Chemistry, 122,* 1 (1984).

144.   D. M. Brouwer, E. L. Mackor and C. MacLean, *Carbonium Ions, 2,* 837 (1970).

145.   R. L. Flurry Jr. and R. K. Wilson, *J. Phys. Chem., 71,* 589 (1967).

146.   R. F. Childs, D. L. Mulholland, M. Zeya and A. K. Goyal, *Solar Energy, 30,* 155 (1983).

147.   R. F. Childs, M. Sakai, B. D. Parrington and S. Winstein, *J. Amer. Chem. Soc., 96,* 6403 (1974).

148.   R. F. Childs and S. Winstein, *J. Amer. Chem. Soc., 96,* 6409 (1974).

149.   I. S. Isaev, V. I. Mamatyuk, L. I. Kuzubova, T. A. Gordymova and V. A. Koptyug, *Zh. Org. Khim., 6,* 2482 (1970).

150.   I. S. Isaev, V. I. Mamatyuk, T. G. Ergorova, L. I. Kuzubova and V. A. Koptyug, *Izvest. Akad. Nauk. S.S.S.R. Ser. khim.,* 2089 (1969).

151.   G. A. Olah, G. Liang and S. P. Jindal, *J. Org. Chem., 40,* 3259 (1975).

152.   K. Ogura and T. Matsuura, *Bull. Chem. Soc. Jpn., 43,* 2891 (1970). B. Miller, *J. Amer. Chem. Soc., 89,* 1690 (1967).

153.   D. M. Brouwer, C. MacLean and E. L. Mackor, *Disc. Farady Soc., 39,* 121 (1965).

154.   S. K. Chadda, R. F. Childs, R. Faggiani and C. J. L. Lock, *J. Amer. Chem. Soc., 108,* 1694 (1986).

155.   For reviews on circumambulatory rearrangements see R. F. Childs, *Tetrahedron, 38,* 567 (1982). F. G. Klarner, *Topics in Stereochem., 15,* 1 (1984).

156.   I. Nabot, F. Tomás and I. Zabala, *J. Catal., 71,* 41 (1981). F. P. Plá, P. M. V. Martin and I. N. Gil, *J. Mol. Catal., 52,* 277 (1989).

157.   R. F. Childs and B. E. George, *Can. J. Chem., 66,* 1343 (1988).

158.   R. F. Childs and B. E. George, *Can. J. Chem., 66,* 1350 (1988).

159.   D. Bryce-Smith, A. Gilbert and H. C. Longuet-Higgins, *J. Chem. Soc. Chem. Commun.,* 240 (1967). G. Tennant, *Annu. Rep. Chem. Soc., 68 B,* 241 (1971).

160.   J. A. Berson and N. M. Hasty, Jr., *J. Amer. Chem. Soc., 93,* 1549 (1971). T. J. Katz, E. J. Wang and N. Acton, *J. Amer. Chem. Soc., 93,* 3782 (1971)

161. L. Kaplan, D. J. Rausch and K. E. Wilzbach, *J. Amer. Chem. Soc.*, *93*, 1549 (1971).

162. For a review of the photochemistry of aromatic compounds see D. Bryce-Smith and A. Gilbert, *Tetrahedron*, *32*, 1309 (1976).

163. H. Kuura and U. Haldna, *Reakts. Sposobn. Org. Soedin.*, *3*, 162 (1966); *Chem. Abstr.* *66*, 119381c (1967).

164. W. von E. Doering, M. Saunders, H. G. Boyton, H. W. Earhart, E. F. Wadley, W. R. Edwards and G. Laber, *Tetrahedron*, *4*, 178 (1958).

165. R. F. Childs and A. Mika-Gibala, *J. Org. Chem.*, *47*, 4204 (1982).

166. R. F. Childs, M. Zeya and R. P. Dain, *Can. J. Chem.*, *59*, 76 (1981).

167. H. Hart and J. D. DeVrize, *J. Chem. Soc. Chem. Commun.*, 1651 (1968). R. Criegee, H. Gruner and D. Schönleber R. Huber, *Chem. Ber.*, *103*, 3696 (1970).

168. G. R. Elia, R. F. Childs and G. S. Shaw, *73rd CIC Congress*, Halifax, Nova Scotia, 1990 Abstr. 903.

169. G. Quinkert, *Angewandte Chemie, Internat Edit.*, *14*, 790 (1975) and *Pure and Appl. Chem.*, *33*, 285 (1973). J. Griffiths and H. Hart, *J. Amer. Chem. Soc.*, *90*, 5296 (1968). W. G. Dauben, L. Salem and N. J. Turro, *Accounts Chem. Res.*, *8*, 41, (1975)

170. H. E. Zimmerman, *Adv. Photochem.*, *1*, 183 (1963). K. Schaffner, *Adv. Photochem.*, *4*, 81 (1966). D. I. Schuster, *Accnts. Chem. Res.*, *11*, 65 (1978).

171. E. C. Friedrich, *J. Org. Chem.*, *33*, 413 (1968). V. P. Vitullo, *J. Org. Chem.*, *34*, 224 (1969).

172. R. F. Childs and B. D. Parrington, *Can. J. Chem.*, *52*, 3303 (1974). S. M. Blackstock, K. E. Richards and G. J. Wright, *Can. J. Chem.*, *52*, 3313 (1974).

173. G. A. Olah and Y. K. Mo, *J. Org. Chem.*, *38*, 353 (1973).

174. J. W. Pavlik and R. J. Pasteris, *J. Amer. Chem. Soc.*, *96*, 6107 (1974).

175. N. Filipescu and J. W. Pavlik, *J. Amer. Chem. Soc.*, *92*, 6062 (1970).

176. H. Hart, T. R. Rodgers and J. Griffiths, *J. Amer. Chem. Soc.*, *91*, 754 (1969).

177. J. Griffiths and H. Hart, *J. Amer. Chem. Soc.*, *90*, 5296 (1968).

178. R. F. Childs, B. D. Parrington and M. Zeya, *J. Org. Chem.*, *44*, 4912

        (1979).

179.   P. Baeckström, U. Jacobsson, B. Koutek and T. Norin, *J. Org. Chem., 50*,
        3728 (1985).

180.   H. D. Becker, *The Chemistry of the Hydroxyl Group*, S. Patai, Ed., Wiley,
        New York, p. 835 (1971).

181.   T. Matsuura, Y. Hiromoto, A. Okada and K. Ogura, *Tetrahedron, 29*,
        2981 (1973).

182.   H. E. Zimmerman and D. I. Schuster, *J. Amer. Chem. Soc., 83*, 4486
        (1961) and *84*, 4527 (1962).

183.   D. I. Schuster and K.-C. Liu, *Tetrahedron, 37*, 3329 (1981).

184.   R. F. Childs, G. S. Shaw and A. Varadarajan, *Synthesis*, 198 (1982).

185.   R. F. Childs and S. Chadda, *Can. J. Chem., 63*, 3449 (1985).

186.   K. E. Hine and R. F. Childs, *J. Chem. Soc. Chem. Commun.*, 145 (1972).

187.   K. E. Hine and R. F. Childs, *J. Amer. Chem. Soc., 93*, 2323 (1971).

188.   K. E. Hine and R. F. Childs, *J. Amer. Chem. Soc., 95*, 6116 (1973).

189.   G. Büchi and E. M. Burgess, *J. Amer. Chem. Soc., 82*, 4333 (1960). D. I
        Schuster and D. H. Sussman, *Tetrahedron Letts*, 1657 (1970). D. I.
        Schuster, M. J. Nash and M. L. Kantor, *Tetrahedron Lett.* 1375 (1964). T.
        Takino and H. Hart, *J. Chem. Soc. Chem. Commun.*, 450 (1970). H. Hart
        and T. Takino, *J. Amer. Chem. Soc., 93*, 720 (1971). H. Hart, *Pure and
        Appld. Chem., 34*, 247 (1973). W. A Ayer and L. M. Browne, *Can. J.
        Chem., 52*, 1352 (1974).

190.   R. Noyori, Y. Ohnishi and M. Katô, *J. Amer. Chem. Soc., 94*, 5105
        (1972).

191.   H. Nozaki, M. Kurita and R. Noyori, *Tetrahedron Lett., 33*, 3635 (1968).

192.   R. Noyori, Y. Ohnishi and M. Katô, *J. Amer. Chem. Soc., 97*, 928 (1975).

193.   H. Hart and A. F. Naples, *J. Amer. Chem. Soc., 94*, 3256 (1972).

194.   R. Srinivasan, *Adv. in Photochem., 4*, 113 (1966).

195.   A. R. Oseroff and R. H. Callender, *Biochemistry, 13*, 4243 (1974).

196.   B. Becher and T. G. Ebrey, *Biophys. J., 17*, 185 (1977).

197.   C. R. Goldschmidt, O. Kalisky, T. Rosenfeld, and M. Ottolenghi, *Biophys.
        J., 17*, 179 (1977).

198.   R. R. Birge, T. M. Cooper, A. F. Lawrence, M. B. Masthay, C. Vasilakis,
        C. F. Zhang, and R. Zidovetzki, *J. Amer. Chem. Soc., 111*, 4063 (1989).

199.   R. H. Lozier, R. A. Bogomolni, and W. Stoeckenius, *Biophys. J., 15,* 955
        (1975).
200.   D. Lukton and R. R. Rando, *J. Amer. Chem. Soc., 106,* 258 (1984).
201.   M. Sheves and T. Baasov, *J. Amer. Chem. Soc., 106,* 6840 (1984).
202.   C. Pattaroni and J. Lauterwein, *Helv. Chim. Acta, 64,,* 1969 (1981).
203.   G. S. Shaw, Ph.D. Thesis, McMaster University, Hamilton, Ontario,
        Canada, 1988.
204.   P. K. Das, G. Kogan, and R. S. Becker, *Photochem. Photobiol., 30,* 689
        (1979).
205.   R. F. Childs and G. S. Shaw, *J. Amer. Chem. Soc., 110,* 3013 (1988).
206.   J. M. Donahue and W. H. Waddell, *Photochem. Photobiol., 40,* 399
        (1984).
207.   K. A. Freedman and R. S. Becker, *J. Amer. Chem. Soc., 108,* 1245 (1986).
208.   R. S. Becker and K. Freedman, *J. Amer. Chem. Soc,, 107,* 1477 (1985).
209.   R. S. Becker, K. Freedman, and G. Causey, *J. Amer. Chem. Soc., 104,*
        5797 (1982).
210.   R. S. Becker, K. Freedman, J. A. Hutchinson, and L. J. Noe, *J. Amer.
        Chem. Soc., 107,* 3942 (1985).
211.   K. Freedman, R. S. Becker, D. Hannak, and E. Bayer, *Photochem.
        Photobiol., 43,* 291 (1986).
212.   R. S. Becker, *Photochem. Photobiol., 48,* 369 (1988). M. Ottolenghi,
        Adv. Photochem., 12, 97 (1980). R. R. Birge, *Ann. Rev. Biophys. Bioeng.,
        10,* 315 (1981).
213.   W. H. Waddell, A. M. Schaffer, and R. S. Becker, *J. Amer. Chem. Soc.,
        95,* 8223 (1973).
214.   M. M. Fischer and K. Weiss, *Photochem. Photobiol., 20,* 423 (1974).
215.   T. Rosenfeld, A. Alchalel, and M. Ottolenghi, *Photochem. Photobiol., 20,*
        121 (1974).
216.   N. Friedman, M. Sheves and M. Ottolenghi, *J. Amer. Chem. Soc., 111,*
        3203 (1989).
217.   R. F. Childs, G. S. Shaw and R. E. Wasylishen, *J. Amer. Chem. Soc., 109,*
        5362 (1987). G. S. Harbison, P. P. J. Mulder, H. Pardoen, J. Lugtenburg,
        J. Herzfeld, R. G. Griffin, *J. Amer. Chem. Soc., 107,* 4809 (1985). G. S.
        Harbison, S. O. Smith, J. A. Pardoen, J. M. L. Courtin, J. Lugtenburg, J.

Herzfeld, R. A. Mathies and R. G. Griffin, *Biochemistry, 24*, 6955 (1985).

218.   E. E. van Tamelen, R. H. Greeley and H. Schumacher, *J. Amer. Chem. Soc., 93*, 6151 (1971).

219.   A. W. Krebs, *Angew. Chem. Internat. Ed., 4*, 10 (1965).

220.   R. K. Lustgarten, M. Brookhart and S. Winstein, *J. Amer. Chem. Soc., 89*, 6350 (1967).

221.   Y. Takahashi, S. Sankararaman and J. K. Kochi, *J. Amer. Chem. Soc., 111*, 2954 (1989).

220.   W. Abraham, J. Epperlein, D. Kreysig and C. Michael, Patent, East Germany DD 135248 (1979); *Chem. Abs., 91*, 1849 (1979). W. Werner, B. Dreher, K. Buck and D. Kreysig, *J. Prakt. Chem., 324*, 925 (1982). W. Werner, B. Dreher, D. Kreysig, N. A. Sadovskii and M. G. Kuz'min, *J. Prakt. Chem., 329*, 569 (1987) S. Ishikawa, M. Mabuchi and T. Koyama, Patents, Japan, JP 8688264; JP 6188264; *Chem. Abs., 106*, 11168u and JP 8691663 A2; JP 6191663; *Chem. Abs., 106*, 25767a.

223.   T. Ikemi, T. Nozoe and H. Sugiyama, *Chem. and Ind. London*, 932 (1960). A. Ter Borg, R. Van Helden, A. F. Bickel, W. Renold and A. S. Dreiding, *Helv. Chim. Acta., 43*, 457 (1960).

224.   L. J. Johnston, J. Lobaugh and V. Wintgens, *J. Phys. Chem., 93*, 7370 (1989).

225.   P. A. Christensen, Y. Y. Huang, A. Meesters and T. S. Sorensen, *Can. J. Chem., 52*, 3424 (1974).

226.   R. Bladek and T. S. Sorensen, *Can. J. Chem., 50*, 2806 (1972).

227.   R. F. Childs and C. V. Rogerson, *J. Amer. Chem. Soc., 98*, 6391 (1976); *100*, 649 (1978).

228.   R. F. Childs and C. V. Rogerson, *J. Amer. Chem. Soc., 102*, 4159 (1980).

229.   L. A. Paquette and O. Cox, *J. Amer. Chem. Soc., 89*, 1969 and 5633 (1967).

230.   G. G. Mihai, P. G. Tarassoff and N. Filipescu, *J. Chem. Soc. Perkin I*, 1374 (1975).

231.   Cf. A. V. El'tsov, Y. K. Levental' and O. P. Studzinskii, *J. Org. Chem. USSR., 12*, 2407 (1976).

232.   A. D. Broadbent and J. M. Stewart, *Can. J. Chem., 61*, 1965 (1983) and *J. Chem. Soc. Chem. Commun.*, 676 (1980).

233. O. P. Studzinskii and A. V. El'tsov, *Zh. Org. Khim.*, *18*, 1904 (1982) and *Zh. Org. Khim.*, *16*, 1100 (1980).

234. C. H. Rochester, Acidity Functions, Academic Press, London, 1970.

235. E. Haselbach, E. Vauthey and P. Suppan, *Tetrahedron, 44*, 7335 (1988).

236. K. H. Meyer, *Justus Liebig's Ann. Chem., 379*, 37 (1911). S. Coffey, *Chem. and Ind. London*, 1068 (1953), A. M. Dawson and A. Johnson, *J. Soc. Dyers Colorists, 82*, 49 (1966).

237. K. Seguchi and H. Ikeyama, *Chem. Lett.*, 1493 (1980).

238. A. Lablache-Combier in *Photochemistry of Heterocyclic Compounds*, Edit O. Buchardt, J. Wiley and Sons, New York, p. 207 (1976). A. T. Balaban, G. Fischer, A. Dinculescu, A. V. Koblik, G. N. Dorofeenko, V. V. Mezheritskii and W. Schroth, *Advn. Heterocycl. Chem. Suppl., 2*, 158 (1982). S. T. Reid, *Advn. Heterocycl. Chem., 33*, 1 (1983).

239. K. Okada, K. Hisamitsu and T. Mukai, *Tetrahedron Lett.*, 1251 (1981). M. J. Climent, H. García, S. Iborra, M. A. Miranda and J. Primo, *Heterocycles, 29*, 115 (1989). F. D. Saeva and G. R. Olin, *J. Chem. Soc. Chem. Commun.*, 943 (1976).

240. J. A. Barltrop, A. W. Baxter, A. C. Day and E. Irving, *J. Chem. Soc. Chem. Commun.*, 606 (1980).

241. J. A. Barltrop, K. Dawes, A. C. Day and A. J. H. Summers, *J. Chem. Soc. Chem. Commun.*, 1240 (1972) and *J. Amer. Chem. Soc., 95*, 2406 (1973). J. A. Barltrop, K. Dawes, A. C. Day, S. J. Nuttall and A. J. H. Summers, *J. Chem. Soc. Chem. Commun.*, 410 (1973).

242. J. W. Pavlik and E. L. Clennan, *J. Amer. Chem. Soc., 95*, 1697 (1973).

243. J. W. Pavlik, A. D. Patten, D. R. Bolin, K. C. Bradford and E. L. Clennan, *J. Org. Chem., 49*, 4523 (1984).

244. J. W. Pavlik and J. Kwong, *J. Amer. Chem. Soc., 95*, 7914 (1973).

245. J. W. Pavlik, D. R. Bolin, K. C. Bradford and W. G. Anderson, *J. Amer. Chem. Soc., 99*, 2816 (1977). J. W. Pavlic and A. P. Spada, *Tetrahedron Lett.*, 4441 (1979). J. W. Pavlik, A. P. Spada and T. E. Snead, *J. Org. Chem., 50*, 3046 (1985).

246. J. W. Pavlik and R. M. Dunn, *Tetrahedron Lett.*, 5071 (1978).

247. J. A. Barltrop and A. C. Day, *J. Chem. Soc. Chem. Commun.*, 177, (1975). J. A. Barltrop, R. Carder, A. C. Day, J. R. Harding and C. Samuel, *J.*

Chem. Soc. Chem. Commun., 729 (1975).

248.   J. A. Barltrop, A. C. Day and C. J. Samuel, *J. Chem. Soc. Chem. Commun.*, 823 (1976). J. A. Barltrop, J. C. Barrett, R. W. Carder, A. C. Day, J. R. Harding, W. E. Long and C. J. Samuel, *J. Amer. Chem. Soc.*, *101*, 7510 (1979).

249.   J. A. Barltrop, A. C. Day and C. J. Samuel, *J. Amer. Chem. Soc.*, *101*, 7521 (1979). J. W. Pavlik and J. Kwong, *J. Amer. Chem. Soc.*, *95*, 7914 (1973). N. Ishibe, M. Sunami and M. Odani, *J. Amer. Chem. Soc.*, *95*, 463 (1973). A. Padwa and R. Hartman, *J. Amer. Chem. Soc.*, *88*, 1518 (1966). P. Yates and I. W. J. Still, *J. Amer. Chem. Soc.*, *85*, 1208 (1963). P. Yates and M. J. Jorgenson, *J. Amer. Chem. Soc.*, *85*, 2956 (1963). E. Paterno, *Gazz. Chim. Ital.*, *44*, 1 and 151 (1914).

250.   D. G. Whitten, The Photochemistry of Heterocyclic Compounds (O. Buchardt, ed.), Wiley, New York, p. 524 (1976).

251.   T. J. van Bergen and R. M. Kellogg, *J. Amer. Chem. Soc.*, *94*, 8451 (1972).

252.   E. F. Travecedo and V. I. Stenberg, *J. Chem. Soc. Chem. Comm.*, 609, (1970).

253.   T. Furihata and A. Sugimori, *J. Chem. Soc. Chem. Comm.*, 241 (1975)

254.   F. R. Stermitz, R. P. Seiber and D. E. Nicodem, *J. Org. Chem.*, *33*, 1136 (1968).

255.   F. R. Stermitz, C. C. Wei, and W. H. Huang, *J. Chem. Soc. Chem. Comm.*, 482 (1968).

256.   H. Nozaki, M. Kato, R. Noyori, and M. Kawanisi, *Tett. Lett.*, 4259 (1967).

257.   T. L. Lowry and K. S. Richardson, Mechanism and Theory in Organic Chemistry (M. Wasserman, ed.), Harper Row Publications, New York, p. 682 (1981).

258.   L. Kaplan, J. W. Pavlik, and K. E. Wilzbach, *J. Amer. Chem. Soc.*, *94*, 3283 (1972).

259.   A. R. Katritzky and H. Wilde, *J. Chem. Soc. Chem. Comm.*, 770 (1975).

260.   P. Bortolus, G. Cauzzo, U. Mazzucato, and G. Galiazzo, *Z. Phys. Chem.*, *63*, 29 (1969).

261.   A. Fozard and C. K. Bradsher, *Tett. Lett.*, 3341 (1961).

262.   A. Fozard and C. K. Bradsher, *J. Org. Chem.*, *32*, 2966 (1967).

263.   D. E. Portlock, M. J. Kane, J. A. Bristol, and R. E. Lyle, *J. Org. Chem.,*
        38, 2351 (1973).

264.   A. R. Katritzky, Z. Zakaria, and E. Lunt, P. G. Jones, and O. Kennard, *J.*
        Chem. Soc. Chem. Comm., 268 (1979).

265.   A. R. Katritzky, Z. Zakaria, and E. Lunt, *J. Chem. Soc. Perkin I,* 1879
        (1980).

266.   E. C. Taylor, R. O. Kan, and W. W. Paudler, *J. Amer. Chem. Soc., 83,*
        4484 (1961).

267.   E. C. Taylor and R. O. Kan, *J. Amer. Chem. Soc., 85,* 776 (1963).

268.   N. J. Turro, Modern Molecular Photochemistry, Benjamin/Cummings
        Pub. Co. Inc., Mento Pak, Calif., p. 456 (1978).

269.   K. Kano and T. Matsuo, *Tett. Lett.,* 1389 (1975).

270.   K. Kano and T. Matsuo, *Bull. Chem. Soc. Japan, 49,* 3269 (1976).

271.   F. Takeuchi, T. Sugiyama, T. Fujimori, K. Seki, Y. Harada, and A.
        Sugimori, *Bull. Chem. Soc. Japan, 47,* 1245 (1974).

272.   A. Castellano, J.-P. Catteau, A. Lablache-Combier, and B. Tinland, *J.*
        Chem. Res., 70 (1979).

273.   J. Joussot-Dubien and J. Houdard-Pereyre, *Bull. Soc. Chim. Fr.,* 2619
        (1969).

4

Photoinduced Hydrogen Atom Abstraction by Carbonyl Compounds

PETER WAGNER

BONG-SER PARK

Michigan State University
East Lansing, Michigan

I. INTRODUCTION

Hydrogen atom abstraction by photoexcited carbonyl compounds has played a central role in the development of a general picture of how photochemical reactions occur. Scaiano reviewed the bimolecular process in 1973;[1] in the mid-70's one of us authored two reviews that covered excited state reactivity[2a] and reactions of photogenerated biradicals.[2b] The past 15 years have seen many new developments in the area of hydrogen abstraction. The emergence of nanosecond and picosecond laser spectroscopy has allowed the direct study not only of the reactive triplet states but also of the radicals, biradicals, and radical-ions that are produced as intermediates during photochemical hydrogen transfer processes. Very accurate rate constants are now known for a variety of hydrogen abstraction processes. The competition between hydrogen and electron transfer has been addressed far more quantitatively than was possible earlier. Environmental effects on hydrogen transfer have been used to determine information about the structures of organized media. Conformational effects both on intramolecular hydrogen abstraction itself and on the reactions of biradicals have been explored more thoroughly than previously. The study of solid compounds with known crystal structures and the arrival of simple molecular mechanics programs have been of great benefit in this regard. The reactions of biradicals, in particular the mechanism for intersystem crossing and product formation , have become a topic of major interest. New biradical rearrangements appear regularly in all of the journals. Synthetic organic chemists have begun to utilize photorearrangements initiated by hydrogen transfers. With the current intense interest in synthetic applications of intramolecular radical reactions, it is almost a certainty that initiation by hdyrogen abstraction will be used more routinely for natural products synthesis.

This review will begin by a brief survey of what we knew 15 years ago and will then cover major aspects of hydrogen abstraction reactions, explaining all major mechanistic features but emphasizing what has been learned since the mid-70's reviews were written. Some of the topics represent updates of what was known earlier while others represent problems that had received little or no study before then. We could not mention every publication on this topic nor dwell in equal depth on every subtopic in this laarge field; we apologize to authors whose work we may have neglected.

What we knew before 1976:

The photoreduction of benzophenone is among the best known of photoreactions, thanks to its inclusion by Fieser in his laboratory experiments. The first work by Hammond on this reaction was seminal in several regards. It established the general reactivity of triplet states in radical-producing reactions. It showed how quantum yield measurements at various substrate concentrations can provide values of relative rate constants for excited state reactions. (In fact, plots of reciprocal quantum yields *vs.* reciprocal substrate concentration should rightfully be called "Hammond plots".) The use of specific triplet quenchers allowed the differentiation of excited singlet and triplet reactions. Study of substituted benzophenones and other aryl ketones soon established that the chemical reactivities of n,π^* and π,π^* triplets differ markedly. The introduction of Stern-Volmer quenching studies allowed generally reliable measurements of triplet lifetimes, from which rate constants for individual triplet reactions could be gleaned. The Norrish type II reaction was shown to involve a 1,4-biradical formed by intramolecular γ-hydrogen abstraction. The competition between cleavage and cyclization of these biradicals was studied for a wide range of structures. That quantum yields are determined almost entirely by the extent to which these biradicals revert to starting ketones was the first dramatic evidence that quantum efficiencies need not be directly related to excited state reactivity. Measurements of hydrogen abstraction rate constants provided some of the first quantitative structure-reactivity relationships in organic photochemistry. Electron-withdrawing substituents near the C-H bond being attacked strongly decrease rate constants; the reactive n,π^* triplet is a very electron-deficient species. Ring substituents on phenyl ketones affect rate constants primarily by their influence on the relative energies of the reactive n,π^* and "unreactive" π,π^* triplets. When the n,π^* triplet lies higher in energy than the π,π^* triplet, rate constants are particularly low because reaction takes place from low equilibrium populations of the reactive upper triplet. The question of possible stereoelectronic requirements for hydrogen abstraction was raised but not settled. Only a few examples were known of intramolecular hydrogen abstraction from positions other than the γ-carbon of ketones. Bimolecular reactions with electron donors such as alkenes, aromatics, and amines were shown to proceed through charge transfer complexes now called exciplexes. Such reactions with stronger donors such as anilines and DABCO were shown to involve electron transfer and radical-ion intermediates. In both cases, actual hydrogen transfer from carbon to oxygen takes on to a lesser or greater degree the character of proton transfer.

All of these basic studies were performed in the 1960-1975 period. Let us now explore what the next 15 years produced in the way of better understanding and new ideas. This review will concentrate on two aspects of hydrogen abstraction: how various structural and environmental changes affect 1) the rate constants for hydrogen abstraction itself and 2) the behavior of the radical or biradical intermediates produced by the initial hydrogen abstraction. Both must be understood if the efficiency of a given photoreaction is to be understood.

II. THE NATURE OF THE HYDROGEN ATOM ABSTRACTION PROCESS.

There are five fundamental questions involving hydrogen atom abstraction by excited states that have been addressed over the years: 1) how does the electronic configuration of the reacting excited state affect the process; 2) what are the orientational and stereoelectronic requirements for the reaction; 3) to what extent do rate constants depend on electronic and steric demand of substituents on both the excited state and the hydrogen donor; 4) to what extent do rate constants depend on the overall thermodynamics for hydrogen transfer; and 5) to what extent does charge transfer compete with or contribute to hydrogen transfer?

A. Electronic Effects on Reactivity

One issue that was well established early on is the strongly electron deficient nature of n,π^* carbonyl triplets. For example, rate constants for triplet state γ-hydrogen abstraction in δ-substituted valerophenones display a textbook Hammett linear free energy relationship when plotted against σ_I values, with a ρ value of -1.85.[3] It is difficult to do such a study for intermolecular hydrogen abstraction; most substituted alkanes have too many different reactive C–H bonds, and toluenes quench most ketones primarily by a charge transfer process.[4,5] Nonetheless, the large deactivating effect that electron-withdrawing groups have on the reactivity of C—H bonds is revealed in the near inertness of acetonitrile[6], which can be used as solvent for photoreductions. (Acetone, acetic acid, and methyl acetate have not been used much but should be equally unreactive solvents.) The reactivity of 2,2-

dimethylpropane is 8×10^3 M^{-1} s^{-1} per methyl group, whereas that of acetonitrile is only 1.3×10^2, a 60-fold deactivation.[6] The 30-fold greater triplet reactivity of valerophenone compared to γ-cyanobutyrophenone provides the corresponding intramolecular comparison.[3] The strong electron demand of the cyano group overcomes its ability to conjugatively stabilize the product radical site. In fact we calculated in 1972 that a 300-fold inductive destabilization masks a 10-fold resonance stabilization of the transition state.[3] Steel has proposed an interesting alternative analysis of this deactivation by noting a linear free energy relationship between the rate constant for triplet benzophenone attack on various R–CH_3 compounds and the oxidation potentials of the product radicals R–$CH_2 \cdot$.[6] Of course, oxidation potentials are known to show linear Hammett relationships, since σ values are one measure of a substituent's electronic demand. No matter which model one uses, the effect reveals that n,π* triplets, like alkoxy radicals,[7,8] have such high electron demand that the transition states for hydrogen transfer are stabilized by charge transfer from the developing radical to the carbonyl oxygen. The more electron-withdrawing R is, the less the transition state for hydrogen transfer can be stabilized by such charge transfer. This effect has been known for over three decades in radical chemistry.[9]

The complementary measurement of substituent effects on the ketone was not performed until recently. As reviewed below, ring substituents on phenyl alkyl ketones alter the nature of the lowest triplet so severely that Hammett plots are not only nonlinear but scattered.[10] Fortunately, most substituted benzophenones retain n,π* lowest triplets.[11] We compared the triplet lifetimes of a variety of mono- and disubstituted benzophenones in cyclopentane, which served as both a hydrogen source and a constant nonpolar solvent.[12] Steady state studies showed that irradiation produces the expected radical coupling and trapping products of hemipinacol and cyclopentyl radicals, confirming that the triplet lifetimes measure rates of hydrogen abstraction from solvent. The total reactivity range is only a factor of 17 going from 4,4'-dimethoxybenzophenone ($1/\tau = 1.4 \times 10^6$ s^{-1}) to 4,4'-dicyanobenzophenone ($1/\tau = 2.4 \times 10^7$ s^{-1}). The ρ value is only 0.59 or 0.43, depending on whether σ or σ^+ values are used. In either case, the inductive effects are weak, but in the expected direction. The fact that the substituents are not directly conjugated with the reaction center, the half-occupied p orbital on oxygen, prevents any large inductive effects.

One interesting anomaly in this study is the superactivating effect of fluoro substituents, which have very small σ constants. In fact triplet pentafluorobenzophenone is twice as reactive as dicyanobenzophenone and triplet decafluorobenzophenone is too reactive to even measure.[12] This enhanced reactivity may denote a change in mechanism from pure one step hydrogen transfer to a two step process involving a charge transfer complex (see below). Another manifestation of this effect is the six-fold greater reactivity of triplet α-fluoroacetophenone relative to acetophenone itself.[13] In this case captodative hyperconjugative stabilization of the product hemipinacol radical may contribute to the rate enhancement.

$$ \underset{Ph}{\overset{\cdot\,OH}{\diagdown}}\underset{\cdot}{C}\diagup CH_2 \!\!-\!\! F \quad\longleftrightarrow\quad \underset{Ph}{\overset{\overset{\cdot}{+}\,OH}{\diagdown}}C\diagup CH_2 \quad F^{-} $$

A final interesting feature of the study involved substituent effects on the tertiary/ primary selectivity that triplet benzophenone displays towards 2,3-dimethylbutane. The two radicals were trapped by CCl_4 as the alkyl chlorides, as introduced by Walling and Gibian[7]. The t/p ratio increased by a factor of 8 in going from 4,4'-dimethoxybenzophenone to 4-cyanobenzophenone. Thus the most electron demanding ketone triplet is most selective towards tertiary C-H bonds. This selectivity dependence is further evidence that there is significant charge transfer in the transition state, since tertiary radicals are most easily oxidized, and lends further weight to Steel's correlation with radical oxidation potentials.[6]

B. Nature of the Excited State–n,π* vs. π,π*

It was evident very early that ketones with n,π* lowest triplets are far more reactive than are ketones with π,π* lowest triplets. The early evidence came mostly in terms of quantum efficiencies[14] and some rate constants. Thus 1-naphthaldehyde and 2-acetonaphthone are not photoreduced in isopropyl alcohol, but are reduced by tributylstannane, a much better hydrogen donor.[15] It is now accepted convention that the singly occupied n orbital on oxygen is the locus of radical reactivity in n,π* states.

When Yang first pointed out in 1967 that substituted acetophenones with π,π* lowest triplets are photoreduced in isopropyl alcohol,[16] the question arose how such triplets derive their reactivity. Since spectroscopists were then measuring the extent of vibronic mixing of excited states, one explanation proposed was that a certain amount of n,π^* character was mixed vibronically into the lowest π,π^* state.[17] Careful study of several systems indicated that the most of the measured reactivity in such systems comes from low concentrations of the n,π^* triplet in equilibrium with the lower π,π^* triplet.[18] Equation [1] describes observed rate constants for hydrogen abstraction, with the n,π^* state providing most or all of the reactivity when ΔE ($E_{n,\pi}$ - $E_{\pi,\pi}$) < 5 kcal/mole.

$$k_{obs} = \chi_{n,\pi}k_{n,\pi} + \chi_{\pi,\pi}k_{\pi,\pi} \qquad [1]$$

$$\chi_{n,\pi} = (1 - \chi_{\pi,\pi}) = e^{-\Delta E/RT}/[1 + e^{-\Delta E/RT}] \qquad [2]$$

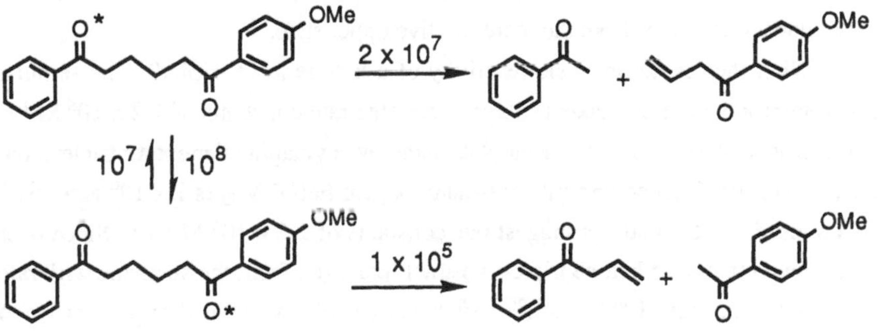

The actual evidence for equilibration was twofold. Rate constants for triplet state γ-hydrogen abstraction by phenyl alkyl ketones are 100 times greater than those for p-anisyl ketones; both are depressed equally by electron-withdrawing substituents on the δ-carbon. Such a result is demanded by Equation [1] if only the electron-deficient n,π^* triplet reacts. However, it is difficult to understand how a π,π^* triplet, even with a few per cent n,π^* character, could be as electron deficient at oxygen as a n,π^* triplet. The constant factor of 100 was interpreted to indicate a 2.8 kcal/mole energy difference between the two triplets of p-methoxyphenyl ketones. This idea of very rapid thermal equilibration of triplet excitation between two states in one chromophore was bolstered by the known efficiency of intramolecular energy transfer in bichromophoric systems,[19] including 1-benzoyl-4-anisylbutane.[20] The diketone reacts from both ends, but primarily (90%) from the benzoyl end, which has the higher excitation energy. Since reaction from each end is quenched with the same efficiency, the two triplets have the same lifetime and thus are equilibrated. The approximate rate constants shown below are predicted by Equations [1] and [2] from the 19:1 product ratio at 25°. Reaction of the anisyl end actually arises from its n,π* triplet 2.8 kcal higher.

The second type of evidence was provided by flash kinetic measurements of activation parameters for hydrogen abstraction by several ring-substituted phenyl alkyl ketones, by Steel for bimolecular reactions[21] and by Scaiano for intramolecular reactions.[22] In both cases activation entropies for a given reaction were almost constant, whereas activation energies increased for ketones with π,π* lowest triplets by amounts which correspond to estimated energy differences between the two triplet levels. Steel points out that reactivity differences arising mainly from varying degrees of n,π* character would appear primarily as activation entropy (not observed), whereas differences based mainly on varying Boltzman factors would appear primarily as activation enthalpy (observed). Scaiano reports five examples, including two for intramolecular hydrogen abstraction, in which triplet reactions of p-methoxyphenyl alkyl ketones have activation energies 2.8 kcal/mole higher than that for the corresponding unsubstituted phenyl ketone. He in fact proposes such measurements as a p-methoxy effect to determine when triplet ketone reactivity arises from the upper n,π* triplet.

It is often assumed incorrectly that π,π* triplets are totally unreactive at hydrogen abstraction. This notion is contradicted by the known ability of alkene triplets[23] and enone triplets[24] to abstract hydrogen atoms, as well as by the observed photoreactivity of some naphthyl ketones. What does seem to be always true is that π,π* triplets have much lower reactivity than do n,π* triplets in "alkoxy radical-like" reactions , such that when n,π* triplets lie only a few kcal above π,π* states, most observed reaction comes from the more reactive upper state.

What then is the intrinsic reactivity of a ketone π,π* triplet? The seminal 1962 Hammond and Leermakers study[15] estimated rate constants of $1\text{-}2 \times 10^6$ M^{-1} s^{-1} for hydrogen abstraction from tributylstannane by acylnaphthalene π,π* triplets, but recently measured quenching rate constants (k_q for Fe(DPM)$_3$ is 1×10^9 M^{-1} s^{-1})[25] and a revised kinetic analysis suggest rate constants of $1\text{-}2 \times 10^5$ M^{-1} s^{-1}. Since later work showed that n,π* ketone triplets abstract hydrogen from the stannane with rate constants approaching 10^9 M^{-1} s^{-1},[26,27,28] the acylnaphthalene π,π* triplets prove to be only 0.01% as reactive as n,π* triplets. The behavior of 1-benzoyl-8-benzylnaphthalene (BBN), which photocyclizes *via* triplet state δ-hydrogen abstraction,[29] provides another comparison. Its ground state geometry is almost perfect for internal hydrogen abstraction, as is the case with o-*tert*-butylbenzophenone (OTBBP).[30] BBN's measured hydrogen abstraction rate constant of 7 x 10³ s⁻¹ is only 0.001% as large as that for the n,π* triplet of OTBBP#, although both reactions are quite exothermic. Hammond and Leermakers provided an explanation for the large difference that still seems reasonable, namely the distinct spin localization in an n,π* triplet as opposed to the delocalized spin in a π,π* triplet. This fact is usually depicted in terms of the n,π* triplet resembling a 1,2-diradical[31] and thus manifesting the chemical reactivity of an alkoxy radical,[7,8] while the π,π* triplets of aryl ketones have little spin density on oxygen. It is very revealing to compare the activation parameters for δ-hydrogen abstraction by triplet BBN (E_a = 4.7 kcal/mole; log A = 7.25) with those for OTBBP (E_a = 2.5 kcal/mole; log A = 10.5). The 10⁵ n,π*/π,π* reactivity ratio reflects 60% entropy and 40% enthalpy differences. It is tempting to suggest that the 10³ variation in A values reflects the low free spin density on oxygen in the π,π* triplet of BBN, in which spin density is located mainly in the ring. This interpretation was not offered by the original authors but is strongly suggested by Steel's analysis.[21]

As mentioned above, several carbon-centered π,π* triplets do abstract hydrogen atoms, especially alkene triplets and enone triplets. Spin is much more localized in alkene triplets than it is in aromatic triplets. However, it must be kept in mind that it is the biradical nature of triplet excited states that induces radical-like reactivity. The much greater propensity for hydrogen abstraction by carbonyl triplets relative to alkene triplets simply reflects the lower kinetic barriers always associated with hydrogen transfer to oxygen relative to carbon. For example, alkoxy radicals are orders of magnitude more reactive than methyl radicals as hydrogen abstractors, despite almost identical thermodynamics.[32]

One of the earliest indications of different n,π* and π,π* reactivity was the pronounced solvent effect on the photoreactivity of p-aminobenzophenone.[33] Although part of the diminished quantum efficiency for photoreduction in polar solvents is due to a combination of low intersystem crossing yield and rapid triplet self-quenching[34], triplet reactivity is very low in polar solvents. The π,π* triplets tend to have enhanced dipoles whereas n,π* triplets typically have diminished dipoles relative to their ground states. Consequently in polar solvents the n,π*-π,π* energy gap is larger, so that $x_{n,\pi}$ and k_{obs} are lower, than in nonpolar solvents. The well studied p-methoxyphenyl ketones, for example, are only 1/10 as reactive in methanol as in benzene.[18,22] Triplet xanthone reacts with cyclohexane 40 times faster in CCl_4 than in methanol; with 2-propanol, 500 times faster in CCl_4 than in 2-propanol.[27] Interestingly, the relatively large solvent effects on xanthone reflect a much greater n,π* reactivity compared to other phenyl ketones. It takes a thiyl or amino substituent to cause benzophenone's π,π* triplet to drop below its normal 68.6 kcal n,π* excitation energy. However, xanthone's n,π* energy is ~74 kcal, comparable to that of its π,π* triplet.[35] Although Scaiano did not directly attribute the high reactivity of triplet xanthone to its high n,π* triplet energy, that and the optimal conjugation in the fully planar radical product seem the most likely reasons.

If the fully conjugated nature of triplet xanthone does explain its high reactivity, then Turro has recently uncovered the opposite extreme.[36] A series of paracyclophanobenzophenones were studied; as the para tether is shortened from 12 to 8 carbons, the benzene rings twist until they are orthogonal to the carbonyl. The absorption and emission properties of the ketones undergo a concomitant change, from those characteristic of lowest n,π* excited triplets to those of some undefined state. Rate constants for reaction of the ketone triplets with 2-propanol also steadily decrease as the carbonyl group becomes less conjugated, ending up only 0.1-1% as large for n=8 as for n=12, which has reactivity comparable to that of 4,4'-dimethyl-benzophenone. Similarly, triplet keto[2.2]paracyclophane has been reported to be very unreactive.[37] Apparently the reactivity of triplet phenyl ketones plummets when the benzoyl group is constrained to be unconjugated. Turro and coworkers suggest that a π,π^* state becomes lowest under such circumstances. However, the lowest π,π^* triplet of normal phenyl ketones corresponds to the L_a transition,[16,18] which is driven to higher energy by loss of conjugation.[38] Therefore the exact nature of the lowest triplet of such "unconjugated" phenyl ketones is not clear. An n,π^* state must still exist at somewhat higher energy, as in aliphatic ketones, so the lowest triplet may involve some unusual interaction of the n-orbital with the benzene ring. The loss of benzylic conjugation in the radical product is not the cause for the low reactivity, since aliphatic ketones have the same triplet reactivity as normal phenyl ketones.[2a,8] Moreover, the loss of reactivity includes charge transfer and energy transfer reactions as well as hydrogen abstraction.[37]

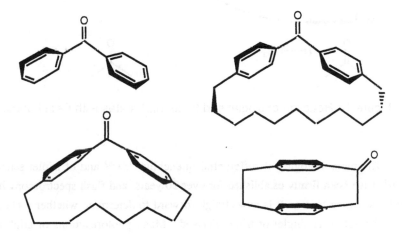

C. Nature of Excited State - Singlet vs. Triplet

There has not really been much in the way of fundamentally new discoveries regarding singlet/triplet reactivity differences. Dewar has performed MINDO calculations on butyraldehyde and concentrated on the differences in reactivity between singlets and triplets.[39] The results support the idea that triplet reactions occur exclusively via biradicals, whereas singlet reactions promote direct radiationless decay and may form products in a concerted fashion. Figure 1 depicts the well known different potential energy surfaces followed by the two states.[1]

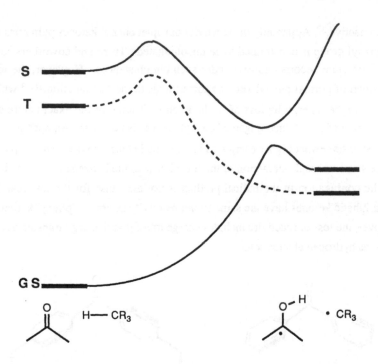

Figure 1. Reaction of singlet and triplet n,π* states with C–H bonds.

The methodologies of differential quenching[40,41,42] and of triplet sensitization[42] have been firmly established for over 20 years, and flash spectroscopy has added a few nice twists. It is now straightforward to determine whether a photochemical reaction is singlet or triplet derived. Many photoreactions of aliphatic ketones involve both singlets and triplets; but quantum yields for product formation from singlets tend to be low. The rapid intersystem crossing in most phenyl ketones usually produces exclusive triplet reactivity in bimolecular reactions. A few *intra*molecular hydrogen abstraction reactions of aryl ketones have been shown to involve excited singlets exclusively or partially, whereas reactiion from both states is common for aliphatic ketones.[8,40] This pattern occurs when n,π* singlet reaction is fast enough to compete with intersystem crossing and triplet reactions are so slow that impurity or self quenching quenches them. Quantum efficiencies for the unquenchable singlet reactions tend to be quite low.[43,44] We shall comment on state differences where appropriate, but mostly triplet reactivity has been studied.

D. Energetics

Scaiano and Previtali suggested that there should be a strong dependence of rate constants on the thermodynamics for hydrogen abstraction.[45] The effects of changing C-H bond strengths in the donors show up in the striking parallel with actual rate constants for hydrogen abstraction by alkoxy radicals.[8] Table 1 lists rate constants for reaction of triplet benzophenone with a representative variety of substrates. Such an energetic dependence is not easily seen when ketones RCOR' of different structure are compared. Various R groups tend to conjugatively stabilize the n,π* triplet and the product hydroxyradical by nearly identical amounts but produce more complicated effects on triplet energies. The reactions of most n,π* triplet ketones even with alkanes are sufficiently exothermic that significant variations of rate constants with triplet energy are rarely observed. Thus the activation energy for attack of both benzophenone and acetophenone triplets on cyclohexane is only 3.5 kcal/mole.[4,21] In this regard it is interesting that Peters has measured by photo-acoustics that γ-hydrogen abstraction by triplet valerophenone is 7 kcal/mole exothermic (72 kcal triplet energy − 65 kcal measured biradical energy).[46] This value nearly equals the difference in bond energies between O–H (104 kcal/mole) and secondary C–H (96 kcal/mole). The assumptions of the earliest thermodynamic calculations, that n,π* excitation effectively breaks the carbonyl pi bond,[7,26,45] have proven to be accurate.

α-Diketones present another picture. The strong through bond coupling of the n orbitals produces relatively low n,π* triplet energies, ~55 kcal/mole. However, α-keto radicals are not resonance stabilized as much as are benzyl radicals. Consequently hydrogen abstraction is less exothermic and rate constants for triplet biacetyl and other diketones are 2-3 orders of magnitude lower than for phenyl ketones.[47] Selectivity towards C–H bonds only doubles, from the 1:24:160 primary:secondary:tertiary ratios measured for phenyl ketones to 1:60:400.[48]

Recently Scaiano has reported a cyclic α-diketone whose triplet has a normal 55 kcal/mole excitation energy but is quenched unusually rapidly by typical hydrogen donors.[49] He has suggested that the triplet may be intrinsically more reactive in a *syn* geometry than in an *anti* because of hydrogen bonding in the transition state. This suggestion is equivalent to simultaneous abstraction by both oxygens, which could well be faster since the half-occupied n orbital is spread evenly over both oxygens.[48]

Table 1. Representative rate constants for bimolecular reaction of triplet
benzophenone with various substrates.[a]

Substrate	k, 10^6 M^{-1} s^{-1}	Substrate	k, 10^6 M^{-1} s^{-1}
CH$_3$CN [b]	0.00013	alkanethiols[e]	14
neopentane[b]	0.04	benzenethiol[e]	260
isobutane[b]	0.6	Bu$_2$S [e]	900
cyclopentane[b]	0.45	H$_2$Se [f,g]	470
cyclohexane[b]	0.75	Bu$_3$SnH [e,g,h]	1300
octadecane[d]	1.3	phenol[e]	1300
toluene[e]	0.5	p-MeO-phenol[e]	4500
p-xylene[e]	3.1	p-CN-phenol[e]	650
cumene[e]	3.7	indole[e]	4000
methanol[c]	0.2	t-butylamine[e]	64
2-propanol[c]	1.9	2-butylamine[e]	230
benzhydrol[b]	7.5	Et$_3$N [e]	3000
pentylsilane[e]	8.8	DABCO[e]	4500

[a] E. A. Lissi and M. V. Encinas in *Handbook of Organic Photochemistry* (J. C.
Scaiano, Ed.), CRC Press, Boca Raton, Vol. 2, Chap. 7, p.111 (1989). [b] in
acetonitrile. [c] neat. [d] in CCl$_4$. [e] in benzene. [f] in THF. [g] acetophenone. [h] ref. 26.

It is not clear that the energy gained by a partial hydrogen bond would be sufficient to
offset the 13 kcal/mole lower triplet energy of diketones relative to benzophenone.

The *anti* conformer is known to dominate the ground state equilibrium of
biacetyl and other acyclic diketones and has been thought to be the reactive triplet
geometry as well. This conclusion derives primarily from the known selectivity of
intramolecular γ-hydrogen abstraction in diketones; the exclusive formation of 2-
hydroxycyclobutanones[50] demands reaction from an *anti* form. If the *syn* triplet
indeed is much more reactive, then its population in acyclic diketones must be
negligible, otherwise type II elimination and 1-acylcyclobutanol formation would also
be observed. This regioselectivity is true for both aliphatic diketones[50] and for α-

keto phenyl ketones.[48] For the latter, it probably reflects n,π* excitation residing primarily on the 1-keto group.[48] It should be noted that hydrogen bonding to both oxygens is unlikely during *intra*molecular abstractions; so it may be asked whether intramolecular reactivity is relevant to intermolecular. However, we were careful to point out in 1976 that the ratio of intra/intermolecular reactivity is the same for diketones as for monoketones. Consequently the conclusion about negligible populations of *syn* conformers in acyclic diketones holds, if indeed it turns out to be generally true that *syn* are intrinsically more reactive than *anti*.

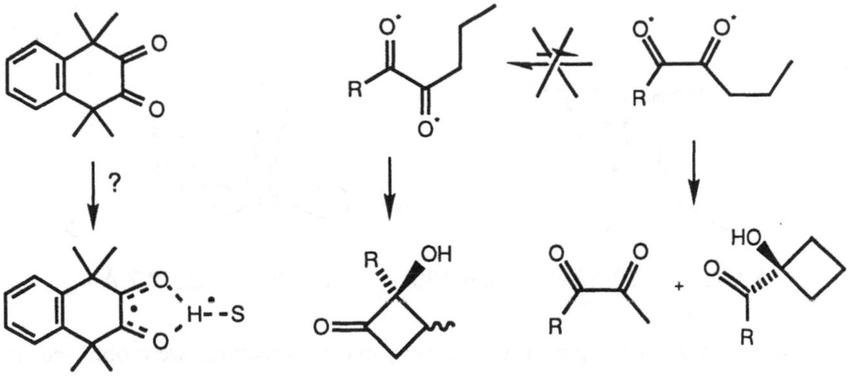

E. Orientational Requirements

Twenty years ago Turro suggested that the hydrogen atom being abstracted should be restricted to lie along the long axis of the carbonyl n-orbital.[51] This suggestion was appealing since it recognized the directionality of p orbitals and the fact that n,π* radical reactivity is centered on this particular half-occupied orbital. It was soon pointed out that many examples were known in which efficient intramolecular hydrogen transfer took place with the developing H--O bond making a fairly large angle with respect to the nodal plane of the carbonyl pi system.[52] It was suggested that reactivity may show a cosine[2] dependence on this angle, that being the electron density function for a p orbital.[2a]

More recently this question has been approached from both theoretical and experimental perspectives. Scheffer has studied a variety of ketones that undergo intramolecular hydrogen transfer in their crystalline states.[53,54] In analyzing their reactivity, he considered the following *ground state* parameters, as depicted in Scheme 1, to be the most important in determining reactivity: d, the distance between O and H; θ, the O–H–C angle; Δ, the C=O–H angle; and ω, the dihedral angle that the O–H vector makes with respect to the nodal plane of the carbonyl pi system. (Scheffer uses τ for this last angle, but we prefer to use another symbol to avoid confusion with lifetimes.) We concur with the validity of Scheffer's comparison of triplet reactivity with ground state geometries because n,π* excitation is known to be so highly localized on the carbonyl group that geometric changes in the rest of the molecule are negligible.

Scheme 1

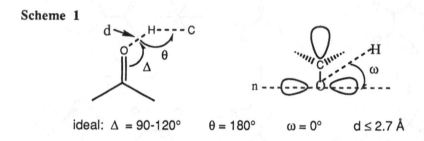

ideal: Δ = 90-120°   θ = 180°   ω = 0°   d ≤ 2.7 Å

In two dozen examples in which x-ray crystal structures were obtained for reactive ketones, the value of d ranged from 2.3-3.1 Å; of θ, from 85-120°; of Δ, from 74-103°; and of ω, from 0-62°, with an average of 43° for 17 γ-hydrogen abstractions. Scheffer suggests the theoretically "ideal" values for these parameters shown in Scheme 1. The sum of the H and O van der Waals radii is 2.7 Å. Scheffer points out that whenever the ground state positions a hydrogen this close to the carbonyl, minimal molecular motion is required for reaction, and minimal motion is all that most crystal lattices allow. Several ketones with longer values of d were found to be unreactive as solids. The fact that a few ketones react with larger values of d probably reflects a varying degree of molecular flexibility in different structures. The value of θ obviously can vary significantly from the linear arrangement thought[55] to be preferable. Likewise, the value of ω can depart significantly from the "ideal" 0°. Both of these "deviations" from ideality have long been known from the reactivity of many steroidal ketones.[2,52] The meaning of Δ is the least clear, since n,π* excitation lengthens carbonyl bonds.[56] The ground state Δ values for reactive ketones often are already less than 90°, whereas the n orbital makes an angle with the C=O bond of 90-120°, depending on orbital hybridization. The reactivity of chlorine atoms does display the expected cosine^2 dependence on the electron density in p orbitals of different rotational states.[57] Consequently it is reasonable to exxpect a gradual decrease in reactivity accompanying the gradual decrease in $\cos^2 \omega$ as ω increases.

There have been several theoretical approaches to answering these orientational questions about hydrogen atom abstraction by n,π* states. Morrison and coworkers reported *ab initio* studies (3-21G basis set) for hydrogen abstraction by triplet formaldehyde from methane.[58] The calculated saddle point for the reaction had the following parameters: $\theta = 176°$, $\omega = 9°$, $\Delta = 109°$, $d = 1.18$ Å. The authors concentrated on why Δ was larger than 90° and compared their d to similar values calculated earlier by other workers. For as yet uncertain reasons, the transition state is 10 kcal higher when Δ is 90° or 130°.

Dorigo and Houk have published extensively on the geometric requirements for intramolecular hydrogen transfers. In particular they have emphasized what factors contribute to entropies and enthalpies of activation.[59] Significant deviation of θ values from 180° and of ω values from 0° is predicted to markedly increase $\Delta H\dagger$ in alkoxy radicals and in ketone excited states, although many steroids with "bad" values of these angles undergo the Barton or the Norrish type II reaction.[2a,52,60]

One failing of the various theoretical contributions is a lack of predicted rate constants or transition state energies for various values of ω, although Dorigo and Houk have just addressed this problem[59]. There have been several experimental observations that suggest zero reactivity when $\omega = 90°$.[61] All of these involve rigid polycyclic ketones or crystalline media, where the reactants are constrained to maintain their ground state geometry. Without careful control experiments, the possibility remains that hydrogen abstraction actually occurs but is followed by 100% efficient reversion of the biradical to starting ketone. The opposite problem is seen in a study of the photochemistry of acetophenone-doped deoxycholic acid crystals [62] Efficient bimolecular hydrogen abstraction was observed even though ω is 90° in the crystal; the authors concluded that there is sufficient molecular mobility in this crystal to allow lower values of ω.

Sauers has addressed this problem by combining theoretical and experimental approaches.[63] He has performed ab initio studies (STO-3G) to calculate the geometries of excited ketones and then developed molecular mechanics parameters that duplicate these geometries. In particular, the C=O bond is lengthened to 1.40 Å and the out-of-plane force constants for the carbonyl group are reduced to allow for the known pyramidalization of the carbons. He has also added a variable parameter V_2 in the torsion equation for the C=O--H grouping that accounts for the calculated barrier to hydrogen transfer when ω deviates from 0°. He then has compared the strain energies calculated for the transition states for intramolecular hydrogen abstraction in a variety of cyclic and polycyclic ketones with the observed quantum efficiencies of photoreactions initiated by such hydrogen transfer. Below are depicted several unreactive ketones that have hydrogens close enough to the carbonyl for reaction but at an angle ω of 90°. The observed nonreactivity is explained by a 7.5 kcal value of V_2, which adds 5.1 kcal/mole of strain energy to the transition state for hydrogen abstraction from 2. The main problem is ruling out other causes for the lack of observed reaction.

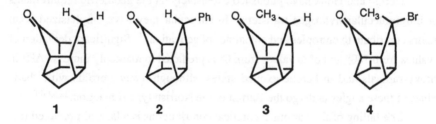

1   **2**   **3**   **4**

The triplet lifetimes of ketones **1** and **2** are only 9-12 nsec, far too short for any known physical decay, yet no reaction products are observed.[64] Scaiano and coworkers point out that such behavior most likely is caused by efficient reversion of a biradical to starting material. They suggest rapid α-cleavage followed by recoupling of the constrained biradical. They reject revertible γ-hydrogen abstraction as the culprit on the basis of a long (3.05 Å) d and the poor angle ω. Such reasoning was admittedly circular. Sauers has provided stronger evidence that hydrogen abstraction does not occur in the cases of **3** and **4**.[63] The former gives no cyclo-pentanol product, which could result from δ-hydrogen abstraction, even when performed in methanol, which is known to enhance the cyclization efficiency of some 1,5-biradicals 20-fold.[30] Equalling compelling is the nonreactivity of **4**; a triplet biradical generated by hydrogen transfer should have eliminated bromine very efficiently. [65] Since neither method of trapping triplet biradicals works, it seems likely that triplet hydrogen abstraction does not occur. There are two possibilities other than zero rate constants for the observed nonreactivity. Since the ketones are aliphatic, excited singlet hydrogen abstraction, which is always highly inefficient,[1,8] may dominate. However, that is not the answer; the fluorescence yields of ketones **1-4** all are similar, and the singlet lifetime of **1** is 12 nsec, the value produced by normal intersystem crossing. Unfortunately, the fast, completely revertible α-cleavage suggested for **1** and **2** may also plague **3** and **4**, especially if hydrogen abstraction at large values of ω is merely slowed down but not eliminated. In fact recent work in which molecular conformation allows α-cleavage to prevail over internal hydrogen abstraction[66] will be discussed below. So we still are left with the strong suspicion that hydrogen abstraction becomes slow as ω approaches 90°, but without any systematic measurements of rate constants. Some recent developments will be discussed later.

III. HYDROGEN TRANSFER FOLLOWING CHARGE TRANSFER COMPLEXATION

The importance of charge transfer in photoreduction has been known for over 20 years. It was first recognized with various amines,[67] then in the photoreduction of easily reducible ketones such as α-trifluoroacetophenone by alkylbenzenes.[68] Its symptoms are rate constants much larger and quantum yields much lower than would be anticipated for normal "radical" reactivity. The abnormal reactivity includes amines being able to reduce π,π^* triplets as well as n,π^* triplets.[69] The basic two step mechanism of partial or complete electron transfer followed by hydrogen (proton) transfer has been accepted for some time, but the exact nature of the presumed exciplex intermediate, especially the extent of electron transfer, has not been well defined. In the case of an exciplex with partial CT stabilization, the ensuing hydrogen transfer can be considered as an internal hydrogen abstraction in a polarized, complexed triplet. In the case of full electron transfer, the ensuing proton transfer is an acid-base reaction of triplet state radical-ion pairs. These questions have received more detailed examination in recent years.

$$K^* + AH_2 \longrightarrow K^{\bullet \delta-} + AH_2^{\bullet \delta+} \longrightarrow KH^{\bullet} + AH^{\bullet}$$

A. Ketones and Substituted Benzenes.

That substituted benzenes quench triplet ketones by a CT process is shown by the linear free energy relationship between quenching rate constants and ionization potentials of the benzenes[68,70]. We undertook a thorough examination of the photoreduction of phenyl ketones by alkylbenzenes in order to establish whether "pure" hydrogen abstraction could be distinguished from the two step process involving prior charge transfer complexation.[5] A series of ring substituted α-trifluoroacetophenones, benzophenones, and acetophenones provided a range of triplet state reduction potentials from 46 to 19 kcal/mole. ($E^*_{red} = E_T + E^\circ_{red}$, where E_T is the 0,0 triplet excitation energy and E°_{red} is the ground state reduction potential.) Several features of the reactions were found to vary monotonically with the value of E^*_{red} (the most positive values of E^*_{red} indicate the most easily reducible triplets):

a) Bimolecular rate constants for reactions with p-xylene vary from 3×10^4 for p-methoxyacetophenone ($E^*_{red} = 19$ kcal) to 3×10^9 for p-trifluoromethyl-α,α,α-trifluoroacetophenone ($E^*_{red} = 46$ kcal). The corresponding rate constants for these ketones reacting with cyclopentane are 5×10^3 and 1.5×10^7, respectively. Since *tert*-butoxy radicals, which usually serve as adequate models for the radical-like reactivity of n,π^* triplets, show the same reactivity towards cyclopentane and p-xylene,[71] the much greater reactivity of the triplet ketones towards p-xylene indicates that hydrogen transfer is being facilitated by some interaction specific to the alkylbenzenes and absent with cyclopentane. A plot of log k_r(xylene) *vs* E^*_{red} gives separate lines for n,π^* and π,π^* triplets, with the π,π^* plot having a slightly larger slope so that it crosses the former at $E^*_{red} = 35$ kcal.

b) In reaction with toluene and toluene-d$_8$, k_H/k_D ratios for both α-trifluoroacetophenone (π,π^*, $E^*_{red} = 38$ kcal) and 4,4'-dicyanobenzophenone (n,π^*, $E^*_{red} > 34$ kcal) are 1; for acetophenone (n,π^*, $E^*_{red} = 24$ kcal), 3.3; and for ketones with intermediate values of E^*_{red}, 1.4-2.5. Ketone triplet lifetimes are the same in benzene and benzene-d$_6$, so the above ratios represent primary isotope effects on hydrogen abstraction from the benzylic C-H bonds.

c) Reaction with p-cymene produces both primary and tertiary radicals. The primary/tertiary ratio ranges from 0.4 to 0.7 for ketones with n,π^* lowest triplets and from 3.9 to 0.5 for ketones with π,π^* lowest triplets, in both cases being highest for the highest values of E^*_{red}.

d) Quantum yields of radical coupling products, extrapolated to infinite alkylbenzene concentration, range from 20% for the ketones with the highest E^*_{red} values to 100% for the n,π^* triplets with the lowest E^*_{red} values and 25-50% for the π,π^* triplets with the lowest E^*_{red} values. It was demonstrated independently that the hydroxyradical intermediates undergo only a small percentage of disproportionation with the benzyl radicals; so low quantum yields must reflect decay to reactant that takes place *before* radical formation.

e) 1,4-Diacetylbenzene and p-acetyl-α-trifluoroacetophenone both show anomalously low rate constants and low amounts of primary radical production from cymene, in terms of their E^*_{red} values.

We have interpreted these results as demonstrating dominant exciplex formation for all of the π,π^* triplets, with the degree of electron transfer in the exciplex increasing as the triplet ketone becomes easier to reduce. For the ketones with n,π^* lowest triplets, direct hydrogen transfer predominates at low E^*_{red} values but gives way to the CT process as E^*_{red} values increase. Figure 2 summarizes these changes. CT complexation of the ketone triplets with the alkylbenzene produces the following effects:

1) It puts a partial positive charge on the alkylbenzene, such that the hydrogen transfer step takes on to varying degrees the nature of a proton transfer. Thus the primary/tertiary selectivity towards cymene shows a variable kinetic acidity effect. This variability is unique evidence that the extent of electron transfer in these exciplexes varies with the thermodynamics.

2) As the degree of charge transfer increases, the transition state for hydrogen transfer becomes more stabilized and the observed rate constant increases. With ketones of intermediate E^*_{red}, exciplex formation is reversible and intermediate isotope effects are observed. With E^*_{red} values above 3.4 kcal, hydrogen transfer gets so fast that exciplex formation becomes rate determining and irreversible, so that the isotope effect disappears. Other unusual rate effects then emerge, such as toluene being more reactive than cumene.

3) As the degree of charge transfer increases, the exciplex becomes more stable, so that radiationless decay to ground state competes better with hydrogen transfer and quantum yields of radical formation decrease. This competition is apparent in the primary isotope effects on quantum yields even when there are no isotope effects on observed rate constants.[68]

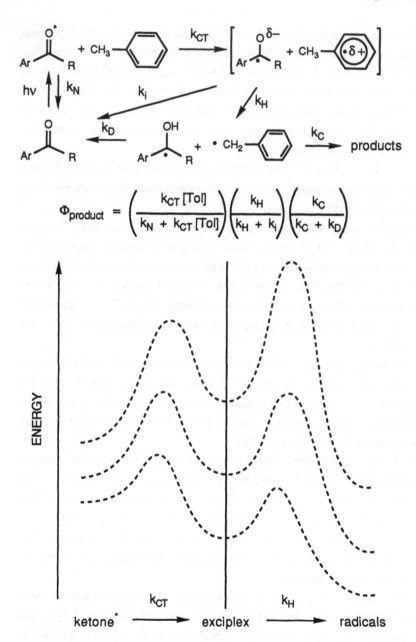

$$\Phi_{product} = \left(\frac{k_{CT}[\text{Tol}]}{k_N + k_{CT}[\text{Tol}]}\right)\left(\frac{k_H}{k_H + k_i}\right)\left(\frac{k_C}{k_C + k_D}\right)$$

Figure 2. Change in rate determining step as the extent of charge transfer decreases from top to bottom.

Various workers have plotted log k_r values for triplet quenching vs. the gas phase ionization potential of the quenchers and obtained reasonably straight plots; occasionally the k_r values for aliphatic substrates such as alcohols and alkanes lie on these plots. It has been suggested that such correlations demonstrate that even these hard-to-oxidize compounds may act as electron donors in exciplex formation with ketones. However, such conclusions are not necessarily warranted. The work just summarized demonstrates that CT slows down as the thermodynamics for electron transfer worsen until hydrogen transfer without prior complexation takes over. The plots of log k vs. E_{red} thus accomodate a change in rate-determing step; their slopes have no simple meaning other than that partial electron transfer occurs.

The behavior of the four acetophenones $PhCOCH_{3-n}F_n$ provides a very telling demonstration of the difference between cyclopentane and the alkylbenzenes as substrates towards triplet ketones.[13] $E*_{red}$ values increase as the number of α-fluorines increases from 0 to 3 and rate constants for reactions of the triplet ketones with toluene or p-xylene also increase uniformly. However, rate constants towards cyclopentane rise more gradually as n goes from 0 to 2 and the value for n=3 then drops to only slightly greater than the value for n=0. The reversal in reactivity is caused by the change in the lowest triplet configuration from $n,\pi*$ for n = 0,1 to $\pi,\pi*$ for n = 2,3. In other words, the standard decrease in reactivity as the equilibrium population of reactive $n,\pi*$ triplets decreases is observed for cyclopentane but not for the alkylbenzenes. Therefore cyclopentane must be reacting solely by a process that depends upon the free radical character of the reacting excited state, i.e. hydrogen abstraction. The alkylbenzenes react primarily by CT complexation, which is not a process unique to radicals, so $\pi,\pi*$ triplets are also reactive.

There do appear to be slight but real differences between $n,\pi*$ and $\pi,\pi*$ triplets in these CT-promoted reactions. As pointed out above, rate constants do not differ more than a factor of 10 but primary/tertiary selectivity with p-cymene differs substantially only for the $\pi,\pi*$ triplets. Singer's original suggestion of quite different geometries for complexation by the two triplets provides a useful explanation.[72] Face-to-face pi-overlap with the $\pi,\pi*$ triplets allows stereoelectronic effects to determine selectivity, whereas the looser overlap of the donor pi-system with the half-vacant oxygen n-orbital does not.

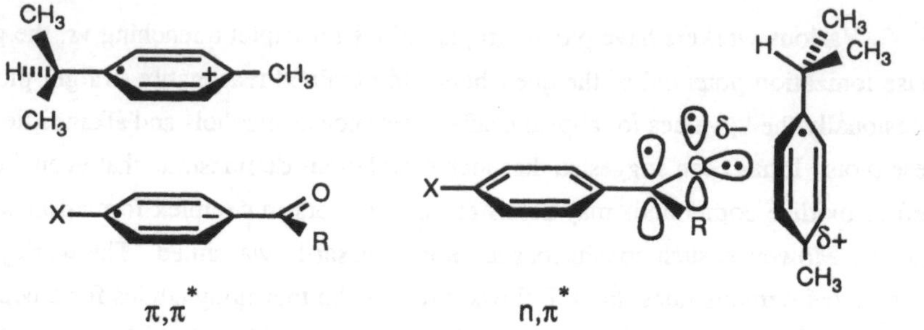

π,π^* n,π^*

The behavior of ketones such as p-diacetylbenzene is quite informative. Their very low reactivity and high selectiivity for tertiary hydrogens probably is due to their symmetric π,π^* lowest triplets, which have no charge separation.[10] With the delocalized half-occupied π and π^* orbitals occupying the same space, there is no concentrated "hole" for the donor orbital to interact with nor any reason for the less stable radical to be formed.

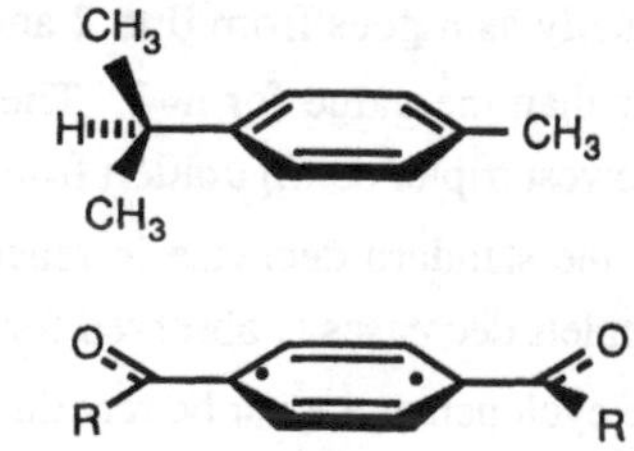

There is much kinetic evidence for a complex with significant charge transfer character that precedes hydrogen transfer in many photoreductions. Apart from the dependence of rate constants on redox potentials and the lack of isotope effects in many systems, activation parameters provide telling evidence. Steel has measured the following triplet E_a and log A values: for benzophenone reacting with cyclohexane, 3.5 kcal/mole and 10.0; for reaction with toluene,1.2 kcal and 8.0; and negative activation energies for quenching of benzaldehyde by toluene.[4] Such kinetics indicate reversible complexation prior to the actual quenching process (hydrogen transfer). There are, however, few reliable reports of spectroscopic identification of such chemically reactive triplet exciplexes except in the case of cycloadditions.[73] Phosphorescence cannot be observed because intersystem crossing and chemical reactions of the exciplex are far too rapid. Weak CT complexes cannot look much different from simple ketone triplets by flash spectroscopy; and the absorption of strong CT complexes is assumed to resemble that of the radical-ions. There has not been a systematic demonstration of a gradual shift from one extreme to the other as the thermodynamics for electron transfer improve, although individual observations may be so interpreted. Scaiano and Wilson have observed a transient absorption in the acetone-durene system that they ascribe to the triplet exciplex.[74] This system does not lead to photoreduction; it remains to be seen whether further work will corroborate this assignment.

An acid promoted dissociation of the triplet exciplexes formed between several substituted benzenes and α-trifluoroacetophenone to hemipinacol radicals (this <u>protonated</u> ketyl radical often is incorrectly called a ketyl radical) has been detected by CIDNP.[75] The different g values of ketyl radical-anions and their conjugate acids provide easy differentiation between the CIDNP produced by radical-ion pairs or radical pairs. Roth has pointed out that higher concentrations of acids are required to intercept exciplexes than free radical-anions.[76] It must therefore also be true that stronger acids are needed to intercept weak CT complexes than for strong CT complexes.

Charge transfer probably is responsible for some rare abstractions of hydrogen from an alcohol O–H bond. This reaction had been reported in 1966 for an intramolecular photoreduction of a hydroxy-isoalloxazine[77] and in 1973 for the photoreduction of several ketones by cyclopropanols, which are oxidized to β-substituted ketones.[78] Similar reactivity has been suggested as responsible for the ketone photosensitized splitting of pinacols and the inefficiency in the photoreduction of acetophenone by 1-phenylethanol[79] In the latter case, the maximum photo-reduction quantum yield with 1-phenylethanol-O-d is 0.70 vs. the 0.50 found for the protonated alcohol. The O–H bond obviously serves as a major contributor to radiationless decay of the expected exciplex. The photosensitized cleavage of pinacols also appears to involve transfer of H from a pinacol to a ketone oxygen. With a stronger electron acceptor as sensitizer, direct cleavage of the pinacol radical-cation would probably occur and produce the same products.[80] However, the thermodynamics for electron transfer from pinacol to ketone probably are too endothermic for this pathway to occur.

In contrast, phenols are relatively easy to oxidize. They react with triplet ketones to furnish phenoxy and hemipinacol radicals[81]; and they quench triplet biacetyl without forming any products, presumably by revertible hydrogen abstraction[82] Scaiano has measured rate constants for quenching of both *tert*-butoxy radicals[83] and triplet ketones.[84] The alkoxy radicals abstract hydrogen from phenol with a rate constant of 3×10^8 M^{-1} s^{-1} in benzene, 2×10^7 in alcohols, and 5×10^6 in pyridine. Phenol is easily one of the most reactive hydrogen donors towards alkoxy radicals. Hydrogen bonding of the phenolic proton to Lewis bases greatly diminishes its reactivity. Towards substituted phenols the alkoxy radical displays it electron-deficient nature with a Hammett ρ^+ value of -0.90. In deuterated alcohols rate constants are only 20-30% as large, since phenol-O-d is the reactant. Arrhenius parameters for phenol itself are $E_a = 2.8$ kcal/mole and log A = 10.7.

Triplet benzophenone behaves very similarly: $k = 1.3 \times 10^9$ in benzene, 8×10^7 in 9:1 acetonitrile:water; $\rho^+ = -0.65$; $E_a = 2.1$ kcal/mole; log A = 10.6. It is far less reactive towards anisole ($k < 10^6$) and even p-dimethoxybenzene ($k = 4 \times 10^8$). Hydrogen transfer clearly is rate-determining. The ketone triplet apparently forms CT complexes with phenol but hydrogen transfer is so facile that it is not accelerated much by prior charge separation. The apparent isotope effect of only 1.3 reflects both the high rate constant and the O to O hydrogen transfer.

Scaiano also studied the reactivity towards phenols of p-methoxypropio-phenone, which has a π,π^* lowest triplet. Its triplet is more reactive than that of benzophenone, showing rate constants $>5 \times 10^9$ M^{-1} s^{-1} towards all phenols and a ρ value of 0. Its enhanced reactivity towards phenol appears to be primarily entropic: $E_a = 3.1$ kcal, log A = 12.0. These two facts suggest a reaction dominated by diffusion rather than chemistry. Apparent k_H/k_D values between 1.3 and 3 9 were measured towards different phenols, again with no obvious explanation. The n,π* triplet is only 1% populated, as mentioned above, so this unusual reactivity must arise from the π,π* triplet. It could be that rapid protonation of the triplet occurs, with electron transfer following. The carbonyl oxygen of such triplets is more basic than in the ground state because of internal charge transfer.[18] With several phenols 55% quantum yields for the formation of phenoxy radicals could be measured, but the value is only 21% for p-iodophenol. This probably represents a heavy atom effect: either enhanced radiationless decay of a very short-lived exciplex; or, more likely given the behavior of biacetyl, enhanced intersystem crossing and in-cage recombination of the initial hemipinacol—phenoxy radical pair.

There is one remarkable case of intramolecular quenching of a π,π* triplet by a phenol which produces a relatively long-lived biradical.[85] The short tether connecting ketone and phenol ensures that proton transfer from phenol to carbonyl cannot occur unless the two benzene rings lie face-to-face. Since the reaction does not occur with the methyl ether, this unusual proton shift validates the idea that exciplexes formed from π,π* triplets indeed involve face-to-face overlap. It is probable that hydrogen bonding further stabilizes such an orientation and leads to the unusually high rate constants for phenol quenching of π,π* ketone triplets.

Shizuka has described the reaction of triplet naphthols with ground state benzophenone.[86] The mechanism involves triplet energy transfer from ketone to naphthol followed by hydrogen transfer from the naphthol to the ketone. Decay of triplet ketone, formation and decay of triplet naphthol, and formation of hemipinacol and naphthoxy radicals in methanol were followed by flash spectroscopy. The rate constant for reaction of triplet 1-naphthol with benzophenone is 1.6×10^6 M^{-1} s^{-1} at 27°. The authors assume that a CT exciplex is first formed. Quenching of the ketone triplet by the naphthol partitions between energy transfer and hydrogen transfer, with the latter having an effective rate constant of 7×10^8 M^{-1} s^{-1}. The 450-fold difference in rate constants for forming probably the same triplet exciplex must reflect the 9 kcal/mole higher triplet energy of the benzophenone. Although products were not isolated, the two radicals apparently disproportionate back to starting materials, given the low rate constants for coupling of phenoxy radicals.

B. Ketones and Amines.

Cohen has reviewed all of the early work that demonstrated the ubiquity of CT interactions between amines and excited states.[87] The fact that rate constants for reaction of excited ketones with amines increase as the oxidation potential of the amines decrease (DABCO > tertiary ~ secondary > primary) provided the first evidence for a photoreduction mechanism involving something other than direct hydrogen atom abstraction, namely CT complexation followed by hydrogen transfer accelerated by the charge separation. Depending on the amine, either a C–H or N–H bond is broken to generate a hemipinacol radical and an aminyl or an α-amino radical.

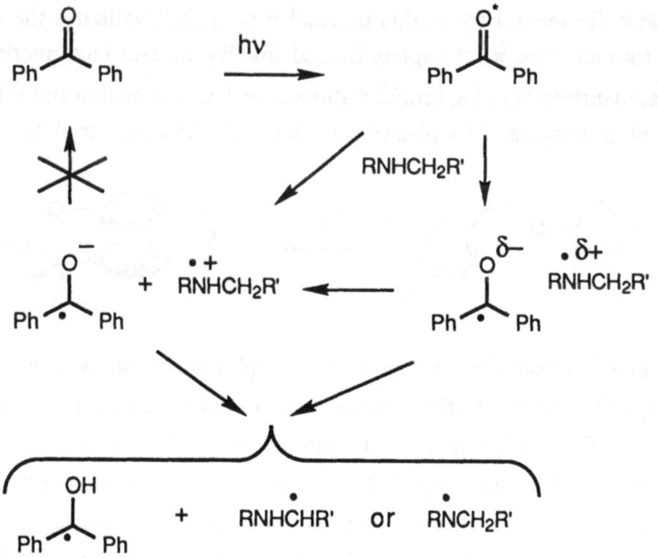

Several points are of interest, including the extent of electron transfer, which step is rate determining, and the overall quantum efficiency. Since the oxidation potentials of amines cover a large range, the exact reaction mechanism can cover the continuum from direct electron transfer being rate-determining to weak CT complexation followed by rate-determining hydrogen transfer. Observed rate constants range from diffusion controlled to something faster than expected for simple hydrogen abstraction. For example, Griller and coworkers have reported a flash kinetics comparison of the reactivity of triplet ketones and of alkoxy radicals towards amines[88]. That the triplet is substantially more reactive in all cases provides excellent confirmation of the realization by Cohen and of Davidson that CT lowers C–H bond energies [67] For tertiary aliphatic amines, rate constants exceed 10^9 M^{-1} s^{-1} and exciplex formation is rate determining. C-H bond strengths do not influence rate constants, as shown by triethylamine and *tert*-butyldimethylamine quenching triplet ketones with the same rate constant.[89] For primary amines, hydrogen transfer must be at least partially rate determining; rate constants are ~10^8 but *sec*-butylamine is four times more reactive than *tert*-butylamine.[88] The former amine contains a very reactive tertiary CH bond α to the nitrogen, whereas the latter amine contains no α C–H bonds and reacts at its N–H bond. In both cases there is sufficient charge separation to make the observed rate constant at least twenty times larger than for reaction of the same amine with *tert*-butoxy radical, which serves as a quantitative model for the purely radical reactions of triplet ketones. Isotope effects bear out these conclusions perfectly.[90] For *tert*-butylamine-N-d_2, k_H/k_D is 1.85, but only 1.25 for 2-aminobutane-2-d. The small isotope effects demand the presence of a reversibly formed exciplex, with exciplex formation and proton transfer being partially rate-determining.

Depending on thermodynamics and solvent polarity, the exciplex may dissociate to a radical-ion pair before radical products are formed. When electron transfer is exergonic, as can be estimated from Weller's equation,[91] radical-ions are formed directly. Both the exciplex and the radical-ions can conceivably revert to ground state reactants by well known radiationless decay and back electron transfer processes. However, the quantum yield for formation of the hemipinacol radical from triplet benzophenone and most amines is unity or higher.[92] Consequently any overall quantum inefficiency arises from subsequent reactions of the radicals. Why is there little or no decay from the exciplex or radical-ion pair? Back electron transfer from a triplet radical-ion pair is spin forbidden and presumably takes at least 100 ns, the lifetime of typical triplet biradicals. In fact shorter lifetimes have recently been measured for radical-ion pairs that form radicals in high yield.[93] The lack of radiationless exciplex decay, which is so important for exciplexes of triplet ketones with alkylbenzenes, presumably reflects the greater intrinsic reactivity of C–H bonds α to a heteroatom as well as greater charge separation in the ketone–amine exciplexes.

Spectroscopic identification of the exciplexes in ketone–amine systems is not expected to be routine when proton transfer to generate radicals is usually so rapid that radiationless decay does not even compete. An intramolecular example may have been observed with a benzophenone tethered to an aniline.[94]

There have been many studies of direct electron transfer from readily oxidizable amines such as DABCO and N,N-dialkylanilines to triplet ketones, including magnetic resonance detection of the radical-ions.[95] Mataga[96] and coworkers were among the first to identify ketyl radical-anions and aniline radical-cations as reaction products by nanosecond flash spectroscopic studies. They concluded that aliphatic amines produce radicals, presumably *via* exciplexes; only amines as easily oxidizable as N,N-dialkylanilines in polar solvents yield radical-ions. The Brandeis group expanded upon these findings by quantitative measurement of the radical yields.[92] Peters reported the first picosecond study of the reaction and was able to follow occurrence of charge transfer at very short times after excitation. With 1 M DABCO in acetonitrile, benzophenone ketyl absorbing at 655 nm has a 20 psec growth time and persists beyond 2 nsec[97]. With 3 M triethylamine in aceto-nitrile, a 610 nm transient appears in 10 ps and converts to the hydroxyradical in 30 psec.[98a] The two amines obviously react differently with the triplet ketone; the more easily oxidized DABCO forms radical-ions that undergo slow proton transfer, whereas triethylamine forms an exciplex that undergoes very rapid hydrogen transfer. It is tempting to suggest that the partial charge transfer which characterizes the exciplex formed from triethylamine is revealed by the lower λ_{max} value than that observed for the ketyl radical-anion formed from DABCO. However, the various workers who have studied this system have not reported consistent results;[98b] so the question of where an exciplex with partial CT character absorbs remains unsettled.

Devadoss and Fessenden recently completed a quite thorough study of the reaction of triplet benzophenone with DABCO.[93] With a combination of pico- and nanosecond flash spectroscopy, they were able to monitor the decay of ketone triplet, the growth and decay of the radical-ion pair, and the quantum yield of hemipinacol radical. A wide variety of other amines also quench the ketone triplet and produce radicals, but only DABCO produces a radical-ion pair detectable on the nanosecond time scale, as Peters reported. Scheme 2 outlines the various reactions that follow initial reaction.

Scheme 2

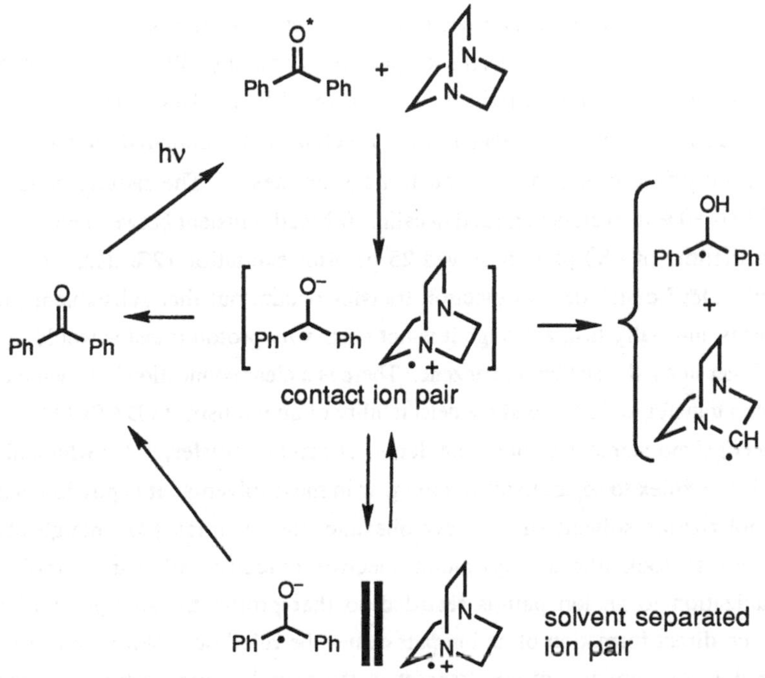

The rate constant for quenching of the triplet ketone by DABCO equals the growth rate of ketyl radical-anion in seven different solvents. In agreement with earlier studies, the rate constant is close to diffusion controlled in all aprotic solvents (hydrocarbons as well as acetonitrile) but only 1×10^9 M^{-1} s^{-1} in methanol and 7×10^7 in trifluoroethanol. As was shown two decades ago, strong hydrogen bonding of the amine lone pair in protic solvents slows down charge transfer quenching substantially.[89,99]

A transient that grows in as the triplet ketone decays was observable in all solvents except saturated hydrocarbons. Since the product of triplet reaction absorbs at the same place as does benzophenone radical-anion (725 nm in benzene, 720 nm in acetonitrile, and 615 nm in methanol), it is characterized as a contact ion pair. Although it often has been suggested that an exciplex *is* a contact ion pair, we prefer to maintain the distinction between exciplexes with variable amounts of charge separation and the radical-ions formed by complete electron transfer.

In low dielectric solvents, the ion pair decays rapidly (30-75 ns) and proceeds in high yield to hemipinacol radicals, as first found by the Brandeis group.[92] The reported quantum yield in benzene is 0.73, which must be compared with a value ≥1 for acyclic primary, secondary, and tertiary amines.[92] The radical yield from DABCO is 80% in cyclohexane and decalin. A broad transient between 650-800 nm with a lifetime of ~80 ps is observed 25 ps after excitation (2% decay of triplet ketone). D&F conclude that electron transfer occurs but that subsequent proton transfer is unusually fast, although it is not clear why proton transfer would be 500 times faster in an alkane than in benzene. There is a clear connection between the rate of proton transfer to ketone and the detectability of an ion pair. In DABCO there is a well recognized stereoelectronic retardation of proton transfer, [85,100] which allows any prior exciplex to separate into an ion pair in most solvents. It is possible that, in a nonpolarizable solvent such as cyclohexane, the exciplex has enough charge separation to look like a ketyl anion spectroscopically and that its molecular reorganization to an ion pair is retarded so that proton transfer predominates. However, direct formation of an ion pair cannot be ruled out. With the less easily oxidized acyclic amines, proton transfer in the exciplex apparently is faster than complete electron transfer.

The contact ion pair is subject to very dramatic solvent effects. All polar solvents increase its lifetime to as long as 3 μsec and decrease the yield of hemipinacol radicals. The presence of just 5% acetonitrile cuts the radical yield to 40%; 1% methanol doubles the lifetime and cuts the radical yield to 4%. There appear to be separate dielectric and acidity effects on the ion pair. Solvents of high dielectric constant allow the radical-ions to diffuse apart, and a solvent-separated radical-ion pair has been identified in the flash spectroscopy. This physical separation prolongs the lifetimes of the radical-ions by preventing both proton transfer and back electron transfer. Hydrogen bonding solvents that are weak acids, such as *tert*-butyl alcohol, have the same effect. However the more acidic alcohols such as methanol produce ion pairs with intermediate lifetimes and provide 50%

quantum yields of hemipinacol radicals. Here protonation by solvent of the ketyl ion apparently catalyzes neutralization of the ion-pair. This effect has already been observed by CIDNP[75,76] and in the intramolecular photochemistry of amino-ketones.[101] Devadoss and Fessenden find no evidence that a solvent separated ion pair can be formed directly from the triplet, as Peters had concluded;[102] they suspect that the profound influence of small amounts of water may be responsible for variable experimental results.

Mataga has reported several examples in which hydrogen transfer from an amine to a triplet ketone occurs before ion-pair formation. Triplet benzophenone reacts with diphenylamine with a rate constant of 1.3×10^{10} M^{-1} s^{-1}, with radical formation occurring directly on triplet decay and *not* from the long-lived ion pair.[103] Whether CT complexation will cause hydrogen transfer may be determined by the geometry of the initial exciplex, which partitions rapidly to either radicals or the radical-ion pair.

Anilines absorb strongly in the near uv, so careful choice of excitation wavelength is required for the study of ketone-aniline photoreactions. Fischer looked at the CIDNP that results from the irradiation of acetophenones with N,N-dimethylaniline.[104] Excitation of the ketone results in triplet radical-ion pairs, whereas excitation of the aniline results in singlet radical-ion pairs. Peters pointed out another crucial difference between singlet and triplet radical-ion pairs, namely the very fast rate of back electron transfer in the former.[97]

The large rate constants for reaction of excited ketones with amines have several quite different and important consequences. On a mechanistic level, they have led to various observations of the "triplet mechanism" for CIDEP and CIDNP. Rapid enough reactions can trap the triplet ketone in its electron-spin polarized state, so that the initial radical or radical-ion pair is also spin polarized, leading directly to polarized EPR spectra.[105] Cross-coupling phenomena then can induce nuclear spin polarization and CIDNP. The best example of this was provided by the reaction of α-trifluoroacetophenone with DABCO. At DABCO concentrations $<10^{-2}$ M, CIDNP emission was observed; but at [DABCO] > 0.05 M, enhanced absorption.[106] The normal radical pair model for CIDNP predicts emission from DABCO radical-cation and ketyl radical-anion pairs. The enhanced absorption at high DABCO concentrations is explained as quenching of spin polarized triplets. It is known that intersystem crossing preferentially populates the lowest energy triplet sublevel of phenyl ketones.[107] On a more practical level, amine-ketone mixtures have been used widely to produce radicals that initiate cross-linking of polymers in various sorts of photo-imaging processes. Large concentrations of vinyl monomers are required for cross-linking; and they are all excellent triplet quenchers. Consequently only bimolecular reactions that are nearly diffusion-controlled can compete with the rapid quenching.

Not much has been added recently to the question of n,π* *versus* π,π* reactivity. Rate constants for reactions with amines appear to be determined primarily by thermodynamics for charge transfer. The acetonaphthones and fluorenone are unreactive.toward primary amines, which have high enough oxidation potentials that hydrogen transfer remains rate-determining.[69] In contrast, they react efficiently with tertiary amines, displaying relative reactivities that correlate with the triplet reduction potentials of the ketones.[1] Consideration of orbital orientations for the two states is presented below.

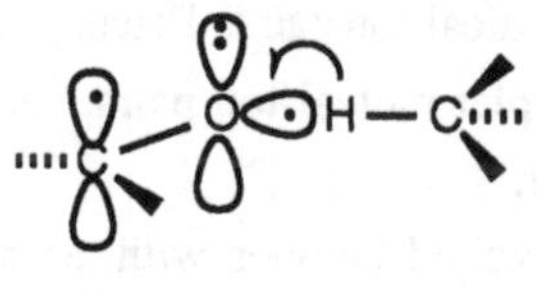

"pure" H-abstraction 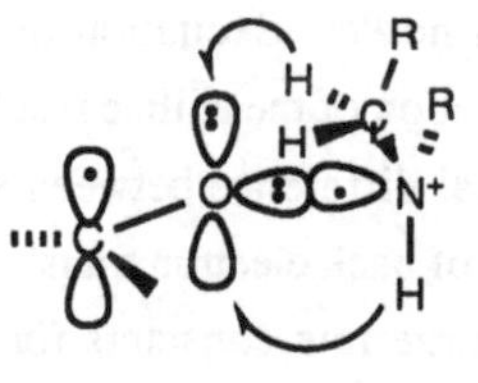   exciplex proton shift

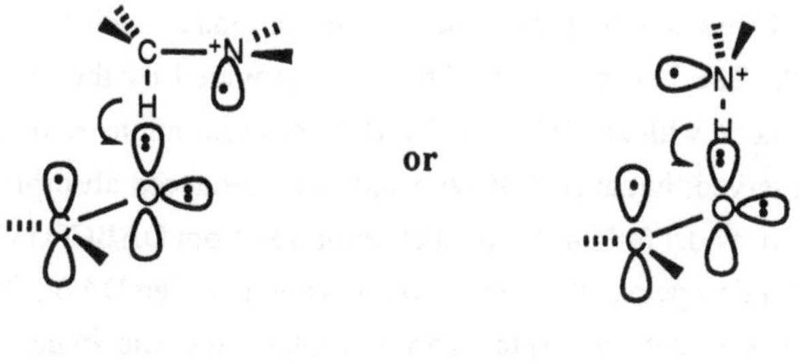

or

proton shift in radical-ion pair

As discussed above, direct hydrogen abstraction by an n,π* triplet with no prior complexation ideally involves approach of a C–H bond along the axis of the n orbital. Hydrogen transfer in an exciplex must occur from a very different orientation, since the nitrogen lone pair orbital overlaps with the carbonyl n-orbital. A C–H bond seems better disposed to interact with a filled p orbital on oxygen than does a N–H bond, but neither can approach a linear arrangement of the three atoms as long as the two n orbitals overlap along their long axes. How can CT complexation speed up H-transfer if it upsets the stereoelectronics? The partial negative charge on the oxygen is not directed; so the hydrogen can probably bond to oxygen at a 45° angle to the two p orbitals. In a radical-ion pair the two ions presumably are free to rotate; the C–H and N–H bonds should be able to line up fairly linearly with a filled p orbital on oxygen. The approach is drawn in the π-plane; but it could also be perpendicular.

There is good evidence from the behavior of aminoketones that tight overlap of the donor lone pair with the half-empty n orbital is necessary and that this overlap enforces restricted orientations in the exciplexes. There is no measurable internal triplet quenching in N-methyl-4-benzoylpiperidines, in which the nitrogen can approach the oxygen only in a boat conformation.[108] Rate constants for internal quenching in ω-dimethylaminoalkyl phenyl ketones are maximum for β– and γ-amino ketones, in which 5 or 6-atom rings can be formed by overlap of the nitrogen lone pair with the half-occupied carbonyl n orbital.[109,110] Internal quenching by sulfur in ω-benzoylsulfides shows similar behavior, attenuated by the long C–S bonds and the different orbital parentage of sulfur lone pairs.[111] These results provide further evidence that triplet ketones react with aliphatic tertiary amines by CT complexation rather than by full electron transfer, since current belief is that intramolecular electron transfer is faster through bonds than through space.[112]

IV. RADICAL REACTIONS FOLLOWING INTERMOLECULAR HYDROGEN TRANSFER

The coupling and disproportionation reactions of the radicals formed by hydrogen abstractions are well known. Nonetheless, several features of photogenerated radical chemistry have been clarified in recent years.

It was generally assumed for decades that the coupling reactions of two different radicals are statistically controlled to furnish 1:2:1 ratios of products, with the total yield of the two homo-coupling products equalling that of the cross-coupling product. This expectation is met if all three coupling reactions have the same rate constant, as in the case of diffusion-controlled reactions. However, Ingold's demonstration of a class of "persistent" radicals showed how slowly sterically congested dimers are formed by radical coupling.[113] If one of the homo-coupling reactions is distinctly slower than the other, then steady state kinetics predict a percentage of cross-coupling larger than 50%. Because the actual rates of homo-coupling must remain equal when they are initially formed in a 1:1 ratio, the steady state concentrations of the two radicals becomes unequal. Fischer has reached similar conclusions;[114] and we have actually observed such results.[13] The best measures of rate constants for coupling of the hydroxy radicals generated by photoreduction are in the 10^8 to 10^9 M^{-1} s^{-1} range, depending on solvent.[115] It seems that solvation of the hydroxy group provides sufficient steric drag to slow down coupling well below the diffusion limit.

$$K^{\cdot} + S \longrightarrow A + B$$

$$A + A \overset{k_{AA}}{\longrightarrow} A{-}A$$

$$B + B \overset{k_{BB}}{\longrightarrow} B{-}B$$

$$A + B \overset{k_{AB}}{\longrightarrow} A{-}B$$

$$\frac{A{-}B}{A{-}A + B{-}B} = \frac{2k_{AB}\,[A][B]}{k_{AA}\,[A] + k_{BB}\,[B]}$$

The ability of hydroxy radicals to reduce ground state molecules is well known. In particular, photoreduction of ketones in alcohols and amines produces radicals that rapidly reduce ground state ketone, such that the quantum yield for ketone disappearance exceeds unity. In the well known case of benzophenone photoreduction in 2-propanol,[116] the equilibrium constant for exchange strongly favors the benzylic radicals so that quantum yields for benzpinacol and acetone formation approach unity.[116] Even when equilibrium constants for hydrogen atom transfer between ketone and hydroxyradical are close to unity, the far greater concentration of starting ketone can produce much the same effect. Schuster has reviewed[117] all of the early evidence for this process and repeated the photoreduction of benzophenone-d_{10} by benzhydrol-d_0, in which the pinacol product is almost entirely d_{20} at early stages of reaction. (Benzophenone-d_{10} is formed in the exchange, so that the labelling gets mixed as the reaction proceeds.) Schuster was not successful in determining the effect of light intensity on the exchange, but Rubin later showed the the quantum yield of benzpinacol formed in aliphatic alcohols drops appreciably at high intensity.[118] As the steady state concentration of radicals increases, radical–radical reactions compete better with exchange and yields of cross-coupled pinacols increase. He even found some para-coupling, as often occurs with benzyl radicals.[119] and which is responsible for the light absorbing transients that often afflict studies of photoreduction.[120]

$$Ph_2CO + R_2CHOH \xrightarrow{h\nu} \left(Ph_2\overset{OH}{\underset{\bullet}{C}} \right)_2 + Ph_2\overset{HO}{\underset{\bullet}{C}}-\overset{OH}{\underset{\bullet}{C}}R_2 + PhCO\text{—}\underset{}{\bigcirc}\text{—}CR_2^{OH}$$

Schuster also provided strong evidence for a relatively low rate constant for coupling of benzophenone hemipinacol radicals. Weiner had shown that camphorquinone is a super scavenger of most hemipinacol radicals.[121] When benzophenone is irradiated in 2-propanol containing the quinone, under conditions in which none of the light is absorbed by quinone, no benzpinacol is formed but the normally ignored cross coupled pinacol $Ph_2C(OH)C(OH)Me_2$ is observed.[121] This product represents the small amount of in-cage radical coupling that competes with diffusion apart. When, however, benzophenone was irradiated with benzhydrol and quinone, no benzpinacol was formed.[116] Therefore there is no measurable in-cage radical coupling. The rate constants measured by Steel for such coupling[122] are low enough that >99.9% of the initial caged radicals are expected to diffuse apart.

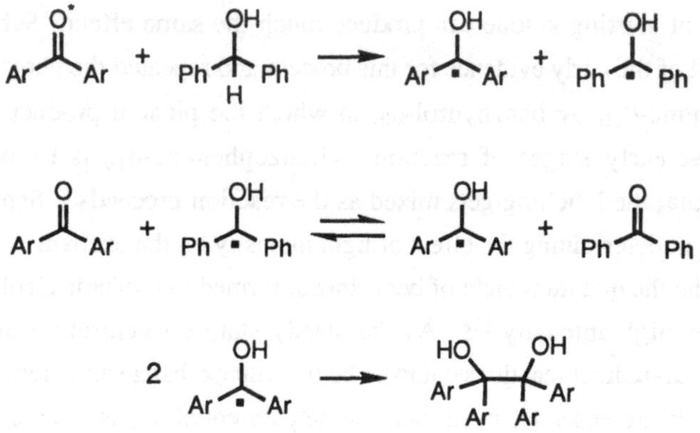

Actual rate constants for these hydrogen transfers have been measured only recently. Steel measured the rate constants for hydrogen transfer to benzophenone from both its own hemipinacol radical and acetone hemipinacol radical as 1.5×10^4 and 3.5×10^4 M^{-1} s^{-1}, respectively.[122] It is curious that the very exothermic transfer and the degenerate transfer have such similar rate constants. The method involved a scrupulous analysis of product ratios from the photoreduction of benzophenone by 2-propanol in acetonitrile as a function of ketone concentration. The paper serves as an excellent review of photoreduction kinetics. We recently finished a similar study in which we measured product ratios in the photoreduction of substituted acetophenones by 1-phenylethanol, in particular the formation of acetophenone.[123] As the starting ketone concentration increases, so does the amount of exchange relative to pinacol, while the pinacol content reflects decreasing amounts of the original alcohol. Exchange is noticeable at ketone concentrations below 0.01 M and is complete by 0.1

M. Equation [3] describes how the quantum yield for exchange depends on ketone concentration. The steady state radical concentrations were determined from the product ratios and actinometry; the rate constants for coupling were measured by flash kinetics. Rate constants for hydrogen transfer from 1-phenyl-1-hydroxyethyl radical to four ketones ranged from 3 to 9 x 10^3 M^{-1} s^{-1}, in good agreement with the Brandeis conclusions. We also measured the exchange equilibrium by measuring pinacol yields from a mixture of two ketones irradiated in 2-propanol. The results indicate that the more electron-deficient radical and the more electron-rich ketone are favored at equilibrium, in agreement with intuition and earlier findings.[124] It is interesting that α-alkyl substitution favors ketone and disfavors radical. This effect probably is more electronic than steric

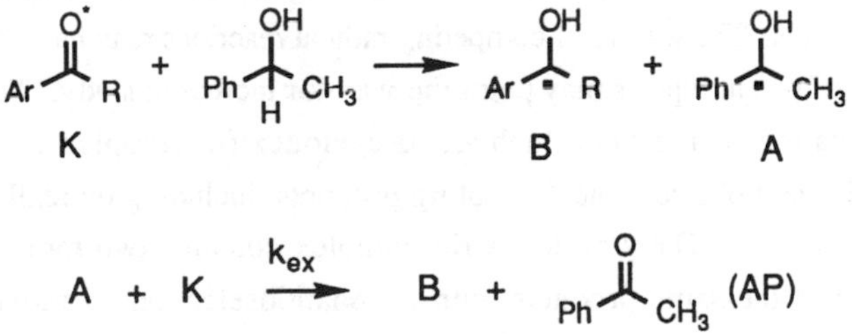

$$\Phi_{AP} = \Phi_{ISC} \Phi^0 \left(\frac{k_{ex}[K]}{k_{ex}[K] + k_{AA}[A] + k_{AB}[B]} \right)$$ [3]

The reactions of aminyl and α-aminoradicals are particularly involved.[87] Like hydroxyradicals, the latter reduce ground state ketone and generate enamines or imines. This process increases the quantum yield for ketone disappearance. The aminyl radicals formed from primary and secondary amines disproportionate with the hemipinacol radicals to regenerate starting materials and to lower quantum yields[125]. Thiols enhance quantum efficiency by intercepting the aminyl radicals.

Benzophenone has been reacted photochemically with a variety of natural products such as aminoacids, nucleic acids, proteins, and lipids in order to test whether oxy-radical reactants show any site selectivity. As one example, Breslow investigated how ω-(p-benzoylphenyl)carboxylates and other surfactant benzophenones attack the various carbons of surfactants such as cetyl trimethyl ammonium bromide and sodium dodecylsulfonate.[126] With mixed micelles, most of the carbons usually were attacked about equally except for those near the charged end of the surfactant. Carbons 5-7 (from the ionic end) were noticeably more reactive than the rest, presumably because the surfactant ketones are looped such that the polar carbonyl group is near the charged groups of the other surfactant molecules. We shall see more evidence for such positioning later. At concentrations well below micelle formation the C-15 position of CTAB underwent 60% of the total hydrogen abstraction. It was concluded that the surfactant is tightly coiled as a monomer in aqueous solution and only its neutral end is available for attack.

Finally, photoreduction has not provided much synthetic utility except for the production of pinacols. The spectre of competing radical reactions may have been too scary. A couple of recent reports may point the way for increased study. Turro has found that irradiation of the benzaldehyde–β-cyclodextrin complex as a solid produces an 80% yield of two radical coupling products, including optically active benzoin with 15% ee.[127] The γ-cyclodextrin complex contains two molecules of aldehyde and gives the same products with no enantioselection. Irradiation of benzaldehyde in solution generates benzil and a variety of radical-radical products. The cyclodextrin obviously holds the two reactants together long enough for the photoproduced radicals to interact with each other *in a chiral environment*. The group at Reims has utilized ketone-amine photochemistry to generate ketyl or hydroxy radicals that then undergo intramolecular cyclizations.[128] The exact mechanism has not be established, but more such efforts can be expected.

56% : [α] = -19°   24%

V. INTRAMOLECULAR HYDROGEN ABSTRACTION

The γ-hydrogen abstraction responsible for Norrish type II reactions provided much of the early structure-reactivity relationships in triplet state photochemistry.[8] Mechanistic interest in this reaction has remained high primarily in terms of characterizing the 1,4-biradical intermediates and monitoring the effects of restricted environments. A large amount of information has become available on other internal hydrogen abstractions. The rate constants for hydrogen abstraction and the behavior of 1,x-biradicals have revealed the strong influence of conformational effects on such intramolecular processes. Since the main fate of most larger biradicals is cyclization, these reactions provide new synthetic possibilities for non-ionic ring formation.

It is worthwhile to summarize our view of how rate constants for intra-molecular triplet reactions reflect conformational effects,[2a,110] as depicted in Scheme 3. We assume that interconversion among the various conformers of a given excited state competes with reaction and decay. One or more conformations may have geometries close to that required for reaction and are dubbed "favorable" or "reactive". There are two types of reactive geometries In one, only nuclear motion along the reaction coordinate is required in order to reach the transition state. (For hydrogen transfer we are talking primarily about the C–H and H–O stretches and rehybridization of the radical sites.) The distance and orientational requirements were presented above. Another reactive geometry is one in which only small amounts of bond rotation or distortion besides motion of the H atom are required to reach the transition state . In these cases, activation energies would include the extra strain. Finally, there are "unreactive" geometries in which the two functional groups are too far apart for reaction; these must be able to rotate into reactive conformations if they are to react rather than decay to ground state. Molecules that react as solids have lowest energy conformations that are reactive.

Scheme 3

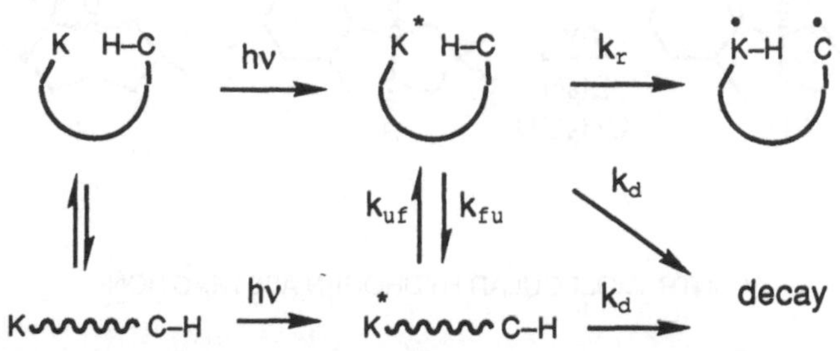

Molecules can be classified both structurally and dynamically in terms of their intramolecular reactivity. There are three structural classes: 1) those with a reactive geometry as the most stable or only conformation; 2) those with a reactive geometry that interconverts with other more stable but unreactive conformations; 3) those with no reactive conformations populated. There are also three dynamic classes of excited states: 1) those in which interconversion among conformers is much more rapid than decay reactions of the individual conformers (k_{uf} and $k_{fu} \gg k_d$ and k_r), such that conformational equilibrium is established before reaction; 2) those in which conformational interconversions are all slower than decay (k_{uf} and $k_{fu} \ll k_d$ and k_r), such that ground state conformational preferences control photoreactivity; 3) those in which unreactive conformers rotate into the reactive conformer, which reacts before it can rotate into an unreactive geometry ($k_r \gg k_{fu}$). In this third case the excited state reaction is subject to rotational control, with rate constant k_{uf}, the intramolecular equivalent of intermolecular diffusion control.

The different dynamic boundary conditions have been adequately reviewed in the past decade.[110,129] The first dramatic example of ground state control of triplet reactivity was provided by Lewis' study of benzoylcyclohexanes.[130] Photoenolization of ortho-alkyl ketones was the first reaction recognized to show rotational control of triplet reactivity.[131] Conformational equilibrium is much more common. We wish to emphasize Equations 4 and 5, which translate the Winstein-Holness principle[132] to photochemistry. For systems in which the excited state establishes conformational equilibrium before decay, there is only one excited state lifetime, which is determined by the weighted decay rates of all conformers. If more than one is reactive, then the observed rate constant for reaction is the sum of possibly different rate constants k_r weighted by their individual equilibrium populations χ_r. ($\Sigma\chi_r + \Sigma\chi_u = 1$) When ground state or rotational control obtains, different excited states have different lifetimes whether they undergo the same or different reactions.

$$1/\tau = \Sigma \chi_u \tau_u + \Sigma \chi_r \tau_r \qquad [4]$$

$$k_H{}^{obs} = \Sigma \chi_r k_H \qquad [5]$$

A complete understanding of the dynamics of excited state decay in conformationally mobile molecules requires an understanding of two molecular properties: the relative energies of various conformations; and the barriers to their interconversion. The first provides values of χ_r for the various conformers; the second, rate constants for their interconversion. In addition, rate constants must be known for intrinsic decay of each excited chromophore in the molecule and for intramolecular reactions as a function of distance and geometry.

Several general comments can be made about different classes of structure. In acyclic ketones, the regioselectivity of hydrogen abstraction reflects different rate constants for cyclic transition states of different sizes, which are closely related to χ_r values. For example, Winnik has shown how the rate constants of remote hydrogen abstraction by benzophenones substituted para with long alkyl tails[133] reflect equilibrium constants for the coiling of polymethylene chains into different sized rings.[129] When the carbonyl and C-H bond are part of a ring or substituents on a ring, the conformational limitations of the ring may either increase or decrease the population χ_r of favorable conformations. Lewis demonstrated how incorporation of rings between the carbonyl and the γ-carbon can increase rate constants for γ-hydrogen abstraction.[134] Scheme 4 shows some ketones in which the activation energy for abstraction of a secondary hydrogen remains the 3.5 kcal/mole that it is for bimolecular reaction[4,21] but the entropy of activation becomes less negative as rings freeze intervening C–C bond rotations. Each such frozen rotation increases the rate constant by a factor of ~8, which corresponds to 4 eu. In all three cases, the major conformation is either a reactive one or of comparable energy and the rings limit the number of *unreactive* conformers. In general, however, *there is no constant factor by which rings change intramolecular reactivity.* They may decrease population of a reactive conformation, as Alexander showed for benzoyl-cyclobutane,[135] limit reaction to one site, as the predominant ε-hydrogen abstraction in cyclodecanone indicates,[136] or prevent population of a reactive conformer, as Turro originally found for cyclohexanones with axial α-alkyl groups[51].

Scheme 4

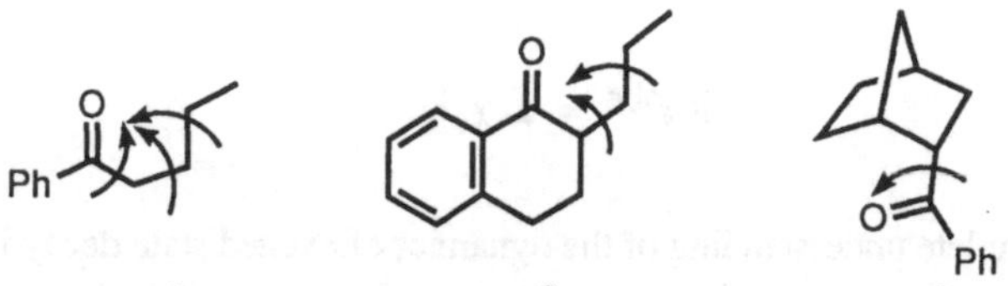

1.3 x 10[8]   6 x 10[8]   7 x 10[9]

A comment about intramolecularity is in order here. The general question of what determines rate constants for intramolecular bifunctional reactions has been of intense interest for years, usually with reference to enzymatic catalysis.[137] The principles described above that have been developed to understand photochemical reactions are in fact general, except for the idea of ground state control. In particular, the idea of conformational equilibrium applies equally well to most excited state and ground state reactions. There does appear to be a need to better blend the ideas that have sprung from photochemical studies with those based on various ground state reactions.

A. 1,5-Hydrogen Transfers (γ-Hydrogen Abstraction)

By 1972 the Norrish type II reaction of ketones[8] was known to involve both singlet and triplet state γ-hydrogen abstraction that generates 1,4-biradical inter- mediates. The triplet reactions of aryl ketones have been the most widely studied. The corresponding singlet reaction is less well understood because of short biradical lifetimes and radiationless decay caused by reaction.[2a,138] The triplet generated biradicals cleave, cyclize, and disproportionate back to starting ketones in proportions that vary widely with structure. Added Lewis bases solvate the hydroxy group of the biradical and suppress its reversion to ketone. It cannot be overemphasized that the quantum yields for such two-step triplet reactions often reflect partitioning of the biradicals more than relative rate constants for triplet reaction. This fact was first demonstrated by the behavior of γ-substituted butyrophenones.[139] A more recent example is provided by the behavior of an acetyl-2-methylenecyclohexane, which photocyclizes primarily *via* what would appear to be the less readily formed biradical. Only traces of product were found from the allylic biradical, which should be formed at least as fast or faster than the the the one leading to the observed product.[140] A similar effect has been found for cyclodecanone; its triplet undergoes both γ- and ε-hydrogen abstraction,but none of the 1,4-biradical cyclizes except in *tert*-butyl alcohol.[136b]

Most of the mechanistic interest in the Norrish type II reaction for the past 15 years has involved four questions: 1) the orientational requirements for γ-hydrogen abstraction; 2) the behavior of 1,4-biradicals, particularly how their lifetimes depend on structure; 3) how the reaction is affected by the environment; and 4) how the reaction may be used to probe competing reactions. We shall begin with a brief update and review of triplet state reactivity.

1. *Excited state reactivity.* In long chain ketones γ-hydrogen abstraction occurs with very high regioselectivity; only about 5% δ-hydrogen abstraction competes[52] and direct β-hydrogen abstraction is very rare. This specificity is a good demonstration of conformational equilibrium determining reactivity, with a six-atom transition state being preferred,[52] and makes the type II reaction an ideal one with which to monitor basic structural effects on rate constants for hydrogen abstraction. Its causes are discussed in the section on δ-hydrogen abstraction. Very few rate constants for γ-hydrogen abstraction have been measured for their own sake in recent years but the effects of several substituents have been added to what were known in the early 70's. Tables 2-4 include k_H values for all of the simple δ-, γ-, and ring substituted phenyl ketones that have been reported. They were all determined from maximum type II quantum yields and triplet lifetimes ($k_H = \Phi^{max}/\tau$).[3] Most of the

Table 2. Rate constants in benzene for triplet γ-hydrogen abstraction by
δ-substituted valerophenones.

δ–substituent	σ_I	k_H, 10^7 s^{-1}	ref
H	0.0	14	3
alkyl	−0.05	18	3
dimethylamino	0.06	7.0	109
phenyl	0.10	8.4	359
alkylthio	0.23	4.8	111
phenylthiyl	0.30	4.7	111
methoxy	0.30	2.6	3
carbomethoxy	0.30	3.8	3
carboxy	0.33	2.6	3
phenoxy	0.38	2.1	3
iodo	0.39	17	65
bromo	0.44	5.6	65
chloro	0.47	2.2	3
azido	0.44	1.8	357
alkylsulfinyl	0.52	12	111
cyano	0.56	1.0	3
alkylsulfonyl	0.60	1.2	111
SCN		1.4	111
NHR2+	0.80	0.50	109

lifetimes come from Stern-Volmer quenching studies, which are quite accurate because of the constancy of k_q values and have often been reproduced by direct flash decay measurements. This mixture of old and recent data provides the most complete picture of the electron-deficient character of triplet ketones. Figure 3 plots the data for δ-substituted valerophenones; it updates the plot first published in 1972. We shall merely emphasize the most important aspects of these numbers.

The effects of most δ-substituents on triplet reactivity correlate well with their σ_I values ($\rho = -1.85$), with three exceptions: I, Br, and sulfinyl. We have suggested that they all provide anchimeric assistance to hydrogen abstraction,[141] as is well known for the halogen atoms in free radical hydrogen abstraction.[142] The effects of γ-substituents must reflect resonance stabilization of the developing radical site as well as inductive deactivation. It was suggested in 1972 that the ρ values for δ and ε substituents could be used to extrapolate a ρ value of -4.3 for γ-substituents. This value can be used to correct the actual hydrogen abstraction rate constants for the inductive effects of the γ substituent.[3] Any residual rate enhancement relative to the value for an alkyl substituent is ascribed to resonance stabilization of the radical center by the γ substituent. Table 3 lists the calculated resonance factors. The relative accuracy of the results is attested to by the fact that the two substituents with no conjugative capability, sulfonyl and ammonium groups, have the expected factors of 1; their large effects on k_H are entirely inductive. It is noteworthy that substituents with available lone pairs provide more resonance stabilization than do those with π bonds and that an aldehydic hydrogen is the most reactive yet measured.

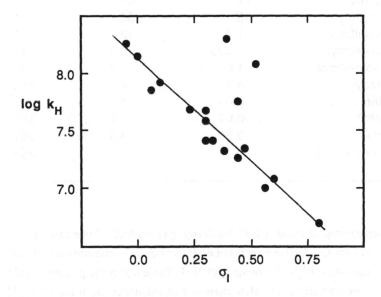

Figure 3. Hammett plot of rate constants for triplet state γ-hydrogen abstraction in δ-substituted valerophenones: ρ = −1.85.

Table 3. Rate constants in benzene for triplet γ-hydrogen abstraction by γ-substituted butyrophenones.

γ-substituent	kH, 10^7 s^{-1}	resonance factor	ref
H	0.80	-	3
alkyl	13-18	(1)	3
dialkyl	50	-	3
phenyl	38	9	3
vinyl	50	9	3
dimethylamino	80	9	109
methoxy	62	98	3
hydroxy	38	37	3
phenoxy	18	2.3	3
acetoxy	1.2	4.8	3
fluoro	0.64	6	151
chloro	2.8	23	3
alkylthio	77	54	111
phenylthiyl	45	58	111
alkylsulfinyl	1.2	13	111
alkylsulfonyl	0.035	0.8	111
carbomethoxy	1.0	1.6	3
benzoyl	3.2	4.0	359
cyano	0.40	11	3
NHR_2^+	0.01	1.0	109
azido	0.50	4.0	357
=O (aldehyde)	90.		263

Positive nitrogens enhance n,π* reactivity over ten-fold;[143] other ring substituents have very small inductive effects but do lower observed rate constants significantly when the lowest triplet becomes π,π*.[12,18] Table 4 lists the percent population of the upper n,π* triplet in valerophenones as calculated from Equation [1]. Values of ΔE can be calculated from Equation [2], but only when kH is decreased by the π,π* triplet being lower. All of the early work on ring substituents involved

Table 4. Rate constants for triplet γ-hydrogen abstraction in benzene by ring substituted valerophenones.

substituent	k_H, 10^8 s^{-1}	% n,π*	reference
H	1.3	99	18
o-alkyl	0.30	25	274
m-alkyl	0.39	35	274
p-alkyl	0.18	15	274
o-OMe	0.03	3	18
m-OMe	0.002	0.2	18
p-OMe	0.006	1	18
p-OAc	0.44	25	144
p-OCF$_3$	1.3	95	145
o-F	1.4	99	18
m-F	1.8	99	18
p-F	1.5	99	18
o-Cl	0.35	?	18
m-Cl	1.6	99	18
p-Cl	0.30	15	18
o-CF$_3$	1.3	>99	18
m-CF$_3$	3.2	>99	18
p-CF$_3$	2.8	>99	18
o-(N)	1.9	>99	18
m-(N)	3.1	>99	18
p-(N)	6.8	>99	18
m-(NH+)	20	>99	143
p-(NH+)	33	>99	143
o-CO$_2$Me	0.36	?	10
m-CO$_2$Me	2.8	99	10
p-CO$_2$Me	0.12	40	10
o-CN	2.3	99	10
m-CN	3.0	99	10
p-CN	0.68	25	10
m-COR	1.4	40	10
p-COR	0.27	10	10
p-SMe	<0.0001	<0.01	18
p-SCF$_3$	0.30	15	145

stabilization of π,π^* triplets by the conjugatively electron-donating alkyl and methoxy groups, which occurs for ortho, meta, and para substitution.[18] Acetoxy[144] and CF_3O[145] groups have only weak electron donor properties and they stabilize π,π^* triplets less. Inductively electron withdrawing groups such as CF_3 stabilize n,π^* triplets more than π,π^* triplets. Conjugating electron withdrawing groups such as carbonyls and cyano[10] stabilize n,π^* triplets inductively and π,π^* triplets conjugatively. The latter effect is somewhat larger so that π,π^* triplets end up being slightly lower in energy than the reactive n,π^* triplets. This effect holds only for para substitution, however. Chloro substitution works the same way because Cl is only a weak conjugative electron donor. The o-chloro and carboethoxy groups are slightly deactivating; the question marks in Table 4 admit that we cannot assess the steric effects of ortho substituents quantitatively.

1. *Orientational requirements.* Simple straight chain ketones normally have their largest α-substituent eclipsing the carbonyl.[146] The most populated conformer thus is in a geometry very close to that required for reaction, requiring only rotation around the β-C—γ-C bond. The ketone can attain a chair or twist-chair transition state geometry quite easily. Although a linear O–H–C arrangement ($\theta = 180°$) presumably is preferred,[59] it is well established that hydrogen abstraction occurs at values of θ closer to $90°$.[52,54]

All of the rate constants in the three tables above include the equilibrium constant for the bond rotation shown below; the χ_r value for the reactive conformation.from Equation [4] must be nearly identical for all the model systems. Additional substitution can affect conformational equilibria in the acyclic ketones. A bulky substituent at the α-carbon might lower χ_r values significantly, while trialkyl substitution at the β carbon is already known to enhance χ_r.[3] As mentioned above, rings reduce conformational freedom such that χ_r values can either increase or decrease. It is only in the last decade that quantitative consideration has been given to the factors that determine both intrinsic rate constants and conformational effects on reactivity.

Scheffer has single-handedly provided most of our knowledge about hydrogen atom abstraction in the crystalline state, in which the geometries of the reacting moieties are known from x-ray studies.[54] We summarized his findings above and shall present some behavior unique to the solid state below. Here we present his comparisons of solution reactivity to crystal structure.

n = 1-5

ring size	d (Å)	ω	Δ	k, 10^8 s^{-1}	E/C[a]
4	3.1	23°	78°	0.3	11.5
5	2.8	31°	96°	1.2	11.5
6	2.6	42°	90°	1.2	0.60
7	2.7	42°	82°	5.7	1.45
8	2.7	48°	77°	6.7	0.22

[a] elimination/cyclization ratio; see below.

Not surprisingly, only compounds with rings that hold the carbonyl and a target C-H bond close together show solid state photoreactivity. Scheffer has studied five α-cycloalkyl-p-chloroacetophenones, all of which react both in solution and in the crystal.[54,147] They give comparable product ratios in both phases, which will be discussed below. Interestingly, four of them must react in a boat rather than a chair geometry, with only the *least* reactive cyclobutyl compound able to attain a half-chair. With the knowledge that organic molecules normally crystallize in their most stable conformations, examination of solution rate constants for hydrogen abstraction revealed no correlation with the geometric parameters for reaction.[147] The data are presented above. The attempted correlation rests heavily on the distance measurements, since the angle differences probably can account for less than a factor of two in rates. The rate constants do correlate well with those for bimolecular attack of various radicals on the different sized cycloalkanes, so the intrinsic reactivities of the different C-H bonds determine relative reactivities. Scheffer concluded that Equation [4] applies, with comparable values of χ for all five. He suggests that these ketones react not from their most stable conformations but rather from higher energy ones that provide a more favorable geometry for reaction. The factor that needs the most adjustment is θ, which averages only 114° for the crystals. It is also possible that they all react from their crystal geometries, in which case χ_r values are all close to one. All of the ketones except the cyclobutyl are considerably more reactive than p-chlorovalerophenone despite their awkward geometries. The rate enhancements probably represent the normal entropic gain associated with the ring.[130] The high observed k values cannot contain a really low value of χ_r; so any common geometric adjustment in solution must involve a relatively low energy conformational change.

2. *Biradical lifetimes and reactions.* The biradical's existence was first established by trapping studies with thiols[148] and by racemization of ketones with asymmetric γ carbons.[26,138] The factors that affect partitioning of 1,4-biradicals among cleavage, disproportionation, and cyclization have been reviewed.[149] We shall merely summarize the main structural features. Cleavage appears driven by the stereoelectronic necessity for overlap of the breaking bond with both singly occupied p orbitals.[139,150] Anything that prevents such molecular alignment retards or suppresses cleavage. α-Substitution by alkyl groups,[151] fluorines,[152] and rings[153] all enhance cyclization. α–Diketones do not undergo any cleavage,[50] most likely because resonance holds the would-be breaking bond perpendicular to the hydroxy-radical site.[48] It is not as clear what affects cyclization and disproportionation, other than that hydrogen bonding to Lewis bases dramatically inhibit the latter and change the stereoselectivity of the former. As an example, we can compare valerophenone and α,α-difluorovalerophenone. The cyclization/disproportionation ratio in benzene for the latter is 1.5:1[152] but only 1:9 for the former.[151]

Another important feature of cyclization is the diastereoselectivity displayed when the prochiral carbonyl carbon becomes tetrahedral. Lewis reported years ago that two different types of selectivity arise, as demonstrated by valerophenone and α-methylbutyrophenone. They both form 1-phenyl-2-methylcyclobutanol; but the former gives a 3.5:1 Z/E ratio while the latter gives only the Z isomer.[150] In α-methylbutyrophenone, the methyl and phenyl groups assume anti orientations in the biradical before it cyclizes. Such steric preferences can involve several kcal/mole. In valerophenone, nonbonded interactions between methyl and phenyl are developed only as the two ends of the biradical begin to bond and obviously produce an energy differential of less than 1 kcal. We shall point out other examples of what may be a general rule, namely that pre-existing conformational preferences can be much larger than those developed only during cyclization. When the nonbonded interactions responsible for diastereoselectivity are developed only during cyclization of the biradical, it is not always easy to predict which mode will prevail, as the following illustrates.[154] We are currently reviewing all of the accumulated evidence on this topic and plan to publish a comprehensive analysis sometime soon.

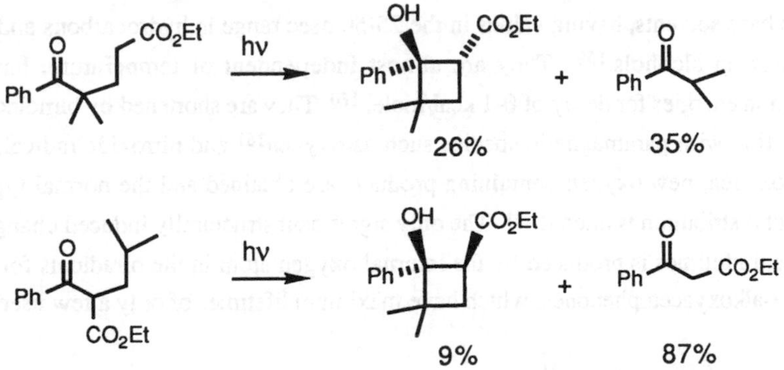

In 1974 we reported that α-allylbutyrophenone forms 4% 2-phenyl-2-norbornanol while undergoing the type II reaction. This fraction of 5-hexenyl radical cyclization in the 1,4-biradical allowed us to conclude that the biradical has a submicrosecond lifetime.[155]

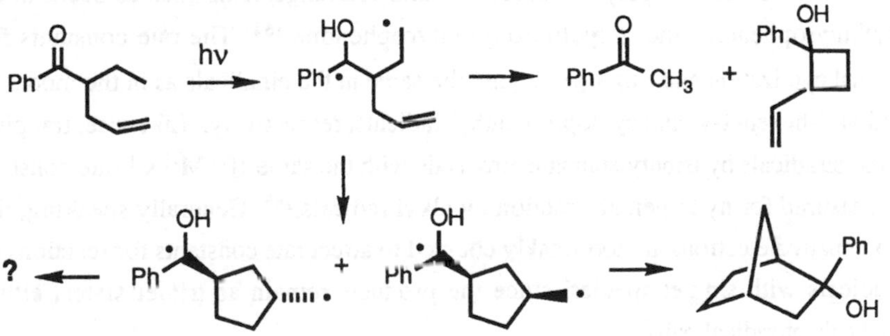

In 1977 Scaiano and Small reported the first flash spectroscopic detection of the 1,4-biradical intermediate from γ-methylvalerophenone, which has a 30 ns lifetime in benzene.[156] Scaiano rapidly developed two independent methods to monitor triplet generated hydroxybiradicals: 1) direct detection of their transient absorption; or 2) indirect detection by following the growth of the strongly absorbing paraquat radical-cation, which is produced when added paraquat oxidizes the hydroxy radical site.[157] He made and has thoroughly summarized several essential observations about biradical lifetimes.[158] They are not very sensitive to substitution at the γ-carbon or on the benzene ring of the ketone.[156] They are lengthened ca. 3-fold in Lewis base solvents, having values in the 25-50 nsec range in hydrocarbons and 75-160 nsec in alcohols.[159] They are almost independent of temperature, having activation energies for decay of 0-1 kcal/mole. [160] They are shortened by bimolecular interaction with paramagnetic species such as oxygen[161] and nitroxide radicals[162]. With oxygen, new oxygen-containing products are obtained and the normal type II product distribution is altered.[161] The only significant structurally induced change in biradical lifetimes is produced by the internal oxygen atom in the biradicals formed from α-alkoxyacetophenones, which have maximum lifetimes of only a few nsec.[163]

It soon became apparent that these biradicals undergo two separate types of reactions: those that are purely monoradical in nature; and those that are triplet in nature. In the first category are intramolecular rearrangements such as occur in α-allylbutyrophenone and γ-cyclopropylbutyrophenone.[164] The rate constants for internal cyclization and ring-opening are the same in the biradicals as in the model 3-methyl-5-hexen-1-yl and cyclopropylethyl radicals, respectively. Likewise, trapping of the biradicals by tributylstannane proceeds with the same 10^6 M^{-1} s^{-1} rate constant as measured for hydrogen abstraction by alkyl radicals.[165] Generally speaking, the two unpaired electrons are too weakly coupled to affect rate constants for reactions of biradicals with singlet species, since the products remain as triplet states, either biradicals or radical pairs.

The two ends of the biradical have characteristic but different radical reactivity. Simple alkyl radical sites abstract hydrogen from several compounds besides stannanes, including thiols.[148] The rate constants with which such compounds react with the biradicals can be measured either by steady state Stern-Volmer quenching or by flash kinetics . In fact quenching of both type II products[148] and biradical absorption[165] provided the first accurate measurements of rate constants for hydrogen abstraction from thiols by alkyl radicals.

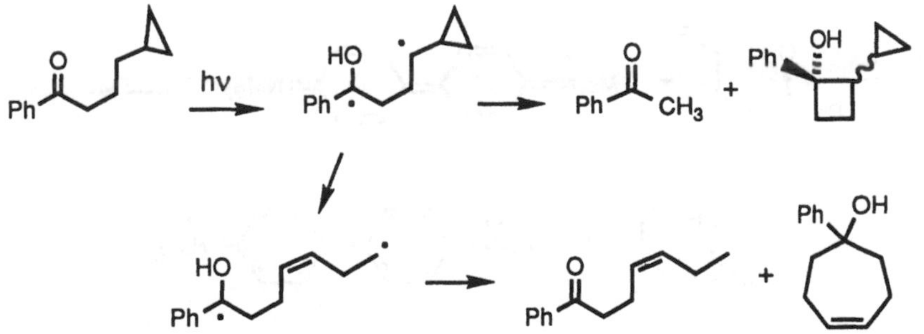

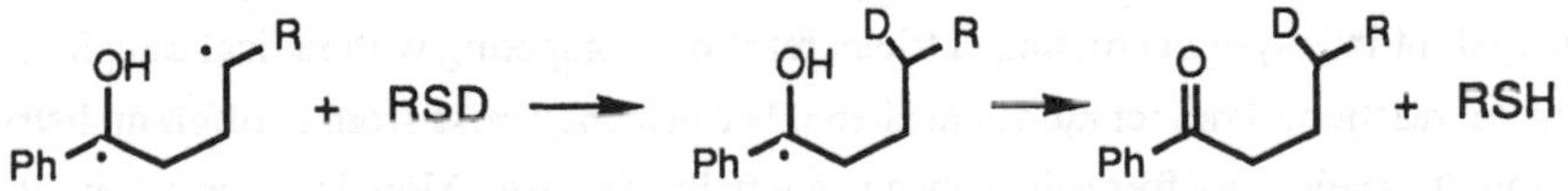

The hydroxy end of the biradical is especially prone to oxidation,[166] just like the hydroxy radicals involved in photoreductions. Thus the paraquat detection technique; rate constants for paraquat quenching[157] are the same for the biradicals as for monoradicals and exceed 10^9 M^{-1} s^{-1}. Other good oxidants also react with the biradicals. One intriguing example is the conjugate acid of 4-cyanopyridine, which quenches the type II reaction of valerophenone and in the process provides a new product.[167] A Stern-Volmer slope of 11 M^{-1} indicates a rate constant of 10^8 M^{-1} s^{-1} for the biradical trapping. This value is much too high to represent radical addition to the pyridine ring and a redox process has been suggested, although the mechanism for coupling at the γ-carbon is not yet clear.

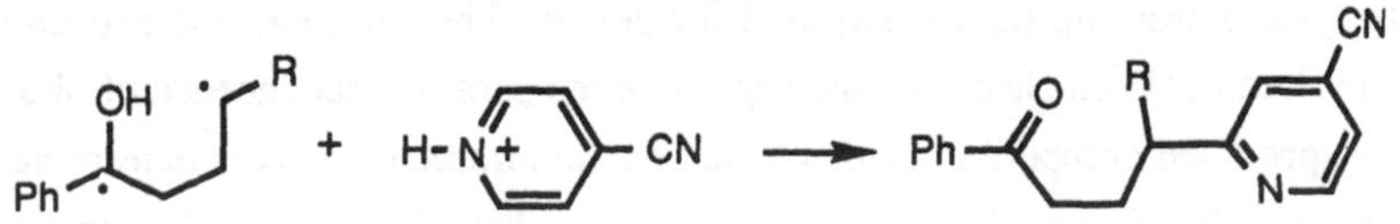

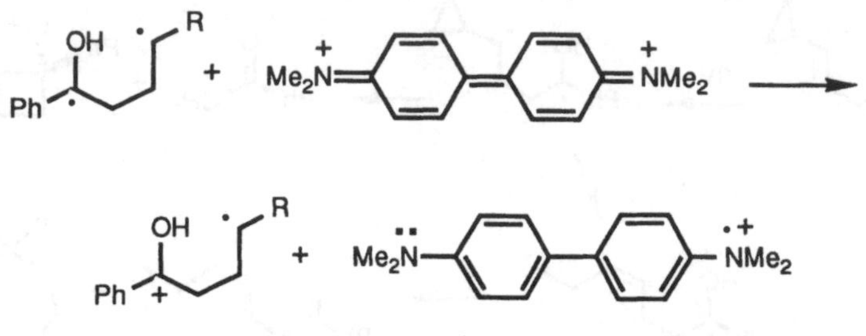

The bimolecular reactions of biradicals with paramagnetic species can involve electron spin, since one allowed spin product of a triplet plus doublet encounter is a singlet plus doublet. The reactions of biradicals with oxygen[161] or NO[168] could be considered simple radical couplings; and appropriate products are indeed found. However, the increase in biradical decay rates are greater than the proportion of these new products would indicate. Moreover, changes in ratios of elimination/cyclization indicate that the oxygen perturbs the normal reactions of the biradical. Nitroxide radicals do likewise.[162] Scaiano has interpreted this behavior as paramagnetic catalysis of intersystem crossing, a triplet reaction, competing with radical trapping, a doublet reaction. Product ratios are altered because they arise from a different distribution of biradical conformations than is normally the case. More isc occurs from the stable anti conformers so more cleavage than normal is observed. This idea brought up the question of what normally determines product partitioning in biradicals.

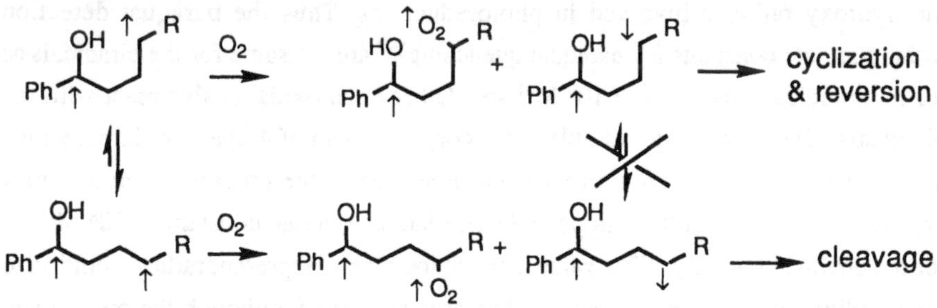

Simple unimolecular biradical decay involves formation of singlet products, so intersystem crossing (isc) is required for decay. The timing of isc and product formation had not been clear until the negligible temperature dependence of biradical lifetimes provided compelling evidence that triplet biradical decay is determined by rates of isc. The chemical reactions of the singlet biradicals are so fast that isc is effectively irreversible. What remains to be described is how singlet biradical reactions can be faster than reverse isc and still show huge variations in product selectivity. Singlet reactions could have activation barriers of several kcal/mole and still outstrip reverse isc, which must have a rate $\leq 3 \times 10^7$ s^{-1} for typical 1,4-biradicals. However, Scaiano has postulated that solvent and structurally induced changes in product composition from type II 1,4-biradicals arise from variations in isc rates of various biradical conformations.[169]. He proposes that singlet biradicals maintain conformational memory of their triplet precursor, which is to say that they undergo no conformational interconversion, because they react so rapidly. The ability of paramagnetic species to shorten biradical lifetimes and change product distributions[158] suggests that <u>externally induced</u> isc can be product determining and supports the possibility that the same holds for intrinsic, internally induced isc. The presumed reason for product discrimination is that the most stable conformation is anti and can undergo only cleavage. The near microsecond lifetimes of most biradicals assures that they attain conformational equilibrium before isc. It is important to note that there are two ways in which isc can occur at different rates from different conformations: 1) even if the different conformers have the same isc rate, singlet conformers will be formed in varying proportions depending on the triplet equilibrium; 2) the conformers may have different intrinsic rates of isc. This latter reason is suggested by the effects of paramagnetic quenchers on product ratios.

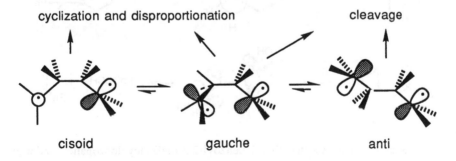

cyclization and disproportionation   cleavage

cisoid   gauche   anti

Scaiano's hypothesis is widely accepted, but it has been difficult to obtain firm experimental support. Several essential bits of knowledge are missing: what determines rates of isc and how do they change with biradical structure? Salem's early discussion of this general question indicated three important factors besides the S–T energy gap that determine isc rates: the distance between the two radical centers;

the relative orientation of the two half-occupied p orbitals; and the relative energy of the singlet zwitterionic state.[170] These all influence the spin orbit coupling that enables isc. Recent work on the triplet biradicals formed by α-cleavage of α-phenyl-cycloalkanones has confirmed that isc in short biradicals is driven primarily by spin-orbit coupling since it depends on the distance between radical centers.[171,172] Several phenyl ketones were designed to generate 1,4-biradicals that may test this feature. Unfortunately, the results have not been particularly encouraging

Caldwell reported that cis-1-benzoyl-2-benzhydrylcyclohexane and γ,γ-diphenylbutyrophenone produce 1,4-biradicals with very similar lifetimes.[173] The ring in the former compound constrains the biradical to a gauche conformation, whereas the biradical from the latter presumably exists mostly in the stretched anti conformation. Obviously differing distances between the biradical ends is not very important here. It might be argued that electron density at the γ-radical site is so delocalized into the two benzene rings that the two geometries really do not differ that much. However, only about 40% of the spin density of benzyl radicals is delocalized over the ortho and para positions;[174] this figure is similar to the per cent decrease in biradical decay rate caused by γ-phenyl substitution.[173]

Scaiano then reported that 1-benzoyl-2,2-dialkylcyclopropanes undergo exclusive photoelimination via biradicals with very short lifetimes, ~20 nsec in methanol.[175] The 5-fold greater decay rate of these biradicals compared to the linear one produced from valerophenone was ascribed to the enforced cisoid proximity of the two radical sites in the cyclopropane derivative. However, Caldwell recently pointed out that these biradicals undergo only cleavage. In contrast, the 2,2-dibenzyl analog undergoes a more normal -30% cyclization and has the same lifetime as the biradical formed from γ-phenylbutyrophenone.[176] He suggests that the unusually short lifetimes of Scaiano's biradicals reflect a rapid, exothermic, *adiabatic* cleavage to triplet enol. This reaction is endothermic in the phenyl substituted biradical and therefore does not compete with normal decay.

The short lifetimes of the biradicals formed from α-alkoxy ketones[171] and their higher than normal yields of cyclization products[177] have both been interpreted as reflecting the greater ease of forming a four-membered ring with oxygen in it, because of the decreased torsional strain brought about by replacement of a CH_2 group. They may equally well reflect stabilization of the gauche or cisoid biradical conformations relative to the anti, but only if the former actually undergo isc faster than the latter. The just cited examples cast doubt on this supposition. We will present an alternative explanation below. It must also be stressed that the wide variations in cyclization/disproportionation ratios, both of which must occur from cisoid geometries, have not been adequately addressed.

Scaiano was particularly interested in solvent effects on biradicals. Lewis bases tend to increase quantum yields of product formation from values of 0.3-0.5 to unity.[178] Since biradical lifetimes increase by comparable factors, it was tempting to equate the lifetime increase with the suppression of the major chemical reaction, disproportionation back to starting ketone, by hydrogen bonding. In one key experiment he showed that the increases in quantum yield and in biradical lifetime caused by different concentrations of pyridine are not exactly parallel.[159] From this result he inferred that solvent effects on quantum yields and on biradical lifetimes represent different interactions. The lifetime lengthening was attributed to destabilization of the cisoid conformations, which are required for back hydrogen transfer and which presumably undergo the most rapid isc. However, the yields of cyclization, which also requires a cisoid geometry, are reduced only slightly by solvation. Moreover, the assumed distance dependence of isc in these 1,4-biradicals has not been strongly demonstrated. We concur with Caldwell's conclusion[179] that no fully satisfactory explanation for solvent effects on biradical lifetimes has yet appeared; we present our own below.

A final caveat is necessary. Scaiano's proposal was based entirely on the behavior of 1,4-biradicals. They are unique in being able to form a product from almost any conformation because they alone can cleave to two olefins. Although it is possible that little if any conformational change can occur after isc in such small, relatively constrained biradicals, larger biradicals that cannot undergo any chemical reaction in stretched conformations must rotate into cyclic geometries before any products can be formed. Therefore the concept of "conformational memory" in singlet biradicals[169], although appropriate for 1,4-biradicals that can form products from most conformations, may not be so relevant for longer biradicals. In fact we have suggested an alternative for 1,5-biradicals, namely that isc and cyclization are coupled.[180] We shall return to this idea below.

Neuman has measured pressure effects on the product ratio formed by type II reaction of valerophenone and found a $\Delta\Delta V^\dagger$ value of only 1-2 cc/mole between cyclization and elimination, with higher pressure favoring cyclization as expected.[181] This small difference in volume of activation is consistent with there being only low barriers for biradical reaction. It can be attributed to larger populations of the more compact cisoid conformations at higher pressures or to different molecular motions along the reaction paths to products.

3. *Environmental effects* . The strong dependence of quantum yields, product ratios, and cyclobutanol stereochemistry on solvent have spurred several workers to check how various restricted environments affect the type II reaction. Both the rate constants for competing triplet reactions and the partitioning of biradical intermediates may be affected by environment. We shall discuss several examples of different media, progressing from the least to the most restrictive.

Several labs found high quantum yields and product ratios that suggest a fairly polar environment for ketones irradiated in aqueous surfactant solution.[182] It is now well accepted that the polar end of the ketones and particularly the biradicals reside mostly near the micelle-solvent interface, called the Stern layer, which has

Lewis base character. Although micelles produce negligible effects on triplet rate constants, they can improve type II/type I ratios by their "super-cage" effect that enhances the recoupling of radical pairs.[183]

The benzoin ethers were a mechanistic puzzle for years, since they were first reported to undergo only α-cleavage to radicals and *no* type II cleavage or cyclization in solution.[184] Since rate constants for γ-hydrogen abstraction approach 10^{10} s^{-1} in simple α-alkoxy ketones,[177] in the benzoin ethers either the rate constant for α-cleavage approaches 10^{12} s^{-1} or the alkoxy groups are held primarily in a conformation unsuitable for hydrogen transfer. The latter must be the case, since triplet lifetimes near 1 nsec have been measured for benzoin and various derivatives.[185] Cleavage of the benzoin ethers themselves cannot be quenched, but the p-methoxy derivative has a triplet lifetime of 5 nsec.[186] Given the normal 100-fold decrease in reaction rates caused by p-methoxy substitution,[18,22] the rate of decay of the benzoin ether triplets must be close to 2 x 10^{10} s^{-1}.

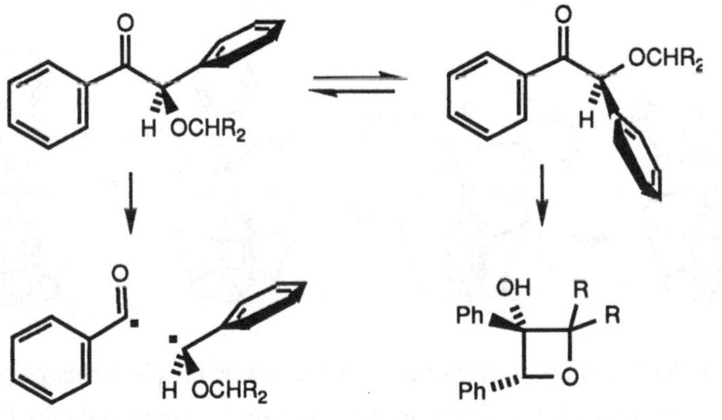

DeMayo reported that ~15% type II reaction does compete with radical cleavage when benzoin ethers adsorbed on silica are irradiated.[187] He points out that the silica must force the ketone into a more reactive geometry than obtains in solution. It also makes α-cleavage mostly revertible because of poor translational mobility on the silica surface. Actually, the use of methanol as solvent also provides 5-10% type II reaction. This result explains the normal lack of γ-hydrogen abstraction, since the carbonyl and alkoxy dipoles presumably prefer to oppose each other except in polar media. Cyclization and type II elimination also occur in micelles.[188]

Ramamurthy has since shown that irradiation of solid complexes of three common benzoin ethers with β-cyclodextrin produces 90% type II reaction.[189] It is a little difficult to understand why the cyclodextrin forces the ketone into what appears to be a single geometry, one suitable for hydrogen abstraction. The scheme below provides a partial picture. There cannot be any significant conformational freedom in the complexed ketones, since the methyl ether gives the same type I/type II ratios as the ethyl and isopropyl ethers, despite its 3-fold lower intrinsic reactivity. There are several other examples of ketones that undergo mainly type I cleavage in solution but mostly or entirely type II reaction when complexed with β-cyclodextrin.[190] These high product selectivities are achieved only when the solid complex is irradiated. Aqueous solutions of ketone and cyclodextrin give more type II reaction than occurs in benzene but radical cleavage still competes strongly. Not only does complexation involve a mobile equilibrium in solution, but one radical fragment sticks out into the solvent and can escape to dimerize.. The decrease in cleavage for the complexed ketones is due to a super cage effect only in the solid state.

Turro looked at the effects of several zeolites on the photochemistry of α,α-dimethylvalerophenone,[191] which undergoes competitive type I and type II reaction in solution.[192] Depending on the cavity size of the zeolite, product ratios (both type I/type II and type II cleavage/cyclization) can either increase or decrease. When mainly type I cleavage is observed, the cavities obviously are too small to allow the conformational changes required for type II reaction. (The cage effect precludes observation of type I cleavage products unless the radical pairs can either dispropor-tionate or rearange while coupling.)

Weiss has explored the effects of liquid crystals on the type II reaction. Their effects on short ketones are variable, whereas their effects on long ketones can be dramatic. His most recent paper describes some very revealing aspects of how guest molecules orient themselves in liquid crystals.[193] He studied the products formed from various p-alkylalkanophenones in n-butyl stearate. As the lengths of the two alkyl chains are varied, the position of the carbonyl group moves from the end to the middle of the ketone. FTIR experiments show that the ketones line up linearly with surrounding butyl stearate molecules, so that the n and m values determine whether the ketone carbonyl and biradical hydroxy group can interact with the solvent carbonyl. Type II elimination/cyclization (E/C) ratios were measured to determine whether hydrogen bonding affects the anti/gauche biradical conformational equilibrium and allows kinked cisoid conformers to exist.

$$m = 0 - 15$$
$$n = 21 - 6$$

When m+n = 21, the length of the ketone molecule matches that of the butyl stearate. The E/C ratio is 3/1 in isotropic phases and in solution but rises to >10 in the smectic and solid phases, with a peak >60/1 (smectic) and >100/1 (solid) when m = 3 and n = 18. Quantum yields in the ordered phases are ~80% as large as in isotropic phases at large and small values of m, but dip sharply with a minimum value of only a few per cent at m = 10. These results can be explained by consideration of how the biradicals align themselves with the solvent. The scheme below shows how four of them would line up with respect to the ester group.

When m = 10, the ketone carbonyl is situated along the middle of the adjoining solvent molecules. It apparently cannot coil into a cyclic transition state for hydrogen abstraction at all, so the quantum yield plummets. What little reaction occurs shows normal product selectivity and probably arises from misaligned or unaligned ketone molecules. When m = 3, the ketone carbonyl is aligned with that of the solvent and the reacting alkyl chain extends into the body of the liquid crystal layer. There is enough conformational flexibility that hydrogen abstraction can occur with normal efficiency. However, hydrogen bonding anchors the biradical to the solvent layer such that the γ-radical site is embedded in the most ordered middle region of the solvent. Since almost no cyclization occurs, this region must be sufficiently ordered to prevent the biradical from coiling into a cisoid conformation. When m = 15, the ketone is situated such that the biradical again is anchored by hydrogen bonding but now with the γ-radical site at the edge of the solvent layer. There is so little order in this region that normal biradical conformational preferences are established. When m = 0, the entire biradical is at the edge of the layer and normal selectivity occurs.

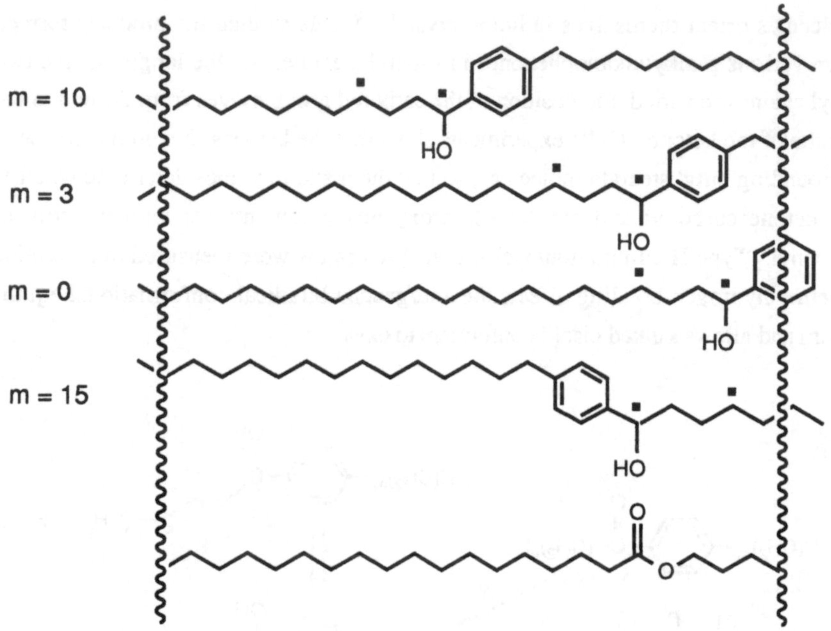

When m is held at 5 and n is varied from 11 to 18, the ketone molecules progress from being too short to too long. The E/C ratio is 10-15 for n<15 but jumps to a maximum >40 for n = 16. Thus the ketone and biradical are subject to the most conformational restrictions when their length matches that of the solvent.

Leigh has provided a very nice example of how liquid crystals can keep a bichromophoric molecule stretched out and prevent the end-to-end quenching that dominates reactivity in solution. This effect occurs in the valerophenone with a phenolic tail mentioned earlier[85] and with β-phenylbutyrophenone.[194]

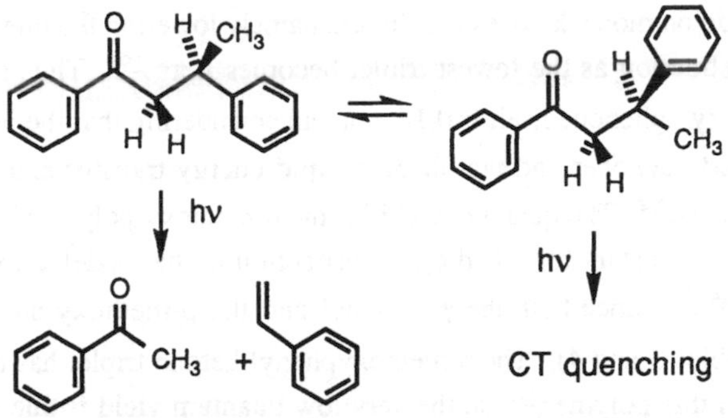

Whitten studied a ketone that is also a surfactant, ω-(p-toluyl)pentadecanoic acid.[195] When prepared as a monolayer film in the C_{20} arachidic acid, this ketone undergoes only a trace of type II reaction. In benzene, it cleaves to p-methylacetophenone with a normal $\Phi = 0.20$; in an aqueous SDS solution, $\Phi = 0.80$. The enhanced quantum efficiency in the micelle was discussed above; it is interesting that the molecule must be looped so that both carbonyls are in the Stern layer. The lack of γ-hydrogen abstraction in the monolayer is ascribed to the linear rigidity of the monolayer environment. The orientation must be something like the m = 0 case of Weiss shown above, yet no reaction occurs; the monolayer would appear to allow much less molecular flexibility than does the liquid crystal.

The photochemistry of polymers containing keto groups has been studied extensively and reviewed by Guillet.[196] The type II reaction has been of particular interest inasmuch as it is partially responsible for the photodegradation of polymers. In this regard, temperature effects on the quantum efficiency for cleavage of ethylene-CO copolymers have been associated with glass transitions that reduce conformational mobility and the eventual freezing out of key rotations.[197] At temperatures above the glass transition, the type II cleavage of both amorphous host polymers and guest ketones proceeds with efficiencies similar to those in solution. Whatever constraints the polymer places on molecular motion are not large enough to prevent the relatively small motions needed for γ-hydrogen abstraction and cleavage.

Ilrdlovic and Guillet have carefully studied the behavior of polyacrylophenones and copolymers of styrene and phenyl vinyl ketone.[198] Guillet has even produced a purposely photodegradable polystyrene copolymer based on type II cleavage of the backbone. It is interesting that p-substituents produce the same

effects as they do on model ketones in solution, namely lowering the rate constant for
γ-hydrogen abstraction as the lowest triplet becomes π,π*.[199] Thus the quantum
yield for polyacrylophenone itself is 0.14, not bad considering that the product vinyl
ketone is a good quencher and that there is rapid energy transfer among chromophores in polymers.[200] The quantum yield for the p-methoxy polymer is only 0.001.
The intrinsic rate constant for γ-hydrogen abstraction in the model ketone is on the
order of 1×10^5 s^{-1}, since both the γ-carbonyl and the p-methoxy slow hydrogen
abstraction (Tables 3 and 4). The p-methoxyphenyl ketone triplet has a lifetime of
several μsec. in this polymer,[201] so the very low quantum yield is due primarily to
quenching by impurities and product.

The crystalline state also provides some unique behavior ; we include three
examples from Scheffer's work. The achiral ketone α-(3-methyladamant-1-yl)-
acetophenone crystallizes in a chiral space group. Irradiation of a large single crystal
results in a 70% yield of a single cyclobutanol in 82% enantiomeric excess, the exact
configuration depending on that of the original crystal.[202] The overall process in the
solid is stereospecific and topologically controlled, whereas in solution a racemic
epimer is the major product.[203] This ketone also has the worst geometric factors:that
Scheffer studied: d = 2.8 Å (okay), ω = 62°, Δ = 77°.

Four symmetric macrocyclic diketones photocyclize with the same sort of
phase dependent stereoselectivity.[204] Cyclohexadecane-1,9-dione and cycloeicosane-
1,11-dione undergo photocyclization in the crystal to primarily cis-cyclobutanols. In
contrast, cyclooctadecane-1,10-dione and cyclodocosane-1,12-dione produce mainly
trans-cyclobutanols in the crystal. The few percent elimination in the crystal increases

to 45-50% in solution and the trans/cis cyclobutanol ratio is 2.5-3/1 for all four ketones. In each compound there are two pairs of symmetry related γ-carbons, one of which has a γ-hydrogen closest to the carbonyl (~2.7 Å) Each additional methylene between the two carbonyls causes the ring puckering to alternate such that, on the carbon attacked, the hydrogen that is *not* abstracted alternates between being syn or anti to the carbonyl oxygen. This hydrogen becomes the bridgehead methine that is either cis or trans to the hydroxy group in the cyclobutanols. To show that solid state chemistry is not foolproof, the stereoselectivity shown by the smallest diketone disappears above 30° because of a solid state phase transition.

| | n = 5,7 | 85% {12%} | 13% {43%} |
| | n = 6,8 | 3% {13%} | 88% {37%} |

Finally, the α-cycloalkyl-p-chloroacetophenones mentioned above show some remarkable chemo- and stereoselectivity; the E/C values noted in the earlier table for the solid are almost the same in solution.[205] The ketones with the smallest two rings undergo primarily cleavage; the next two split nearly 50:50; the largest ring gives mainly cyclobutanol. In the last case, the 18% elimination gives mainly *cis*-cyclooctene, the cis/trans ratio being 9:1 in benzene and 72:28 in the crystal. However, over 85% of the cyclization gives the *trans* ring junction. Scheffer provides an intriguing explanation by noting that the p orbital on the ring in the biradical must twist 90±45° to become parallel to the p orbital at the hydroxy radical site and achieve the geometry required for both cleavage and cyclization. (Actually they both can occur at an overlap angle somewhat greater than 0°.) Rotation in the direction that gives cis products is favored for the two smallest rings; rotation in the opposite direction, which favors formation of trans products, is favored for the largest ring; the two rotations are equally likely for the 6- and 7-rings. Because the three largest ketones yield mostly trans-fused cyclization products but cis-cycloalkene

cleavage product, Scheffer concludes that the pre-cis biradical undergoes mainly
cleavage and the pre-trans biradical undergoes mainly cyclization. He further
proposes that the phase-independence of the results indicates that these preferences
are intrinsic to the molecules' structures. The diagram below depicts the proposed
biradical geometries and indicates the product ratios as a function of ring size.

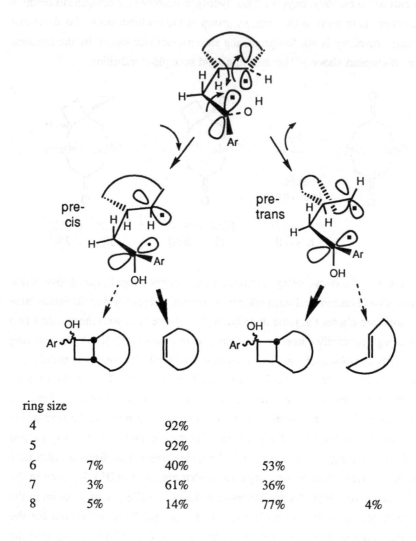

ring size				
4		92%		
5		92%		
6	7%	40%	53%	
7	3%	61%	36%	
8	5%	14%	77%	4%

Scheffer did not explain why the pre-cis biradical prefers cleavage and the pre-trans biradical prefers cyclization. We presume that the latter cyclizes preferentially because cleavage would form a strained double bond and that the former prefers cleavage because a cis ring junction causes too much torsional strain in the larger rings. In fact, the whole idea that product stability influences biradical partitioning ignores Scaiano's postulate about product-determining isc. (Who ever said that biradicals would be easy to understand?). We do note that Scheffer ignored anti biradical geometries. Given the dubious evidence for any conformation dependence of 1,4-biradical lifetimes, it is unlikely that these biradicals are so short-lived that even in solution their reactions are subject to the least motion restrictions that hold in the solid. Although type II biradicals in the solid state generally stay in their nascent cisoid form until they react, in solution they usually live long enough to equilibrate with anti conformers. The α-rings should cause there to be little energetic difference between cisoid and anti conformers , so the two probably are about equally populated. Since the anti conformer can undergo only cleavage and the pre-cis anti form should still be preferred for the larger rings, the presence of an anti geometry would not upset Scheffer's basic conclusions.

Wamser has studied the photochemistry of cyclohexylacetyl groups bonded by Friedel-Crafts acylation to polystyrene beads, which provides an α-cyclohexyl acetophenone structure.[206] The parent ketone reacts with a triplet rate constant of 6 x 10^8 s^{-1} in pentane, in excellent agreement with Scheffer's measurement (after correction for the four-fold deactivation imparted by a p-chloro group). The polymer bound ketone reacts with almost the same quantum efficiency as the free ketone in pentane, 0.5, but is unreactive in ethanol, which does not swell the polymer and allow the molecular motion necessary for reaction.

Every rule needs an exception; Mohr has reported that the spirodiketone below undergoes type II photoelimination both as a solid and in solution.[207] As indicated above, most α-diketones undergo only cyclization and no cleavage at all. This rigid cis diketone can undergo γ-hydrogen abstraction only one way; the spiro structure prevents the biradical from cyclizing but allows both p orbitals to overlap with the central bond such that cleavage is stereoelectronically possible.

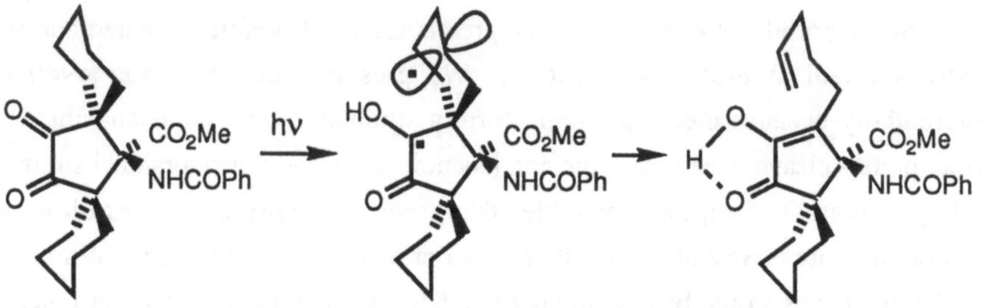

4. *Deconjugation of α,β-unsaturated carbonyls.* The photorearrangement of α,β-unsaturated ketones and esters to their β,γ-isomers is well known.[208] Careful study of several cases indicates that the reaction occurs from ^{1}n,π* states, which undergo γ-hydrogen abstraction in competition with intersystem crossing.[209] These compounds tend to have π,π* lowest triplets that usually do not undergo hydrogen atom abstraction. Acid or base catalysis regenerates a mixture of α,β- and β,γ-unsaturated compounds from the dienols. The group at Reims has shown that the use of either enantiomer of ephedrine induces as much as 30% ee in several unsaturated esters and lactones.[210]

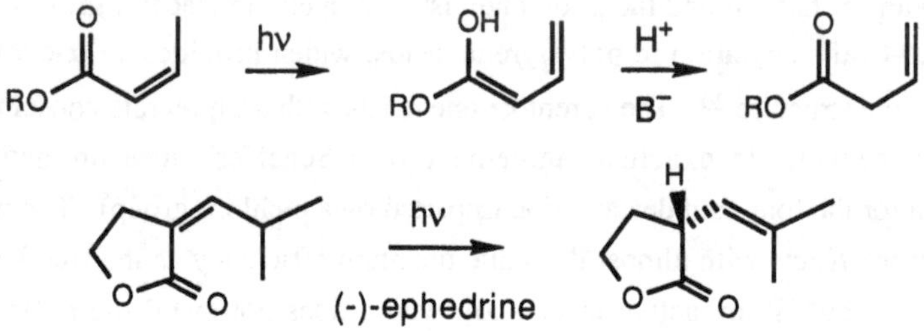

5. *Photoenolization of ortho-alkylphenyl ketones.* γ-Hydrogen abstraction from an ortho-alkyl group is identical to the deconjugation reaction and provides one of the simplest forms of photochromism. The transient enols are highly reactive in the Diels-Alder reaction, having been trapped with a variety of dienophiles.[211] Early flash kinetic studies provided bimolecular rate constants for some of these trapping reactions.[212] The enols undergo facile acid/base catalyzed reketonization. We have just discovered that they generally cyclize to benzocyclobutenols when not trapped. We shall concentrate first on triplet reactivity, then on the fate of the biradical

intermediates, and finally on the enols. Scheme 5 summarizes all of the important competing reactions.

Scheme 5

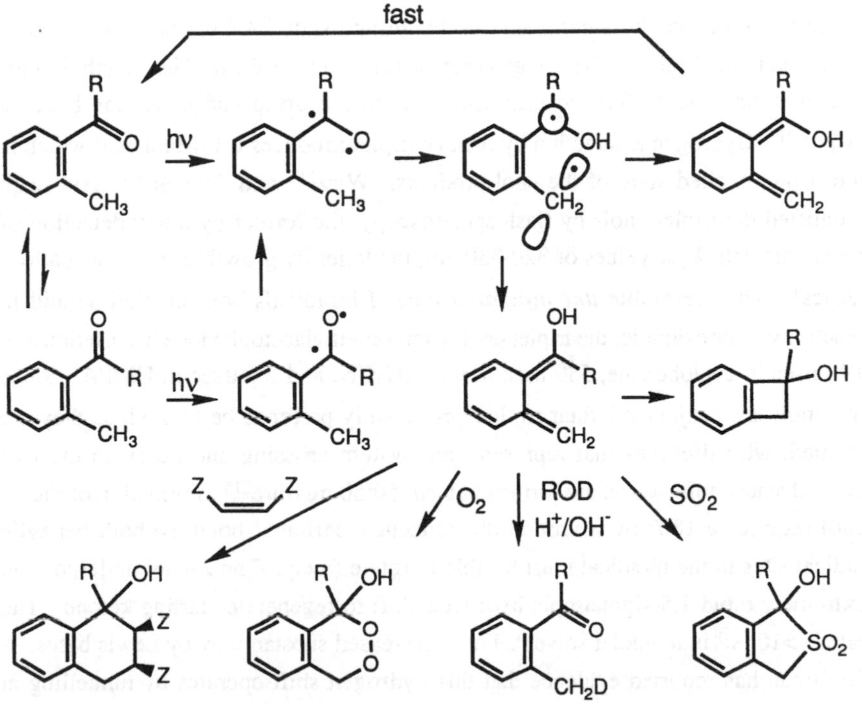

The initial hydrogen abstraction reaction is known to be a rare example of rotational control of triplet reactivity.[131] In o-alkylphenyl alkyl ketones, some reaction may occur from n,π* singlets; in o-alkylbenzophenones, the reaction occurs only from triplets.[213] Because of slow rotation around the benzene—carbonyl bond, the *syn* and *anti* rotamers become kinetically distinct triplets. The *syn* triplet undergoes γ-hydrogen abstraction even from a methyl group with a rate constant $>10^9$ s^{-1}, far faster than rotation into the *anti* geometry. The longer-lived *anti* triplet has time to undergo competitive reactions,.since it must first rotate into a *syn* geometry in order to enolize. This rotation has solvent dependent rate constants $\sim 10^7$ s^{-1} that determine the triplet lifetime of the *anti* rotamer and its rate of enolization. Scaiano has observed by flash spectroscopy two triplets for o-methylbenzaldehyde and o-methylacetophenone,[214] the latter having lifetimes of 0.5 and 34 nsec, nearly identical to those estimated by quenching studies.[130]

The actual geometries of the two distinct rotamers are not known, so it is not possible to discuss the distance and angles for hydrogen abstraction. The benzoyl chromophore certainly is not planar in the ground state[38,213] but probably is more nearly so in the triplets. The *syn* geometry is favored over the *anti* in o-methylaceto-phenone; and the preference increases as both the ortho and α groups become larger.[213] Hydrogen abstraction by the *syn* triplet produces a 1,4-biradical which is the triplet excited state of the enol products. Wirz[215] and Scaiano[214] have each identified the triplet enols by flash spectroscopy, the former by direct detection of transients with λ_{max} values of 320-330 nm, the latter by growth of paraquat cation-radical. They resemble *unconjugated* type II biradicals both in lifetime and in reactivity. For example, the triplet enol from o-methylacetophenone has lifetimes of 190 nsec in cyclohexane, 580 nsec in wet CH_3CN, and 2.6 μsec in HMPA. Since the enols are conjugated, their triplets presumably prefer to be twisted, as they are formed, with lifetimes that represent intersystem crossing and decay to the two ground state enols, which are formed in a *ca.* 50:50 mixture[215]. Formation of the E-enol requires a 180° twist around the benzene—carbonyl bond, so both benzylic radical sites in the biradical must be able to rotate freely. The Z-enol undergoes an extremely rapid 1,5-sigmatropic hydrogen shift to regenerate starting ketone. The rate is $>10^7$ s^{-1} in nonpolar solvents but is decreased substantially by Lewis bases.[215] Grellman has reported evidence that this hydrogen shift operates by tunnelling at lower temperatures.[216] Only the E-enols live long enough to be trapped by dieno-philes.[215] They also undergo facile catalyzed H/D exchange and regeneration of starting ketone.

For a given ketone, its *anti* triplet, its enol triplet, and its Z-enol all have similar lifetimes, depending on solvent.[215] Consequently understanding of the overall mechanism proved elusive for years. Study of cyclic ketones such as 8-methyl-1-tetralone was particularly helpful, since they exist only in a *syn* conforma-tion, have only one short-lived triplet[131], and form only the Z-enol[215]. The 2,6-dialkyl ketones appear to provide the same benefit; but their photochemistry remains to this day confusing, especially the triplet decay kinetics. Matsuura first noted that 2,4,6-trimethyl ketones.form benzocyclobutenols as products[217] and that enols are involved, since they can be trapped by oxygen.[218] We found that several 2,6-

dimethylacetophenones and valerophenones form benzocyclobutenols, with the quantum yields enhanced by Lewis bases[213] just as in other hydroxybiradical processes. In the valerophenones, type II reaction occurs from an easily quenched triplet but cyclobutenol formation occurs from a short-lived triplet. Scaiano could detect a long-lived (300 nsec) triplet by sensitization but no biradical-like intermediate.[214] For years it was assumed that cyclobutenols are formed *only* when the corresponding enols would be too sterically crowded; an extra ortho group prevents the HO–C–R group from lying in the plane of the benzene ring. It was not clear whether cyclobutenols or enols are formed first and which then rearranges to the other.[211] Heating the cyclobutenols causes them to revert to E-enol and then to ketone.[219] In studying this process, the group at Nancy found that several [2,3]benzocycloalkenones with ring sizes of 9 to 14 also photocyclize to benzocyclobutenols in quantum yields less than 0.10.[220]

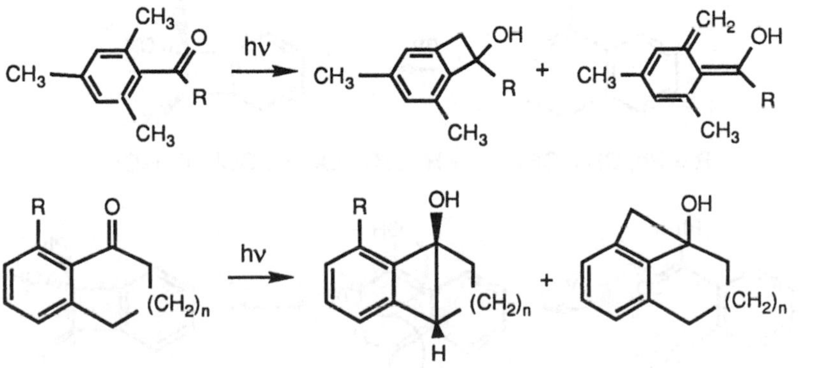

R = H ; n = 4 - 6   R = CH₃ ; n = 3 - 6
R = Me ; n = 5 - 9

The product distribution from o-alkyl ketones seemed predictable until three years ago when Wilson reported that 1-phenylbenzocyclobutenol is formed in good yield from normal solution irradiation of o-methylbenzophenone.[221] Enol is formed also and can be trapped.[222] He suggested that cyclobutenols escaped detection in all the earlier studies of simple o-alkyl ketones because they revert to starting ketone during gas chromatographic analysis. Sammes also had reported a low yield of benzocyclobutenol itself from irradiation of o-methylbenzaldehde.[223] We have recently confirmed and expanded these findings by observing benzocyclobutenols as the major or only products from several o-alkyl ketones.[224] They all revert to ketone when heated above 80°. Therefore we must now suggest that cyclobutenols are always formed but are most stable (*i.e.*, survive GC) when buttressed by an extra ortho substituent or when opening would create an unstable medium-sized ring. What is most surprising is that all of the cyclobutenols that we have observed are formed 100% as the E isomer. This is the opposite stereoselectivity from most other biradical cyclizations. Stereoselective formation of the less stable product suggests a concerted process. As suggested by Matsuura,[218] the conrotatory thermal closure of the more stable E-enol would seem to be the most likely mechanism. We have confirmed this idea by showing that cyclobutenol formation from both 2-alkyl and 2,4,6-dialkylbenzophenones is quenched by acid and by SO_2, both of which trap enol but merely cause slow E→Z isomerization of the benzocyclobutenols.[224]

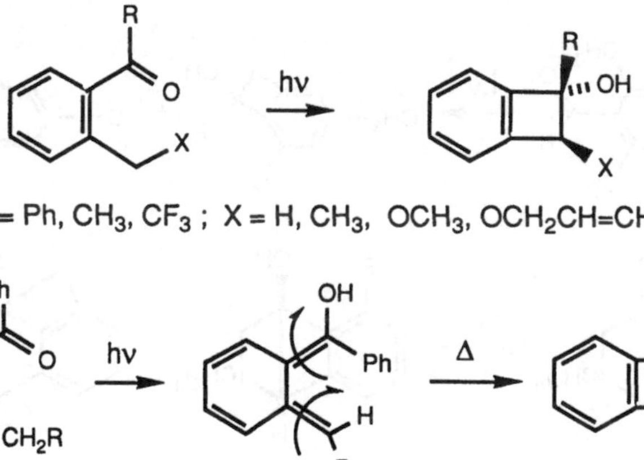

R = Ph, CH₃, CF₃ ; X = H, CH₃, OCH₃, OCH₂CH=CH₂

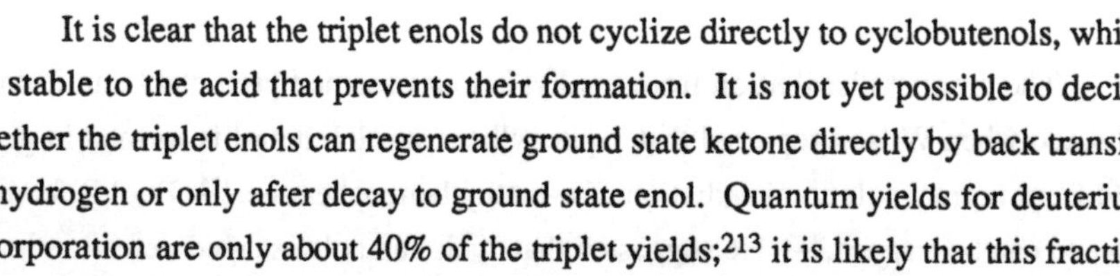

It is clear that the triplet enols do not cyclize directly to cyclobutenols, which are stable to the acid that prevents their formation. It is not yet possible to decide whether the triplet enols can regenerate ground state ketone directly by back transfer of hydrogen or only after decay to ground state enol. Quantum yields for deuterium incorporation are only about 40% of the triplet yields;[213] it is likely that this fraction represents E-enol formation, with the Z-enol reverting to ketone before it can be trapped. However, Wirz has found that irradiation of metastable enols at low temperature causes them to revert to ketone.[215] The acetophenone and benzaldehyde derived triplet enols have lifetimes around 1 μsec, five times longer than typical type

II biradicals, and show similar large solvent effects; those from o-methylbenzo-
phenones live only 24-30 nsec.[214,215] Thus the more sterically congested enols are
formed faster than the less congested ones. The added delocalization of spin density
afforded by the extra benzene ring would not be expected to enhance intersystem
crossing rates. Since one double bond of the ground state enol can twist and still
remain conjugated to the extra benzene ring, the faster decay in the benzophenone
system may reflect decreased steric strain in the enol. We shall see more of this effect.

The timing of photoenol and cyclobutenol formation has special significance
to a mechanistic puzzle that has persisted for 20 years. Hamer first reported that 1-(o-
tolyl)propandione photocyclizes to 2-hydroxy-2-methyl-1-indanone.[225] He
suggested that normal γ-hydrogen abstraction occurs to form a cyclobutenol that then
rearranges thermally to the product. It is noteworthy that the overall reaction shows
the same diastereoselectivity that we have just described for cyclobutenol formation.
Ogata then showed that acetylenedicarboxylate forms a normal Diels-Alder adduct and
suggested that it is the dienol that rearranges.[226] The same diastereoselectivity would
be expected. Hamer later showed that thermal generation of the dienol produces the
hydroxyindanone but still claimed that rearrangement occurs from the cyclobutenol.[227] Of course neither of these workers realized that the dienol rearranges
thermally to the cyclobutenol; either can presumably rearrange to the indanone. We
shall indicate another possible mechanism later.

Matsuura has thoroughly studied the photochemistry of various 2,4,6-
trialkylbenzophenones, all of which form stable benzocyclobutenols.[218] The

importance of steric crowding in promoting cyclization is indicated by the quantum yields for the trimethyl, triethyl, and triisopropyl ketones: 0.008, 0.14, and 0.60, respectively. These are subject to 25% enhancement by added pyridine; and the triisopropyl ketone reaches a maximum value of 0.77. The authors assume no other reaction; but it seems very likely that the remaining quantum inefficiency is a measure of competing Z-enol formation and perhaps competing decay of E-enol. We have found that the triethyl ketones form only the E cyclobutenol,[224] in contrast to the early Kyoto report of solvent dependent Z/E mixtures,[218] which probably were caused by trace acid.

2,4,6-Triisopropylbenzophenone (TIPBP) has been studied the most carefully[228] Of particular interest is the unusual effect of p-substituents on the otherwise unsubstituted benzene ring. Electron-donating groups *raise* triplet rate constants for γ-hydrogen abstraction while electron-withdrawing groups lower them, there being a 30-fold difference between p-methoxy and p-cyano, with TIPBP itself having a k of 9×10^6 s^{-1}. These inductive effects are exactly opposite to those found for simple hydrogen abstraction by substituted benzophenones[12] and the rate constants are two orders of magnitude lower than those ascribed to *syn* o-methyl ketones.[131,214] Because of various observations including a primary isotope effect of only 1.5, Ito and coworkers suggested that the rate determining step for triplet decay is not hydrogen abstraction but coupled rotation around the two benzene–carbonyl bonds. The ground state of TIPBP has the triisopropylphenyl ring twisted 86° out of conjugation with the carbonyl.[229] The authors believed that the hydrogens are too far from the oxygen and stereoelectronically too poorly oriented to react from the ground state geometry. Therefore they proposed that reaction takes place only from a conformation in which the triisopropylphenyl ring is more nearly coplanar with the carbonyl. In support, they found that 1-isopropylanthrone has only one subnanosecond triplet; however, it does not form any cyclobutenol.[230] They explained the odd inductive effects by proposing that the more a para substituent stabilizes the n,π* transition, the larger a barrier there must be to the rotation that deconjugates the benzoyl moiety. This kind of rotation-controlled reaction is very similar to what we originally proposed for the *anti→syn* rotation, with one serious difference. Dynamic NMR studies showed that the rotations are very slow in the ground state, $k_{rot} = 14$ s^{-1}.[231] Since conjugation is more important in the excited states, it is difficult to understand how rotation rates can be 10^6 times faster than in the ground state.

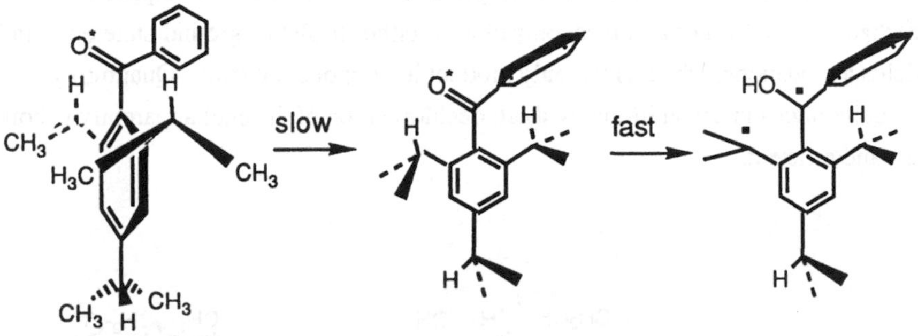

The above picture definitely requires reconsideration. Recently Ito and Matsuura found that TIPBP and its derivatives also undergo photocyclization in the solid state, with an efficiency that depends on para substitution in the same way as in solution.[229] Therefore the odd inductive effects of para-substituents need not involve rotation rates. X-ray analysis showed that with the 86° ring twist, the target benzylic hydrogens are 2.8 Å away from the oxygen and make an angle ω of 55°. The authors propose that reaction occurs from a π,π^* triplet in the solid state. Given all of Scheffer's results, we see no reason to invoke any reactive triplet other than the n,π^*. In fact, the results show that hydrogen abstraction can take place efficiently from the ground state geometry, contrary to what was assumed when the solution results were interpreted. In fact, a two order of magnitude decrease in rate constant seems too large for the deviations from "ideal" orientation measured for TIPBP. We suspect that other factors are involved, such as rotation of the isopropyl groups and the effect that changing conjugation of the nonreacting benzoyl group has on conformations.

The trialkylacetophenones behave quite differently from the benzophenones. The trimethyl, triethyl, and triisopropyl ketones all cyclize in low quantum yield, ~0.01.[218] The trimethyl compound cyclizes ten times more efficiently in dioxane and cyclization appears to take place from a short-lived triplet.[213] We presume that this triplet has a more nearly coplanar benzoyl group than is needed in the benzophenone derivatives. It also forms a long-lived triplet that may well be a more twisted form. It is not clear why the two triplets are kinetically distinct.

Bergmark has examined how α-chlorination changes the course of photo-enolization.[232] Both internal and external trapping take place. The latter likely involves an S_N2' attack on the enol to yield the benzyl ether and happens only in methanol. The former involves ionization of either triplet or ground state enol and yields an indanone, which is the only product in nonpolar solvents. Quantum yields are quite high and triplet lifetimes short; dechlorination of the enol apparently is both fast and efficient.

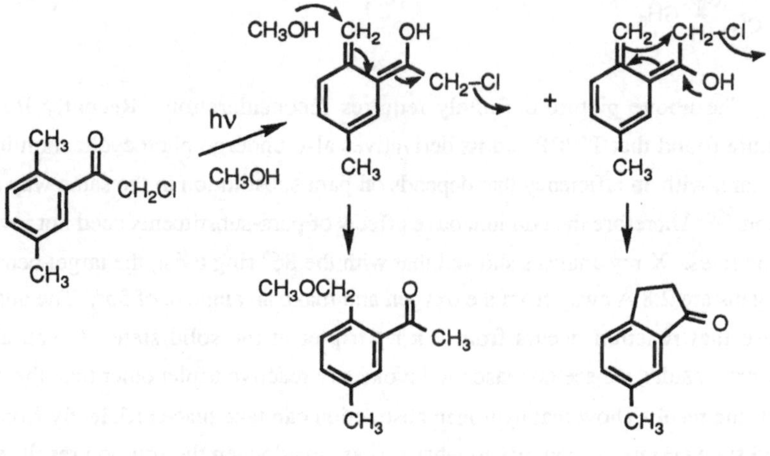

Agosta has studied intramolecular trapping by a triple bond and proposed biradical addition to yield a carbene that either rearranges or is trapped by methanol.[233] A multinational effort revealed no direct evidence for a carbene and found that enols are the main product.[234] The triplet enol has normal 38-45 nsec lifetimes and forms predominantly the less crowded Z enol. This decays rapidly to ground state and shows a k_H/k_D value of 7.3, an interesting primary isotope effect in view of Grellman's conclusion that this reaction occurs by tunnelling.[216] It is curious that no benzocyclobutenol was formed. Could it rearrange to the indanone? Or could its absence indicate that E-enol is trapped by the triple bond, despite the authors' arguments against this pathway?

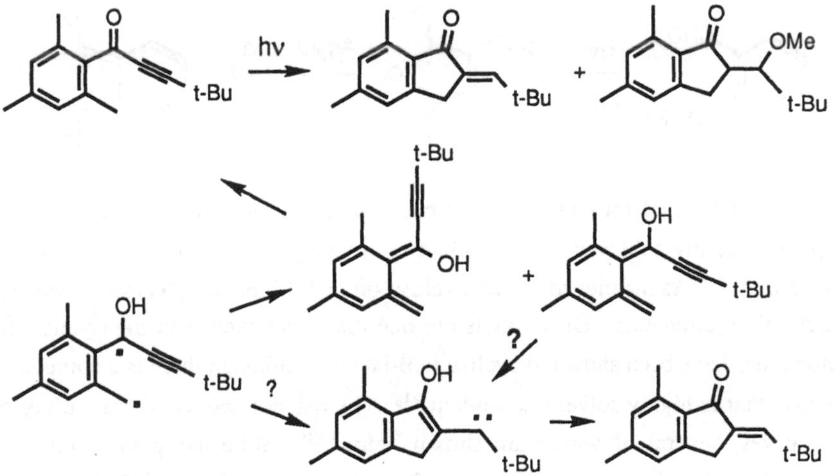

6. Synthetic features.

Photoenolization has been quite useful synthetically in the construction of naphthalene and anthracene skeletons. A couple of typical Diels-Alder reactions are shown below.[235] Charlton and Durst added a useful twist to the synthetic arsenal by showing that the photogenerated enols are readily trapped by SO_2.[236] The cyclic sulfones can be subjected to functional group manipulation before being heated to release SO_2 and the enol. Benzocyclobutenols also are useful compounds since they are converted to the enols both by heat and by light; they may become even more useful now that the generality of their formation has been recognized.

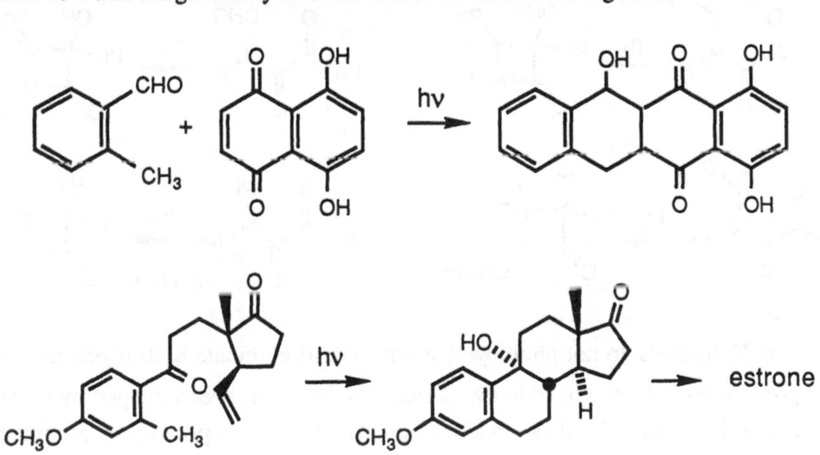

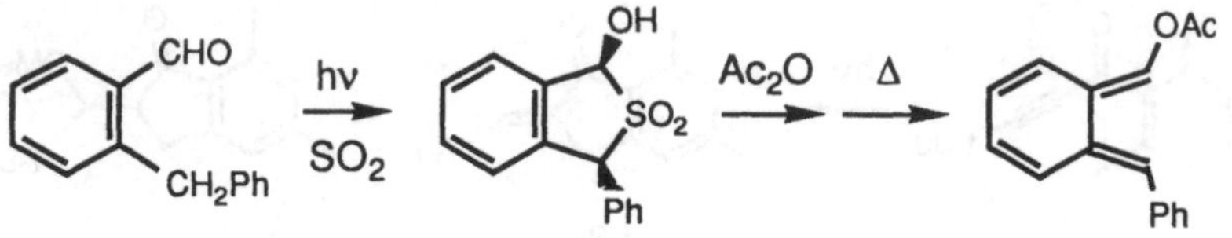

Cyclobutanol formation has not been applied often synthetically, although Paquette recently achieved a stereoselective trans cyclization in the synthesis of punctatin A.[237] As mentioned above, cyclization of 1,4-biradicals has some specific structural requirements. Dicarbonyls are one class that cyclize in high yield. β-Ketoamides have been shown to cyclize to β-lactams, although there is a competing reaction that is highly solvent dependent.[238] The reaction works with a variety of derivatives, several of which are shown below.[239] Shim has pointed out that restricted rotation about the nitrogen can influence triplet reactivity.[240] Toda reported that chiral crystals of the diisopropylamide (X = O) afford optically active lactams in 93% ee.[241]

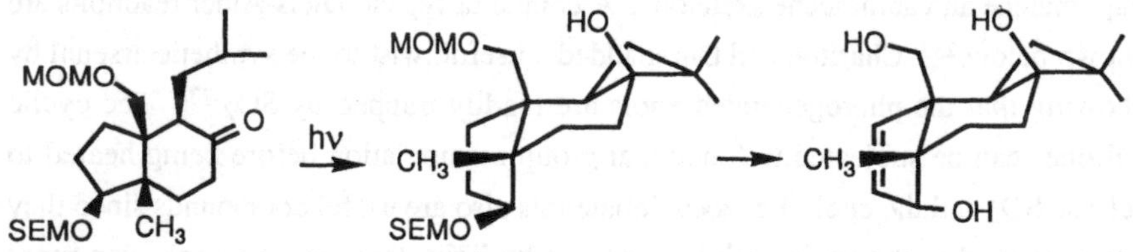

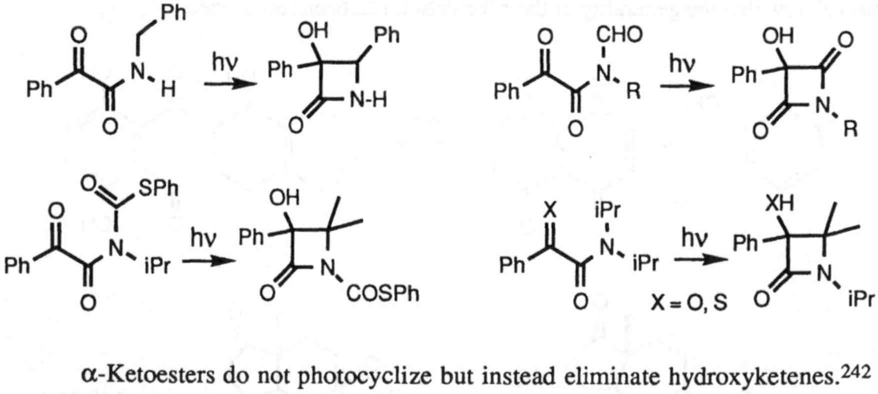

α-Ketoesters do not photocyclize but instead eliminate hydroxyketenes.[242] This process has been exploited in two ways. The ketene has been trapped by imines to generate β-lactams.[243] If one makes the ketoester from a primary or secondary alcohol, the photoelimination results in an oxidation at that site. This reaction offers the potential for selective oxidations in sensitive molecules.[244]

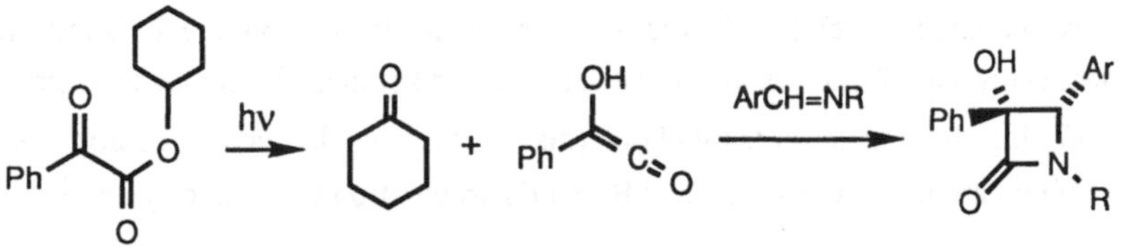

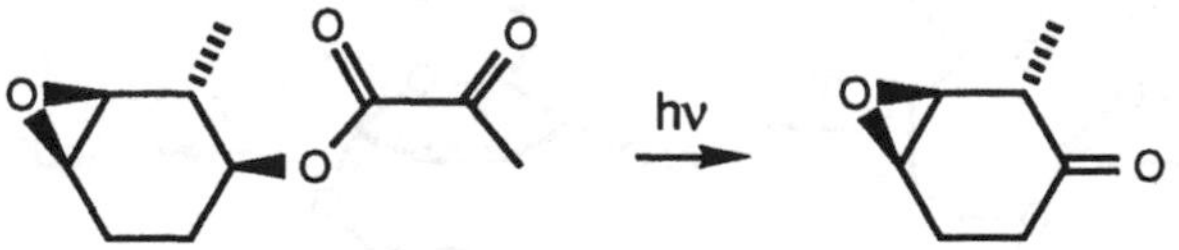

α-Alkoxy ketones photocyclize to oxetanols in good yield.[245] They also cleave, especially in aliphatic ketones, resulting in net oxidation of the alkoxy group. Descotes has exploited this process to both reduce and oxidize the anomeric carbon of cyclic acetals and sugars.[246] It is noteworthy that γ-hydrogen abstraction competes so well with type I cleavage. Binkley has oxidized the primary alcohol site of sugars this way, although epimeric oxetanols were byproducts.[247]

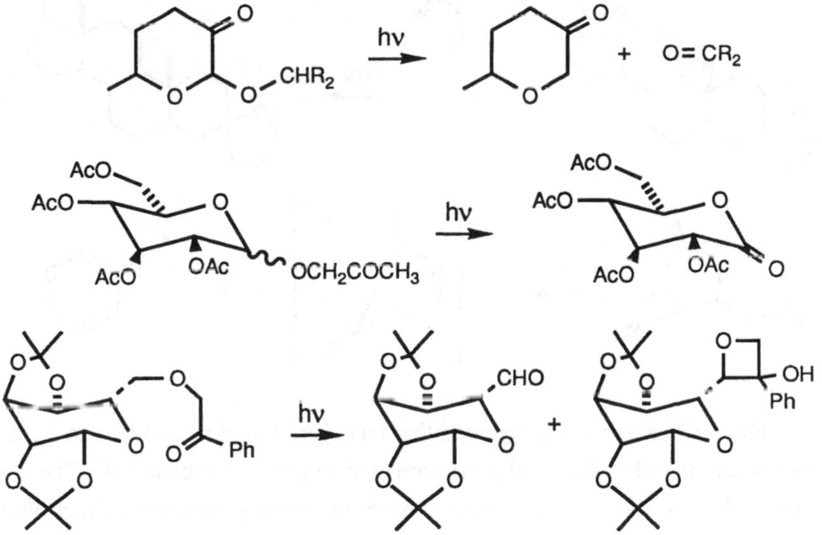

Simple type II elimination also has been put to use in several other ways. Glycosides made from γ-hydroxyketones undergo photoelimination to give O-vinyl glycosides, which can undergo a Claisen rearrangement in certain deoxy sugars.[248] Properly alkylated bicyclo[4.2.0]octan-2-ones formed by 2+2 photocycloaddition of cyclohexenones to alkenes cleave to δ,ε-unsaturated ketones.[249] Scharf has removed the C-17 side chain in steroids by photolysis.[250] Perhaps the most bodacious use of type II elimination was Goodman and Berson's preparation of meta-xylylene.[251]

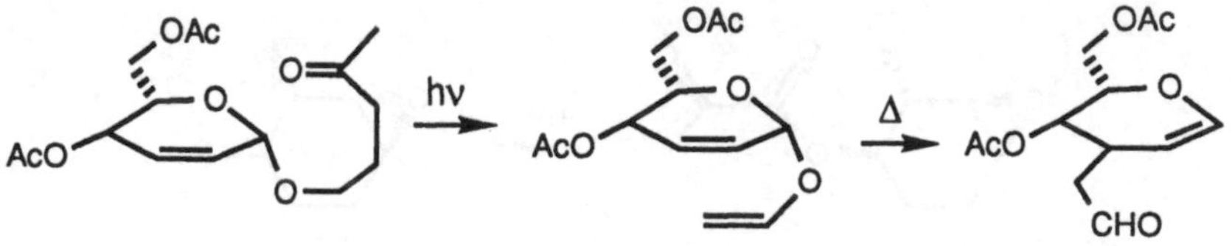

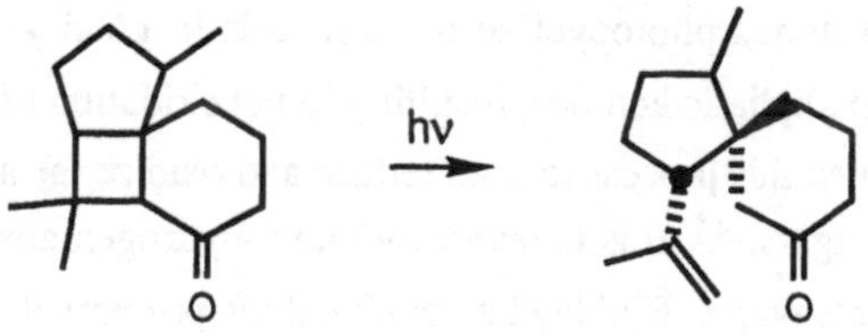

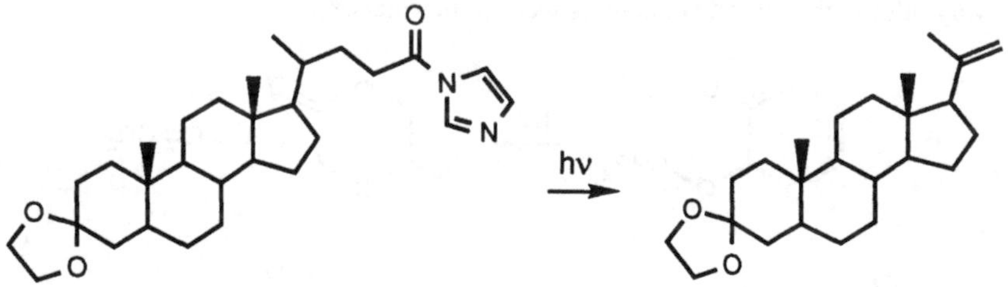

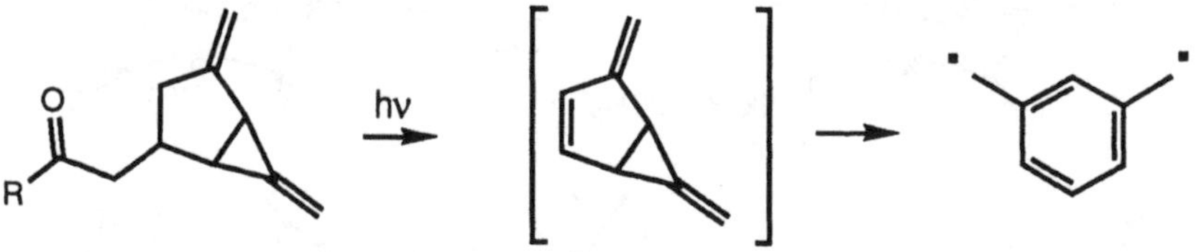

Finally, Scaiano has recognized that incorporation of o-tolyl ketone units in polymers can provide substantial protection against photodegradation.[252] The rapid and thermally reversible enolization of this chromophore provides excellent photoprotection, the efficiency of which is augmented by rapid energy transfer from higher energy chromophores in the polymers.

B. 1,6-Hydrogen Transfers (δ-Hydrogen Abstraction)

Before 1975 there had been only a handful of scattered examples of this reaction that produces 1,5-biradicals, mostly dealing with the photocyclization of various o-alkoxyphenyl ketones.[253,254] Pappas conducted a thorough study of o-alkoxyphenylpyruvate esters;[255] O'Connor reported the cyclization of one o-t-butyl-phenyl ketone;[256] and a Rochester contingent reported a unique example in the photocyclization of 1-benzoyl-8-benzylnaphthalene[29]. All of these cyclic compounds are characterized by an absence of γ C–H bonds, as are β-alkoxy ketones[257], γ,γ-dimethylvalerophenone[52], and 5,5-dimethyl-3-hexen-2-one.[258], all of which photocyclize. In fact, it was a common belief that excited ketones attack *only* γ C–H bonds if any are present, until we showed that δ-methoxyvalerophenone undergoes both γ- and δ-hydrogen abstraction with comparable rate constants.[259] The δ-methoxy group activates the δ C–H bonds and deactivates the γ C–H bonds by the inductive effects mentioned above. From these results it could be concluded that the intrinsic γ/δ selectivity in triplet ketones is 20:1, as it is in alkoxy radicals.[260]

The conformational restraints imposed by cyclic systems are nowhere better illustrated than in Paquette's synthesis of dodecahedrane, which relied on several δ-hydrogen abstractions by cyclopentanone units.[261] The polycyclic frame holds δ-hydrogens close to, and γ-hydrogens away from, the carbonyl group. Our group at Michigan State has studied δ-hydrogen abstraction in three types of ortho substituted phenyl ketones that impose a form of endocyclic restriction[262] on the photochemistry: o-*tert*-butyl ketones; o-alkoxy ketones; and α-(o-tolyl) ketones.[180] First we shall consider noncyclic systems.

1. *Acyclic systems.* The efficient photocyclization of γ-benzoylbutyralde-hyde[263] provides the only simple example other than δ-methoxyvalerophenone of an acyclic ketone that undergoes efficient δ-hydrogen abstraction in competition with γ-hydrogen abstraction, again because of a highly reactive δ C-H bond and a deactivated γ C–H bond. Houk has published several theoretical studies of intra-molecular hydrogen transfer that accurately reproduce the preference for 1,5- *vs* 1,6-hydrogen transfers.[59] It is noteworthy that much of the preference is entropic, as

expected, but that there is little enthalpy difference between the two modes. A chair-like transition state was originally suggested for γ-hydrogen transfers.[52] The calculations predict a stronger preference for a linear C—H—O arrangement (θ = 180°), which is more easily accommodated in a 7-atom transition state than in a 6-atom one. Thus relative enthalpies of activation reflect both ring size strain and nonlinear orientation.in the transition state.

Since our early studies indicated that 1,5-biradicals often do not cyclize very efficiently,[52] we scrutinized the photocyclization of β-ethoxypropiophenone in order to better characterize the behavior of acyclic 1,5-biradicals.[264] Its triplet abstracts a δ-hydrogen relatively slowly, k = 2 x 10^7 s^{-1}; the resulting biradical forms the two diastereomeric oxacyclopentanols. The Z/E product ratio depends strongly on solvent, as expected with hydroxybiradicals; but polar solvents do not maximize quantum yields as they do for 1,4-biradicals. Study of the α-deuterated ketone revealed *higher* cyclization efficiency and migration of deuterium to the δ-carbon. The 1,5-biradical efficiently disproportionates to form the enol of starting ketone, which rapidly tautomerizes; hydrogen bonding by the hydroxy group to solvent suppresses disproportionation to ketone but cannot suppress enolization . The results indicate a primary isotope effect of 3 on enol formation. Type II 1,4-biradicals presumably are too short to achieve the orientation required for this form of hydrogen transfer. Scheme 6 illustrates how the biradicals partition themselves in different solvents.

Scheme 6

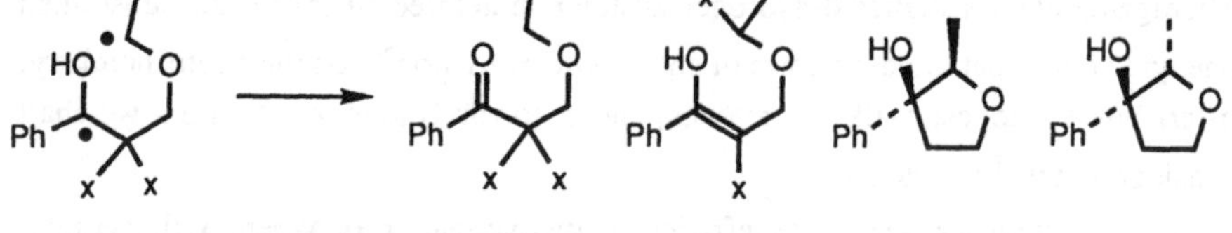

x = H	benzene	16%	35%	40%	9%
	t-BuOH	0	71%	13%	16%

Descotes has applied this reaction to modify the structure of various sugars.[265] When 4-hydroxy-4-methyl-2-pentanone is the alcohol part of an acetal group, irradiation causes hydrogen abstraction only at the acetal δ–hydrogen and not at the γ-methyls. This selectivity is produced by the strong deactivation of the γ C–H bonds by the β-oxygen and by the intrinsic reactivity of an acetal C–H bond. A simple pheremone synthesis has utilized this photocyclization.[266] Likewise, glycosides made from 4-hydroxy-2-butanone undergo photocyclization at the anomeric carbon, often with retention of configuration at that carbon.

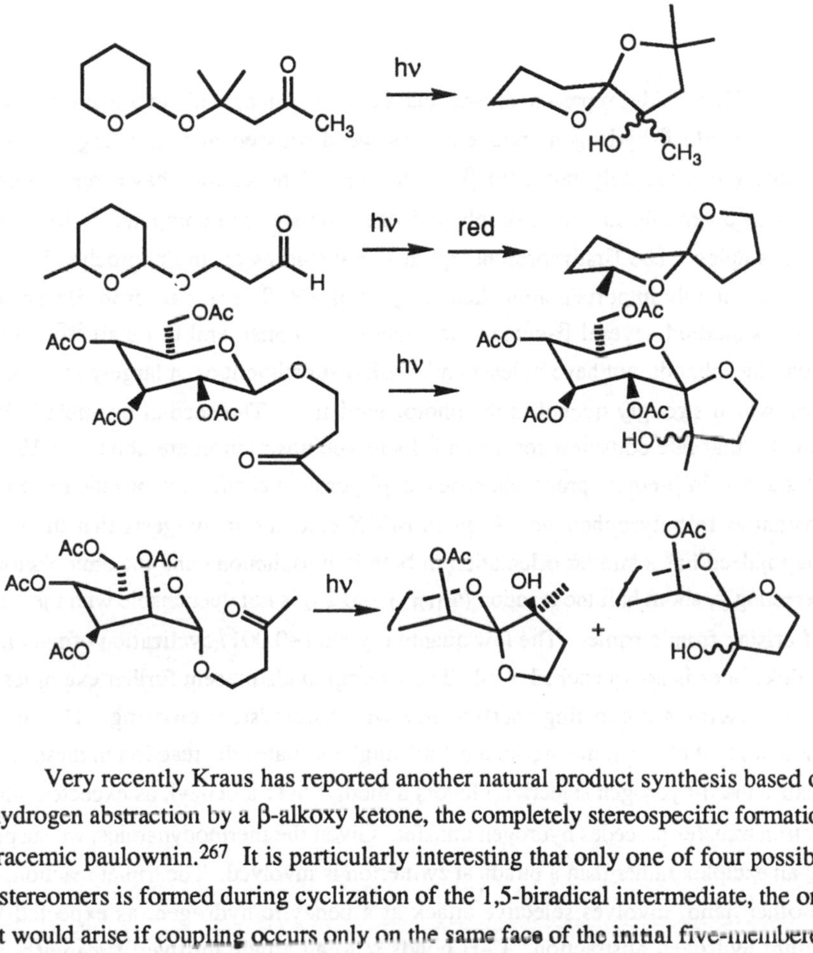

Very recently Kraus has reported another natural product synthesis based on δ-hydrogen abstraction by a β-alkoxy ketone, the completely stereospecific formation of racemic paulownin.[267] It is particularly interesting that only one of four possible diastereomers is formed during cyclization of the 1,5-biradical intermediate, the one that would arise if coupling occurs only on the same face of the initial five-membered ring as hydrogen abstraction took place and with the substituents at the δ-radical site maintaining the same relative configuration they had before hydrogen abstraction. There are no other examples of such photocyclizations of substituted cyclopentanones with which to compare this selectivity.

β-Ketoamides represent another class of compounds with no γ-hydrogens that undergo facile δ-hydrogen transfers. As we discussed above, charge transfer quenching is extremely rapid for β-ketoamines. The amides have much higher oxidation potentials such that simple hydrogen transfer can compete. Below are a few examples. The first represents quite a few studies on the photochemistry of pharmaceutically important aminoketone systems .[268] The rest are from Hasegawa, who has studied several β-ketoamide systems in considerable detail.[269] Those ketoamides that do not have at least one α-alkyl substituent exist largely in the enol form, which strongly quenches the photochemistry . The second example below indicates that rate constants for triplet δ-hydrogen abstraction are about 2×10^7 s^{-1} (the same as in β-ethoxypropiophenone), if γ-hydrogen abstraction has the same rate constant as in butyrophenone. A quick MMX calculation suggests that the ethyl group indeed has a similar orientation in both butyrophenone and the amidoketone. Interestingly, about half the reaction ($\Phi_{TOTAL} \sim 0.25$) is not quenchable with the other half arising from a triplet   The low quantum yield (~0.001) cyclization of naphthyl amidoketones is not quenched at all. These compounds present further examples of singlet reactions competing inefficiently with intersystem crossing. Hasegawa concludes that electron transfer to the n,π* singlet initiates the reaction in these cases because the δ-hydrogen selectivity favors a methyl over a benzyl, as expected when electron transfer precedes hydrogen transfer. Given the thermodynamics, we suspect that an exciplex rather than a biradical zwitterion is involved. The triplet reaction, on the other hand, involves selective attack at a benzylic hydrogen, as expected for straight hydrogen abstraction. C–H bonds α to an amide nitrogen apparently are activated towards hydrogen abstraction about as much as those α to an oxygen, although the planar amide group undoubtedly affects conformational preferences.

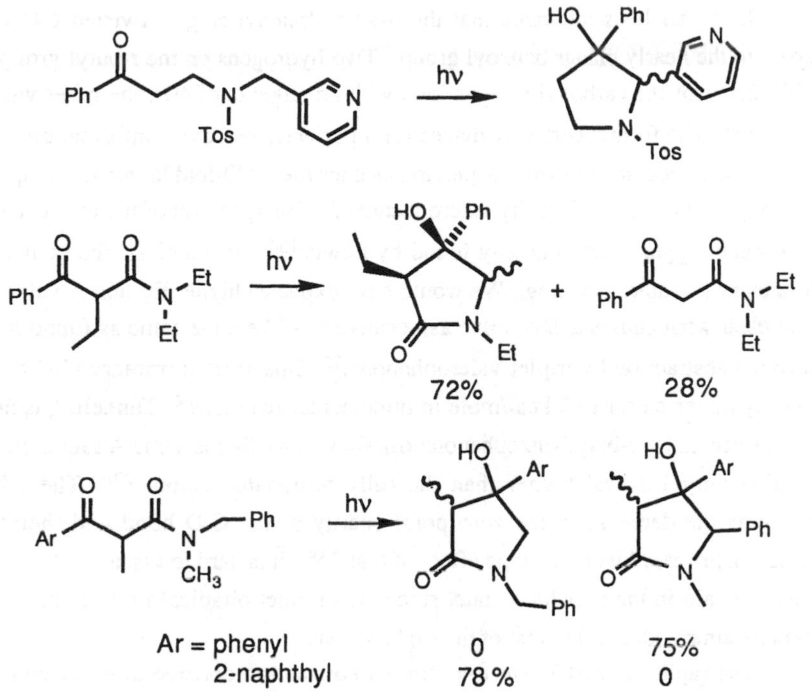

Ar = phenyl
2-naphthyl

| | 0 | 75% |
| | 78 % | 0 |

2. *Ortho-tert-butylphenyl ketones.* Ortho-*tert*-butylbenzophenone (OTBBP) is an unusually well behaved compound, undergoing clean photocyclization not only in solution but also in the crystalline state and in glasses at 77° K.[30] Its n,π* triplet undergoes δ-hydrogen abstraction with a rate constant $>10^9$ s[-1] at room temperature, $\sim 10^4$ at 77° K[270] . The triplet decay shows an E_a value of 2.3 kcal/mole, log A = 10.6. A 1,5-biradical was observed by flash kinetics; its lifetime depends on solvent basicity, varying from 43 ns in methanol to ≤4 ns in toluene. It shows a parallel chemical partitioning, proceeding to the indanol product in quantum yields of 1 and 0.04, respectively. Both oxygen and nitroxide radicals double the quantum yield in benzene, characteristic paramagnetic interception of triplet biradicals.

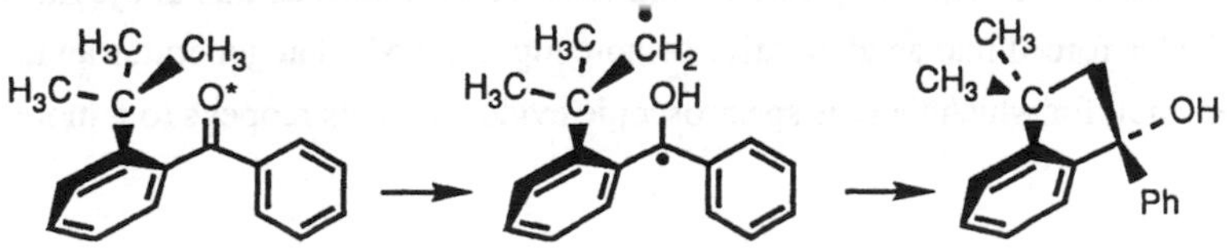

X-ray analysis indicates that the o-*t*-butylphenyl ring is twisted 69° with respect to the nearly planar benzoyl group. Two hydrogens on the *t*-butyl group lie within 2.7 Å of the carbonyl oxygen, one with an angle ω of 40°, the other with ω nearly 90°. The former certainly makes for a perfectly reactive conformation. The rapid kinetics seem to confirm this picture, as does the 5000-fold larger k_H compared to the acyclic analog γ,γ-dimethylvalerophenone.[52] Interposition of the benzene ring has a much bigger effect than any found by Lewis.[134] Nonetheless, the activation parameters are rather puzzling. We would have expected higher E_a and A values. It is not clear what causes a ΔS† value as negative as -12 eu, the same as found for γ-hydrogen abstraction by triplet valerophenone.[22] Unactivated primary C-H bonds show E_a values closer to 5 kcal/mole in bimolecular reactions.[4] Tunneling is not a factor, since 2,5-di-*t*-butylbenzophenone-d$_{18}$ shows exactly the same A factor and an E_a value only 1.0 kcal higher than the fully protonated ketone.[270] The 1 kcal represents the decrease in the zero point energy of the C-D bond and therefore provides a primary kinetic isotope effect of 6 at 25°. It is hard to visualize any relief of steric strain in the transition state, since the product biradical presumably has a geometry almost identical to that of the triplet ketone.

The rapid but inefficient δ-hydrogen abstraction corroborates Bergmark's explanation for the photostability of some o-*t*-butyl ketones.[271] However, the situation is much more complicated with o-*t*-butylphenyl alkyl ketones. 2,4,6-Tri-*t*-butylacetophenone photocyclizes to an indanol in low quantum efficiency.[272] However, 2,4-di-*t*-butylacetophenone provides only 10% indanol, giving an internal redox product in low quantum yield as the major product.[273] The triplet reacts in 26 nsec, more slowly than OTBBP because of its π,π* lowest triplet.[274] However, no known radiationless decay would compete with such rapid hydrogen abstraction. Tri-*t*-butylacetophenone resembles OTBBP in having a very twisted *t*-butylphenyl ring both in the triplet and in the biradical. Coupling to indanol is facile from this geometry. We have suggested that the 2,4-di-*t*-butylphenyl ring is more conjugated with the carbonyl in the triplet and with the benzyl radical center in the biradical. Since indanol formation requires a bond twist that diminishes this conjugation, the biradical is forced into an alternative ortho-coupling mode that generates an unstable spirotrienol, for which there is spectroscopic evidence. This reopens to a more stable bis-tertiary biradical that then can disproportionate to the observed unsaturated alcohol. Pericyclic rearrangements of the spirotrienol could also lead to products.

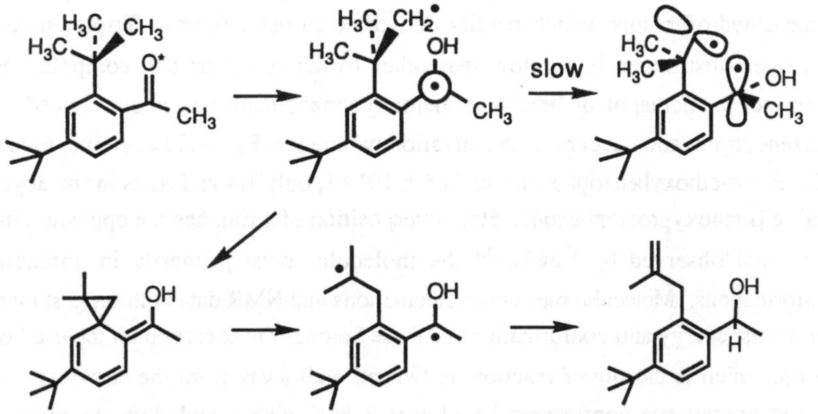

The *t*-butylanthraquinones are another interesting group that can undergo triplet state δ-hydrogen abstraction.[275] The 1,5-biradical undergoes both cyclization and the redox/rearrangement first seen in the acetophenones. The photochemistry shifts from hydrogen abstraction to benzene ring isomerization as additional *t*-butyl substitution forces one benzene ring out of planarity.

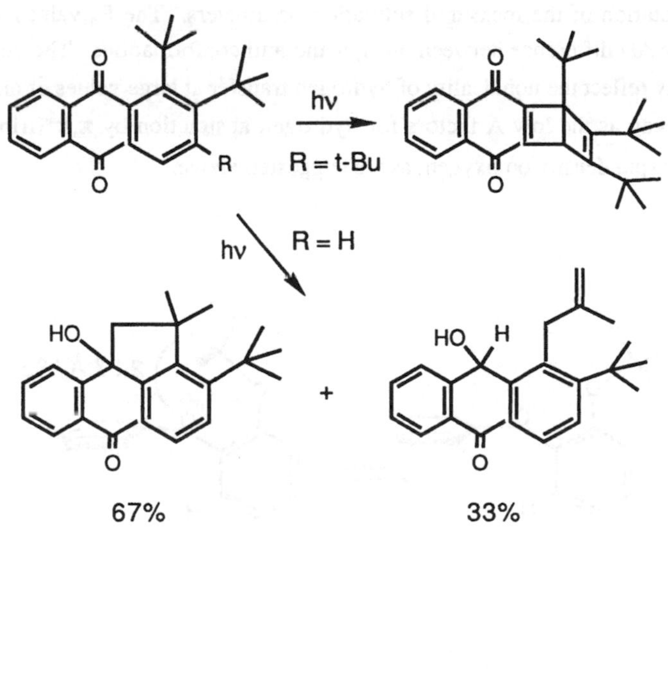

67%   33%

3. *Ortho-alkoxyphenyl ketones.* We have also examined more fully the photocyclization of o-alkoxyphenyl ketones.[276,277] As indicated by early work,[253] the triplet states of o-alkoxybenzophenones react in high quantum efficiency to form benzodihydrofuranols, which readily dehydrate to benzofurans. Triplet state δ-hydrogen abstraction is so slow that other triplet reactions can compete. For example, in methanol or hexane o-methoxybenzophenone is photoreduced. In benzene, cyclization occurs, with activation parameters E_a = 4.2 kcal/mole, log A = 9.2. For o-ethoxybenzophenone k_H is 5 x 10^6 s^{-1}, only 1/4 as fast as in the acyclic analog β-ethoxypropiophenone. Here interposition of a ring has the opposite effect from that observed by Lewis;[134] the molecules exist primarily in unreactive conformations. Molecular mechanics calculations and NMR data both suggest that in the lowest energy anti conformations of these ketones the α-carbon on the o-alkoxy group, which is the site of reaction, is twisted well away from the carbonyl. The lowest energy syn conformers lie almost 2 kcal higher and thus are only 5% populated at equilibrium, which certainly is attained in the microsecond lifetimes of the triplets. This energy difference is minimized by significant twisting of the o-alkoxyphenyl ring out of conjugation with the acyl group. The syn conformers can be termed reactive in terms of calculated d values of 2.6-2.9 Å. However, the target hydrogen atoms define angles ω of 66-85°, hardly "ideal". This observation suggests an interpretation of the measured activation parameters. The E_a values appear to contain the ΔH difference between the syn and anti conformations. The very low A factors may reflect the nonideality of hydrogen transfer at large values of ω, in much the same way as the low A factors for hydrogen abstraction by π,π* triplets may reflect low spin density on oxygen, as we suggested above.

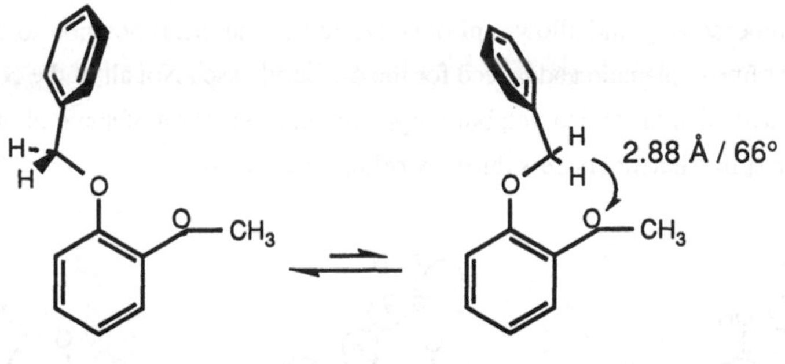

In 2,6-diacylphenyl ethers, there is no syn-anti difference possible and k_H values are ten times larger than in the monoketones. This observation certainly supports our syn/anti argument for the monoketones, but a direct comparison cannot be made. The C–O bond in the alkoxy group of the diketones is twisted perpendicular to the central benzene ring, a quite different geometry from that in the monoketones. Nonetheless, similar d and ω values are calculated, although Δ values are much smaller in the diketones.

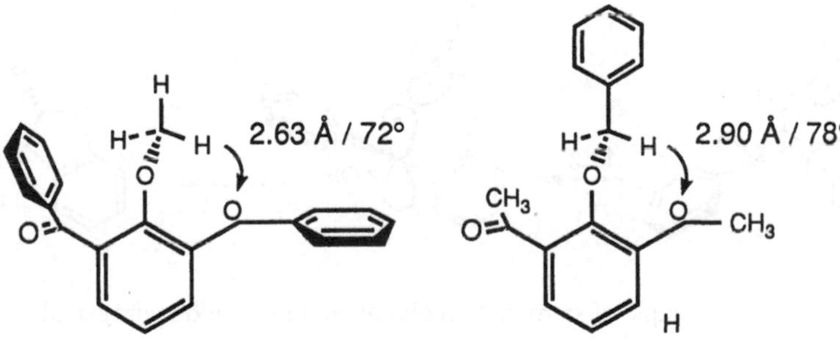

The situation gets even uglier in the o-alkoxyphenyl alkyl ketones. The triplets are subject to the same conformational effects as the benzophenones but are even less reactive with π,π^* triplets. O-methoxyacetophenone is essentially inert; o-benzyloxyacetophenone has a k_H value of only 2×10^6 s^{-1}; o-benzyloxyvalerophenone undergoes only type II elimination. Moreover, cyclization quantum yields are quite low even with added Lewis bases. The 1,5-biradicals, like those formed from the o-t-butylacetophenone, have difficulty cyclizing and form other products such as o-benzoylacetophenone from o-benzyloxyacetophenone.[278] We have suggested that restricted rotation around the benzylic radical site impedes coupling to a five-membered ring and allows spirocyclization at the ortho position to compete. This is the same explanation advanced for the o-*t*-butyl case. Not all of the competing products have been identified yet, but it appears that oxidation of some photoisomer such as the spirotrienenol occurs during workup or analysis.

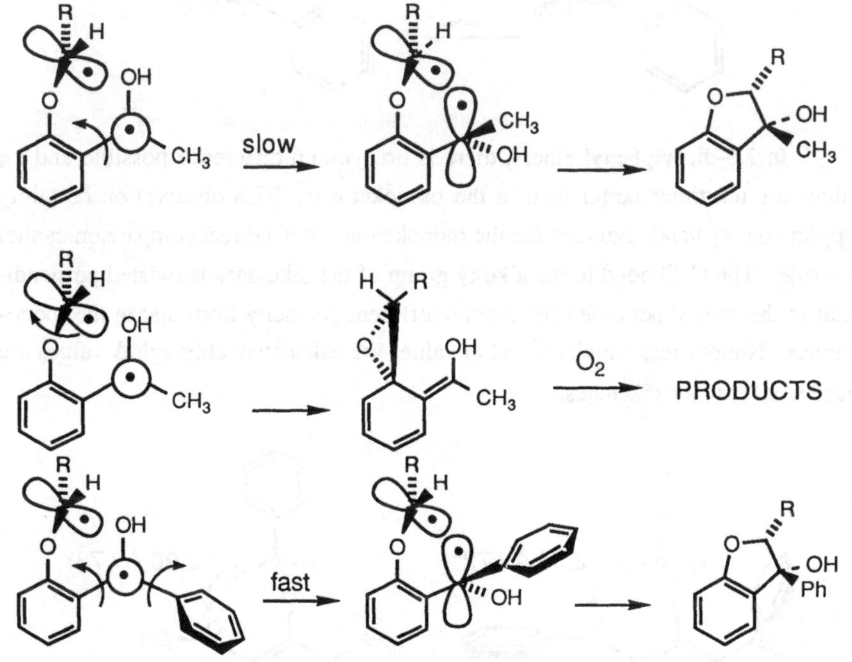

An earlier example of spiroenol involvement, in which δ-hydrogen abstraction competes with α-cleavage of a phenoxy radical, is shown below.[279] Rearranged alcohol was isolated from two separate reactants. o-Allyloxyacetophenone is another example of a molecule that primarily cleaves and cyclizes only in low yield.[254]

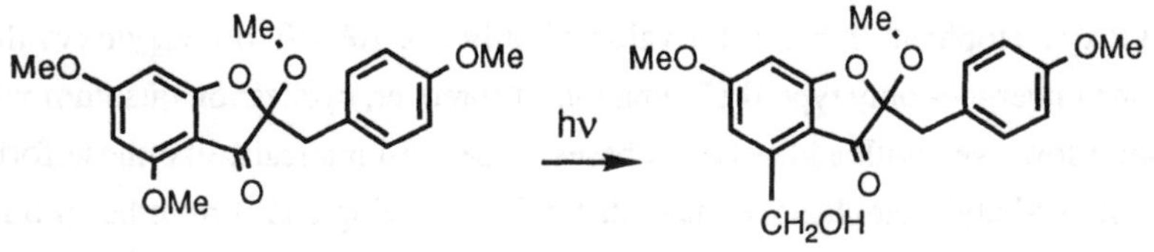

Another case of biradical disproportionation is provided by the photochemistry of o-isopropoxybenzophenone, which forms a benzofuranol and a 1,3-dioxane in a 2:1 ratio.[253] The original paper did not offer a mechanism for formation of the latter. We have determined that biradical disproportionation generates a hydroxy enol ether, which undergoes acid catalyzed internal addition to form the 1,3-dioxane.[277] In this and the above case, decreased coupling of the *bis*-tertiary biradicals cannot be blamed on restricted rotations; the normal proclivity of tertiary radical centers to disproportionate rather than couple must be involved.

The 1,5-biradicals formed from the o-alkoxy ketones all have very short lifetimes, as is typical of biradicals that have an oxygen atom as part of the skeleton.[164] They are undetectable by nanosecond flash spectroscopy, although a weak 13 nsec transient was observed for o-benzyloxybenzophenone in the presence of 0.1 M diene and 0.5 M pyridine.[276] Moreover, no opening of the cyclopropane ring occurs in the photocyclization of o-(cyclopropylmethoxy)benzophenone.[277] With a cyclopropylethyl rearrangement rate of 2×10^7 s^{-1},[280] the maximum lifetime of the biradical is 1 nsec. It is interesting that o-allyloxybenzophenone photocyclizes without any allylic rearrangement;[277] several β-allyloxy ketones[281] behave similarly. In contrast, γ-vinylbutyrophenone forms more cyclohexenol than allylcyclobutanol.[3,282] The behavior of the 1,5-biradicals may not be a simple preference for five *vs.* seven-membered rings, since α-vinylcycloalkanones open and recouple to larger cycloalkenones over a range of ring sizes.[283] We suspect that the short biradical lifetime may not permit the far end of the allyl-radical site to get within bonding distance of the hydroxy-radical site.

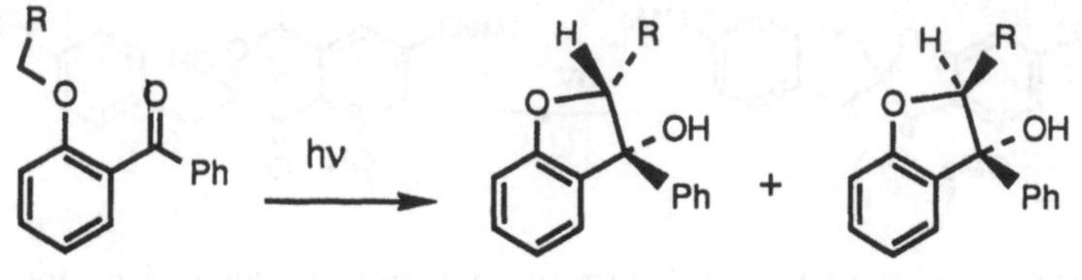

R = Me	92%	8%
R = Ph	88%	12%
R = c-C3H5	86%	14%
R = CH2CH=CH2	86%	14%

The 1,5-biradicals do show significant diastereoselectivity when cyclizing in hydrocarbon solvents, as shown above. However, all of the Z/E ratios approach unity in Lewis base solvents such as alcohols or pyridine, as is the case for almost all hydroxybiradical cyclizations known.[284] It is difficult to determine to what extent this selectivity is determined by biradical conformational preferences set before coupling and by nonbonded interactions created during coupling. The scheme below suggests some of the likely conformational changes involved in this reaction. We presume that the O-benzyl bond becomes perpendicular to the central benzene ring in the biradical and that some of the Z stereochemistry is set by the phenyl rings pointing away from the rest of the biradical. Kirmse has found similar diastereoselectivity in the coupling of 1,5-biradicals generated by internal hydrogen abstraction in ortho substituted phenyl carbenes.[285]

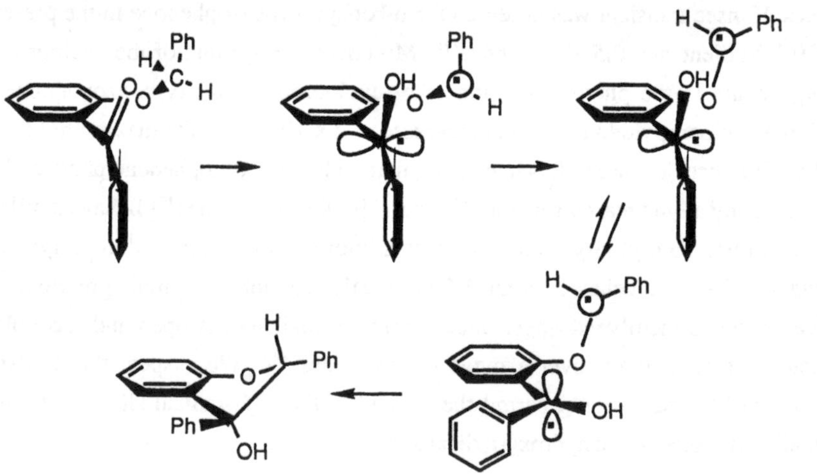

Descotes has irradiated glycosides made from o-hydroxyacetophenone to make the same anomeric ketals as described above.[286] Kraus recently pointed out a strategy using o-alkoxy ketone photocyclizations to synthesize aflatoxins.[287] In particular he showed that 2,6-dialkoxy substitution is required for reaction when there is a 4-methoxy group. Since 2,6-disubstituted phenyl ketones exist with their benzoyl chromophore highly twisted, Kraus made the reasonable suggestion that such twisting destabilizes the π,π^* triplet and allows the n,π^* state to react.

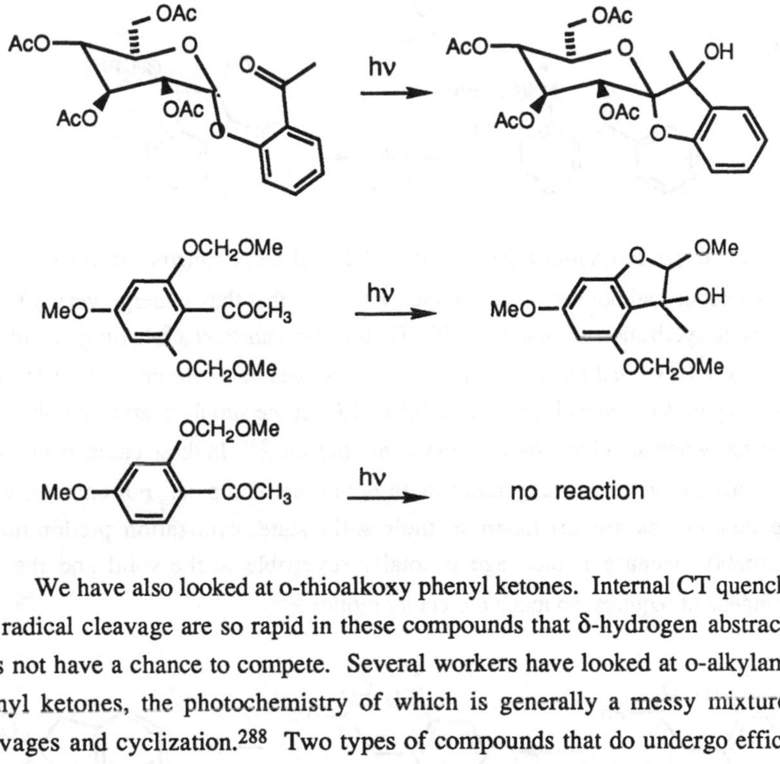

We have also looked at o-thioalkoxy phenyl ketones. Internal CT quenching and radical cleavage are so rapid in these compounds that δ-hydrogen abstraction does not have a chance to compete. Several workers have looked at o-alkylamino phenyl ketones, the photochemistry of which is generally a messy mixture of cleavages and cyclization.[288] Two types of compounds that do undergo efficient triplet δ-hydrogen abstraction and cyclization are o-benzoyl trialkylanilinium salts and the conjugate acid of o-benzoyl N,N-dibenzylaniline.[289] The positive nitrogens cannot act as electron donors, so hydrogen abstraction is the only triplet reaction. The rate constant for the trimethylanilinium ion is 1/30 that for the isoelectronic OTBBP, an interesting compromise between inductive activation of the triplet ketone and deactivation of the C–H bonds by the same positive nitrogen. The photo-

chemistry of the conjugate acids is complicated by the low basicity of the free
o-benzoylanilines, such that free aniline exists in equilibrium with its conjugate acid
and acts as both an internal filter and a quencher.

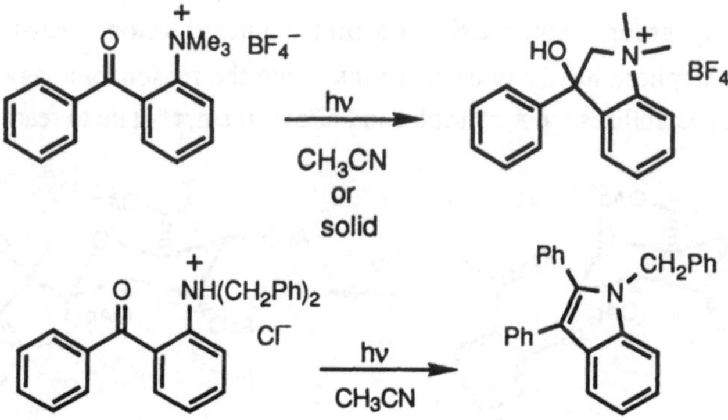

5. α-(o-Tolyl)acetophenones. The photochemistry of α-(o-tolyl)-
acetophenones had not been studied until we found that they undergo very efficient
triplet state cyclization to 2-indanols.[290] Their triplets abstract a δ-hydrogen with rate
constants $>10^8$ s^{-1} and the 1,5-biradical intermediates have lifetimes of 20-50 nsec.
Quantum yields of indanol range from 0.5 to 1.0 for the simple α-arylacetophenones
but plunge when additional α-substituents are present.[291] In these cases, α-cleavage
to radicals and a 1,3-rearrangement to an enol ether compete. Fortunately, when
these compounds are irradiated in their solid state, cyclization predominates,
presumably because α-cleavage is totally revertible in the solid and the 1.3-
rearrangement requires too much molecular motion.[292]

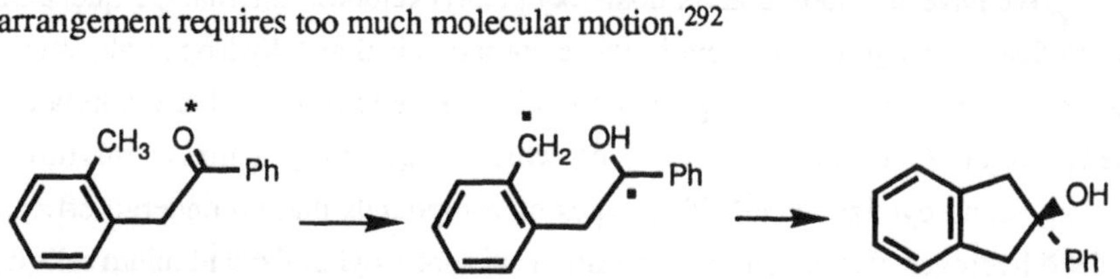

        Extra α-substitution affects both triplet reactivity and the efficiency of
biradical cyclization. For example, α-(o-tolyl)propiophenone undergoes primarily α-
cleavage; and α-(o-tolyl)valerophenone undergoes primarily type II elimination.[293]
For both ketones the rate constant for δ-hydrogen abstraction is only 1/50 what it is in
α-(o-tolyl)acetophenone itself. An extra α-substituent on α-mesitylacetophenone

lowers $k_{\delta\text{-H}}$ by a factor of only 3-4. These changes in rate constant reflect dramatic changes in the most stable conformations of the ketones. X-ray and NMR analysis as well as MMX calculations indicate that the various molecules have the geometries shown below. The conformer with an eclipsed α-aryl group is in a "nearly" reactive geometry that requires only ~1 kcal to rotate to a conformer with a C-H bond within 2.7 Å of the carbonyl at an angle ω of ~25°. The tolyl group can assume a slightly more stable but unreactive geometry by the same rotation, so tolylacetophenone is only 30% as reactive as α-mesitylacetophenone.

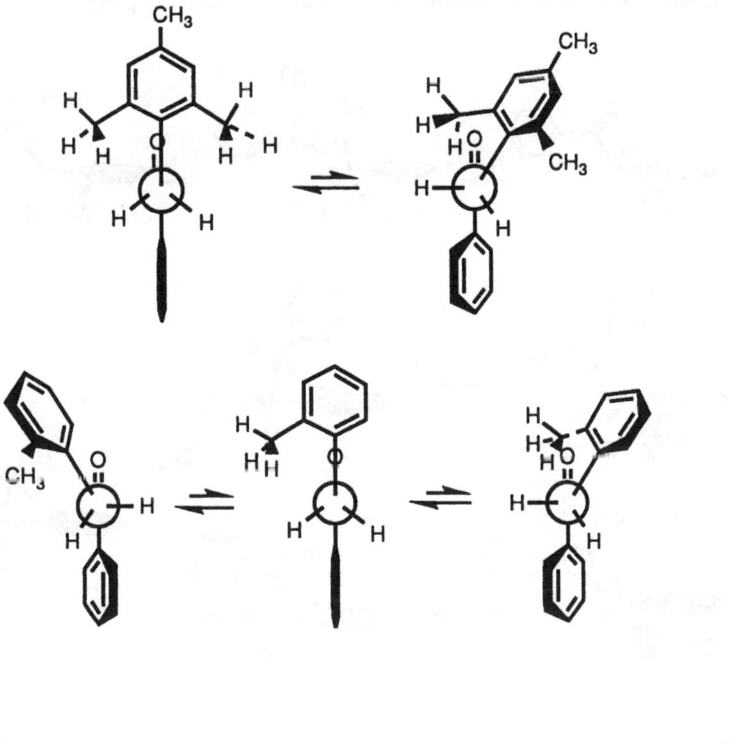

An additional α-substituent makes the α-aryl group a substituted cumene, such that the tolyl or mesityl group twists and rotates away from the carbonyl. The eclipsed geometries become too high in energy to be populated within the short triplet lifetimes and reaction can occur only from the highly twisted geometry, in which d = 2.7 Å but ω = 61-70°. The fact that reaction occurs in the solid proves that reaction from the same "nonideal" geometry is possible in solution. We suggest that the small drop in $k_{\delta-H}$ for the mesityl ketones represents a rare measure of the effect of ω on a rate constant. The much larger decrease for the tolyl ketones reflects the fact that the tolyl methyl group is mainly twisted away from the carbonyl.

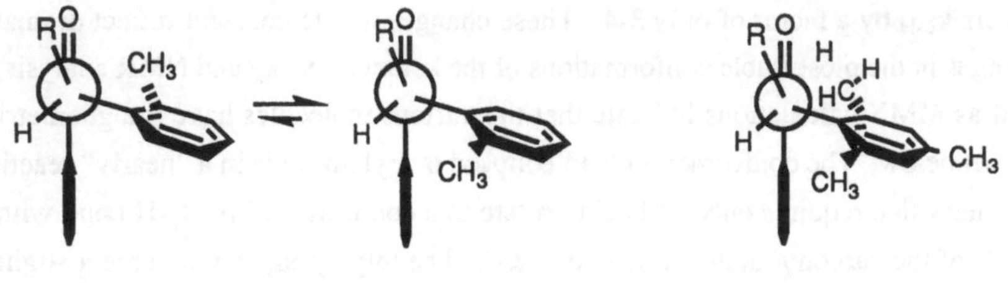

Guillet has shown that the photocyclization of α-(o-tolyl)acetophenone proceeds with exactly the same rate and quantum efficiency in poly(ethylenevinyl acetate) beads as in benzene.[294] This fact makes this reaction an excellent actinometer for studying photochemistry of and in polymers. It also reaffirms the conclusion from our studies that this molecule does not require much molecular motion after excitation to react.

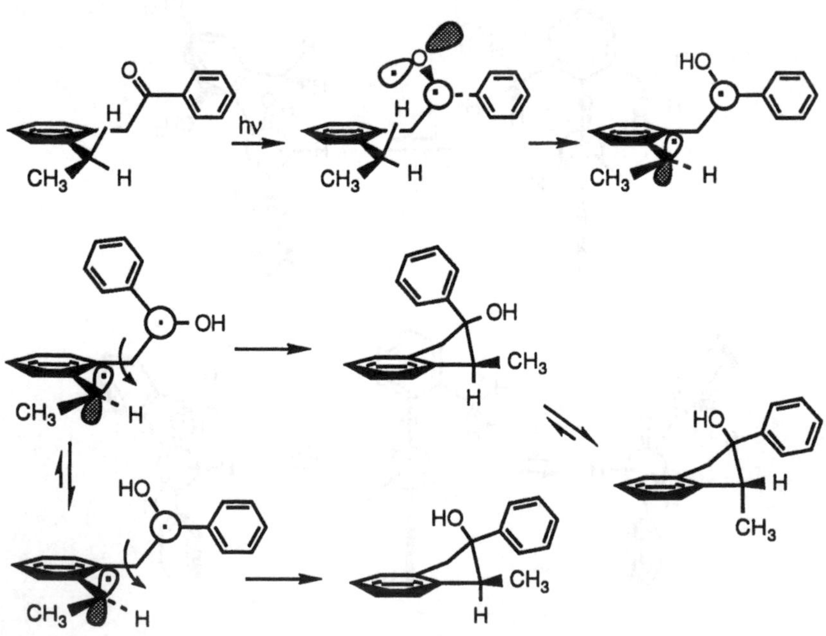

These photocyclizations show the highest diastereoselectivity yet recorded. α-(o-Ethylphenyl)acetophenone produces a 20:1 Z:E ratio of indanols.[295] Analysis of the possible biradical conformations indicates that in these particular biradicals the stereochemistry is set prior to cyclization, as shown below. Both the methyl and the phenyl groups are held predominantly away from the rest of the molecule so that they end up trans to each other when the two biradical ends rotate together.

These particular 1,5-biradical intermediates are intriguing in that they cyclize in high efficiency even when unsolvated, just the opposite of the 1,5-biradical from OTBBP. Apparently a relatively uncrowded five-membered ring is easier to form than the 7-atom transition state required for disproportionation at oxygen. However, α-(triisopropylphenyl)acetophenone forms both the indanol and kinetically stable Z and E enols in yields that depend strongly on solvent polarity, the enols being favored in dioxane.[296] Here is another example of a *bis*-tertiary biradical that prefers disproportionation at carbon over cyclization.

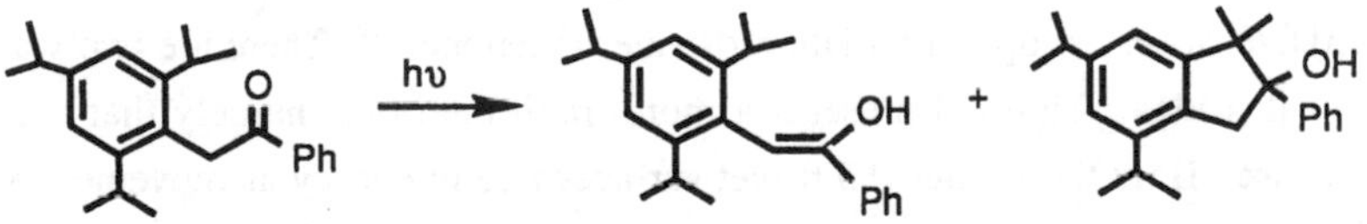

The potential for steric congestion in these ketones produces some strange behavior that could not be predicted unless the molecular geometry were first determined. For example 1,2-dimesitylethanone is locked into a conformation in which neither γ- nor δ-hydrogens are close enough to the oxygen for hydrogen abstraction to take place; only α-cleavage occurs.[66] Although no low temperature NMR line-broadening is detectable, rotations about the mesityl–carbonyl bond and around the α-carbon-carbonyl bond apparently are not fast enough to compete with the rapid radical cleavage that takes place when both developing radical centers are fully parallel to the benzene rings.

C. Biradical Lifetimes

Biradical lifetimes. The three classes of cyclic ketones that undergo δ-hydrogen abstraction form three distinct types of biradicals. Those from α-tolyl-acetophenones have 20-50 nsec lifetimes that depend only slightly on solvent; they cyclize in high quantum yields that are enhanced only slightly by Lewis bases. Those from o-*tert*-butyl benzophenones have ~4 nsec lifetimes that are increased 10-fold by Lewis base solvents. They cyclize in only 5% efficiency in nonpolar solvents but 100% in alcohols. Those from o-alkoxy ketones have lifetimes too short to be measured. Their cyclization efficiency varies; for benzophenones it is near unity; for acetophenones it is low but markedly enhanced by Lewis bases, as is the lifetime. In all cases there is a strong parallel between solvent effects on lifetimes and on product partitioning.

As we pointed out above, similar parallel solvent effects on lifetimes and product partitioning were noted for 1,4-biradicals. Since they were not *exactly* parallel, they were once dismissed as coincidental; but now that there are four types of biradicals that show parallels, reconsideration seems appropriate, as Caldwell has recently concluded.[179] It has been shown that solvent effects on lifetimes are unique to hydroxybiradicals and are presumably due to hdyrogen bonding.[297] We have advanced an alternative explanation to Scaiano's, namely that isc and reaction of triplet biradicals are coupled.[180] Kaptein first presented the overall idea in explaining the CIDNP observed upon irradiation of valerophenone.[298] There are two ways to look at this idea. Figure 4 presents a more traditional one, namely that reaction induces isc. Both the singlet and triplet surfaces rise in energy as movement along the reaction coordinate develops electronic and steric strain, but the singlet surface is soon stabilized by the developing bond, such that the two surfaces cross at points that represent very low observed activation energies. We presume that isc at this point benefits from large spin-orbit interactions[172] as well as state degeneracy. The crossing occurs with only a minimal increase in energy, so activation energies are very low. The small diastereoselectivities developed during cyclization demand barriers to coupling of a few kcal/mole. Since crossing presumably occurs before the transition state is reached, there are two possible explanations for the apparent irreversibility of isc. Either momentum along the reaction coordinate carries the molecule across the small barrier or the molecule stays trapped in a conformational well until it reacts because the barriers to bond rotation are larger than that for coupling. Closs also has suggested that reaction and isc are coupled, with rates proportional to the percentage of singlet character in the biradical.[299] There may be little real difference between the two ideas.

If there is no separate isc, merely different amounts of singlet-triplet mixing at different biradical geometries, one needs to be able to calculate the extent of such mixing as a function of geometry in order to understand structural effects on biradical lifetimes. If isc occurs discretely, but at the early stages of reaction, solvent and structural effects can be understood as entropic. For example, the major reaction of several hydroxybiradicals is disproportionation back to ketone. In all such cases, Lewis bases suppress this reaction and increase the biradical lifetime proportionately. Since a major pathway for reaction and isc has been eliminated, the biradical has a lower probability of finding an isc path. Therefore isc is disfavored entropically and lifetimes lengthen without acquiring any activation energy. There remains a lot of basic theoretical understanding that must be developed in order to predict isc rates at surface crossings and as a function of spin-orbit mixing.

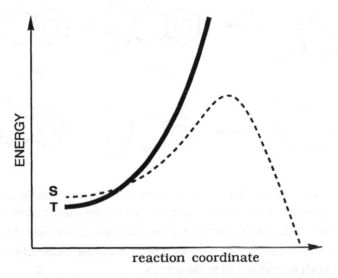

Figure 4. Intersystem crossing of a biradical coupled with reaction.

D. More Remote Hydrogen Abstraction.

There have been several examples of more remote hydrogen abstraction over the past two decades, ranging from ε-hydrogen abstraction competing with γ-hydrogen abstraction in cyclodecanone[136] to the cyclization of benzophenones with long alkyl tails attached para.[133] Only some of these have been studied sufficiently that rates of triplet state hydrogen abstraction can be correlated with structure.

We have looked at two examples of ε-hydrogen abstraction, both of which represent extensions of the α-(o-tolyl) ketone idea. The photocyclization of α-(o-benzyloxyphenyl)acetophenone proceeds in a quantum yield of only 0.05 pretty much independent of solvent.[300] The 3×10^7 s^{-1} triplet decay was assigned as the rate for ε-hydrogen abstraction because at the time there was no known physical decay that rapid. It is likely that CT quenching by the α-alkoxyphenyl group has a much lower rate constant, ~2×10^6 s^{-1}.[185] Thus it was concluded that the low product quantum yields are primarily due to very efficient disproportionation of the 1,6-biradical intermediates to enol of starting ketone.

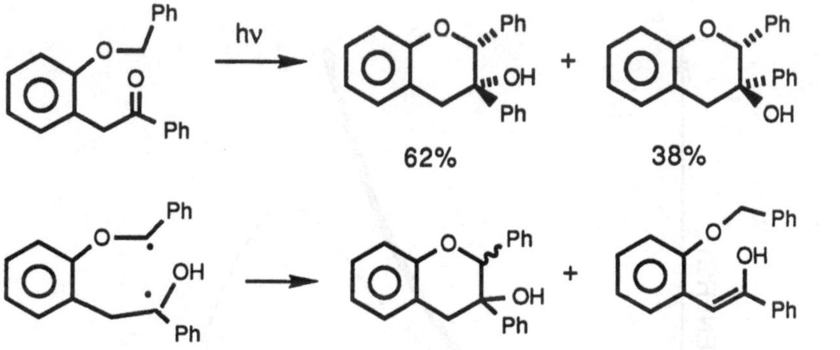

The rate constant for hydrogen abstraction is comparable to that for o-benzyloxybenzophenone and 1/15 that for α-(o-ethylphenyl)acetophenone. Given the known conformational preference for the α-aryl group to eclipse the carbonyl, that order of magnitude decrease in rate must reflect a *syn-anti* rotational preference as in the α-(o-tolyl) ketones as well as lost entropy.

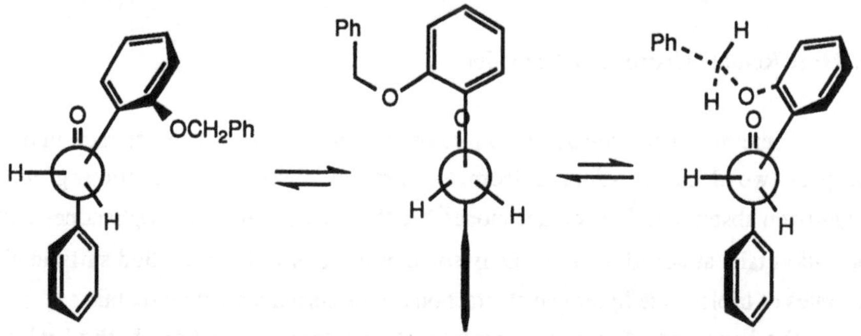

We have also studied the β-(o-tolyl)propiophenones, which photocyclize to 2-tetralols in good chemical yields but very low quantum yields.[301] Rate constants for ε-hydrogen abstraction are quite low, 10^5-10^6 s^{-1}, and must compete with rapid CT quenching by the β-aryl group. The added methylene group lowers k_H over three orders of magnitude from its value in the α-arylacetophenones. None of the lowest energy ground state conformations have a ε-hydrogen close to the carbonyl, so the molecule must distort into a higher energy geometry to react. Interestingly, methyl substitution at the α-carbon increases k_H in these compounds by destabilizing what is calculated to be the most stable unreactive conformation, thus increasing population of the geometries in which an ε-hydrogen is at least within 4 Å of the carbonyl.

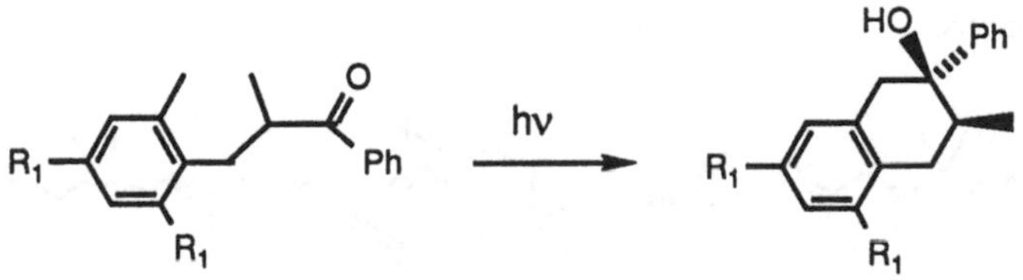

Neither polar solvents nor α-deuteration enhances the cyclization quantum yields of these ketones, so the 1,6-biradicals apparently do not disproportionate to either the keto or the enol form of starting ketone. Of course coupling to a six-membered ring must be a highly favored process in competition with dispropor-tionations that require alignment of C–H bonds as well as p-orbitals.

Probably the first example of ε-hydrogen abstraction, the photocyclization of some naphthoquinone methides reported by Barton,[302] also involved a β-aryl ketone. Wirz showed that the basic chromophore undergoes normal photoreductive dimerization and pointed out the likelihood of a 1,6-biradical in the Barton compound.[303] It is not clear why this intramolecular reaction requires acid.

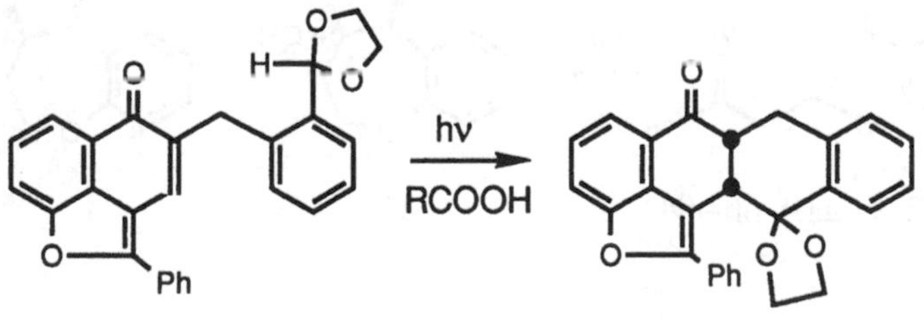

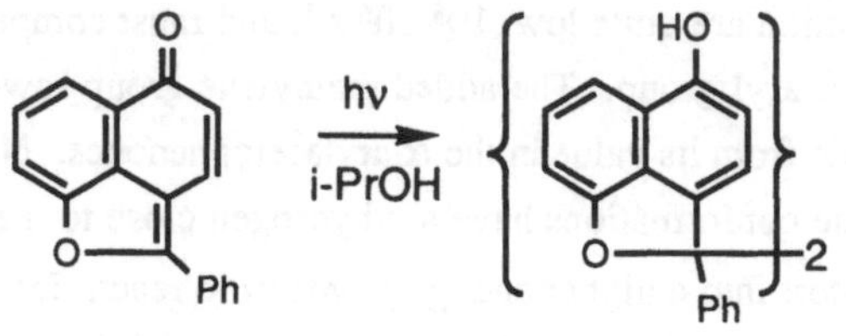

Carless studied several 5-allyloxy-5-methyl-2-hexanones that photocyclize in.45-60% yields to 2-allyl-3-hydroxy-tetrahydropyrans.[304] The 5,5-dimethyl substitution is necessary for ε-hydrogen abstraction to occur; otherwise only Norrish type II reactions occur. No kinetics were reported; but it seems certain that these are triplet reactions.

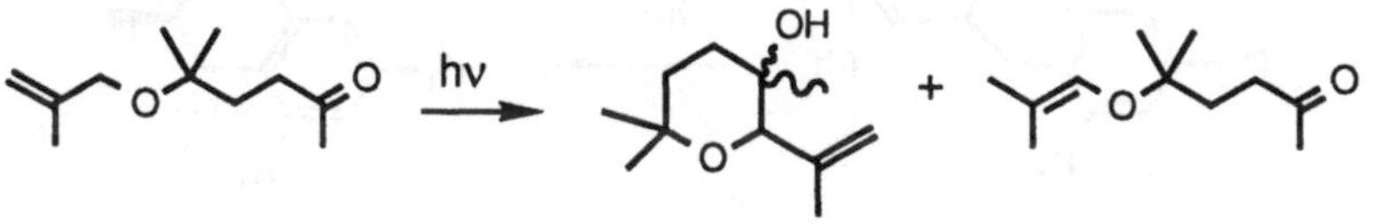

45% (1:1 Z/E)   trace

Carless also reported several examples of ζ-hydrogen abstraction by excited β-(o-alkylphenoxy)propiophenones.[305] Chemical yields were moderate (21-52%) and quantum yields quite low. Again no rate constants were measured but they must be very low since bimolecular photoreduction competes and causes the low yields; triplet reactivity seems certain. Cyclization of the 1,7-biradicals shows little diastereoselectivity but the Z/E ratios reported were unusual in being less than 1. No allylic rearrangements occurred and no disproportionation products were found for the o-isopropylphenoxy ketone, although the enols of starting ketone could not have been detected.

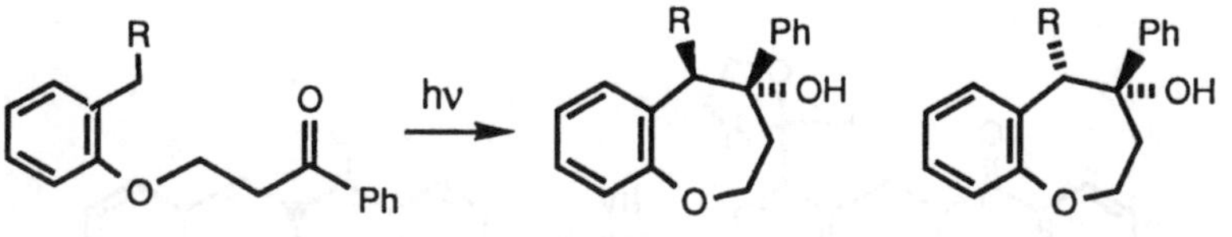

R = allyl, phenyl   2   1

A site specific ζ-hydrogen abstraction by a gibberellin oxo ester produces the corresponding lactone.[306] The authors calculate that the hydrogen abstracted lies only 2.5 Å from the carbonyl oxygen.

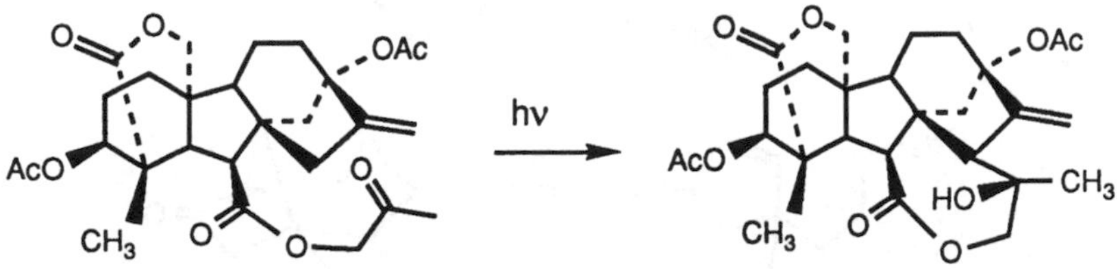

The hydrogen abstraction from the long alkyl tails of n-alkyl p-benzoyl-benzoate esters[133] led to the concept of biomimetic reactions, in this case oxidation of an otherwise unactivated methylene group.[307] Breslow has shown that considerable site selectivity occurs in intramolecular hydrogen abstraction from steroids.[308] Oxidation of long chain alkyl groups can be done with better site selectivity than in the original esters. Myristate esters attached to p-benzoylbenzoic acid via catechol or 1,2-cyclohexanediol undergo hydrogen abstraction and cyclization 85% at carbons 7-10.[309]

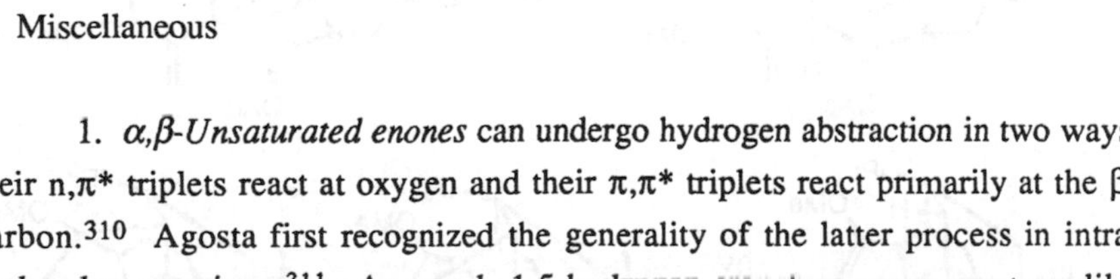

E. Miscellaneous

1. α,β-Unsaturated enones can undergo hydrogen abstraction in two ways: their n,π^* triplets react at oxygen and their π,π^* triplets react primarily at the β-carbon.[310] Agosta first recognized the generality of the latter process in intramolecular reactions.[311] As usual, 1,5-hydrogen transfers occur most readily, although 1,6-transfer also occurs.[312] The structures in the two first examples strongly favor reaction at the β-carbon. Therefore a more symmetric ketone was constructed. Deuterium labelling proved that only the β-carbon abstracts hydrogen.[313]

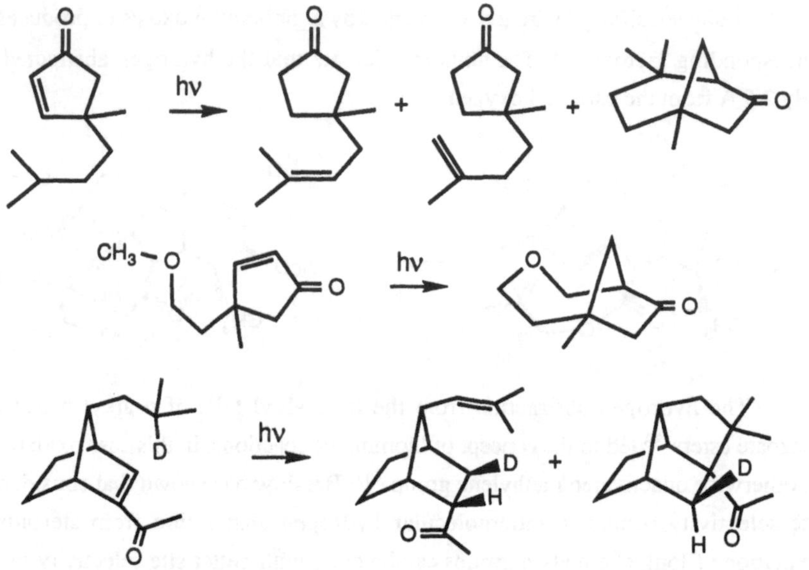

There are examples of polycyclic enones in which hydrogen is transferred at least formally to the α-carbon. In taxinine, the remote hydrogen is held close to both carbons.[314] Schuster has suggested[310] that the bridged steroidal enone[315] reacts at oxygen from its n,π* triplet to generate the enol of one product and at the β-carbon from its π,π* triplet to generate the normal product.

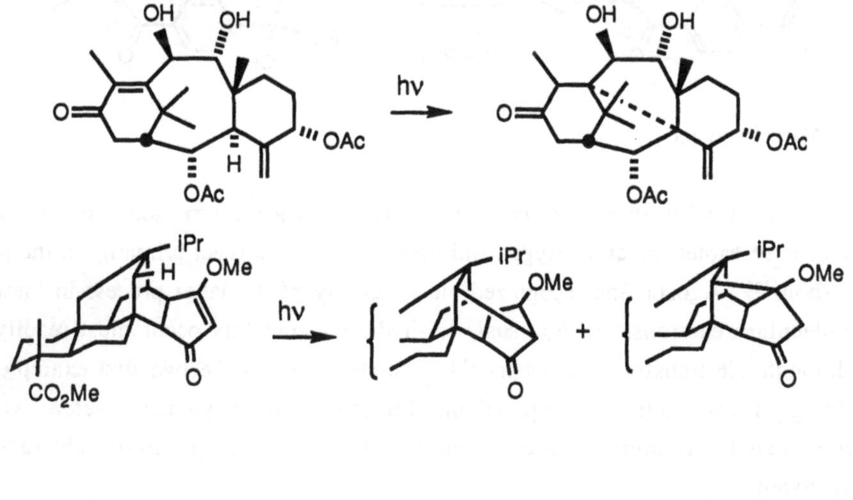

Mehta and Subrahmanyam have applied the photocyclization of o-toluyl-cyclopentene reported by Agosta and Smith[316] to demonstrate that the spiroindane skeleton of fredericamycin can be formed photochemically.[317]

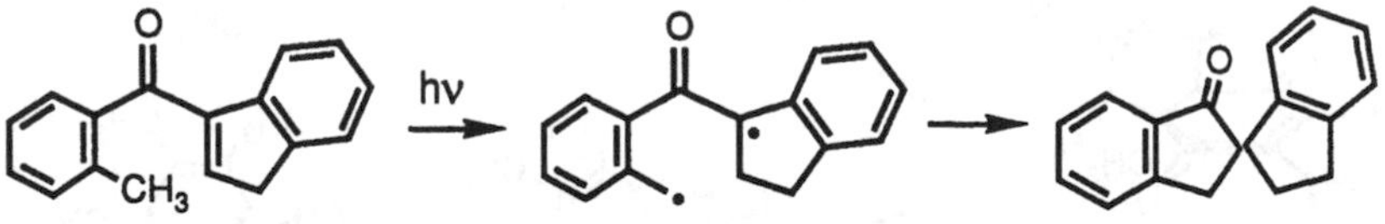

It should be stated that the nature of triplet enones continues to be less than fully understood. One feature that is well understood is the importance of the free rotor effect on triplet lifetime. Triplets of enones that can twist easily do not live long enough to undergo hydrogen abstraction.[318]

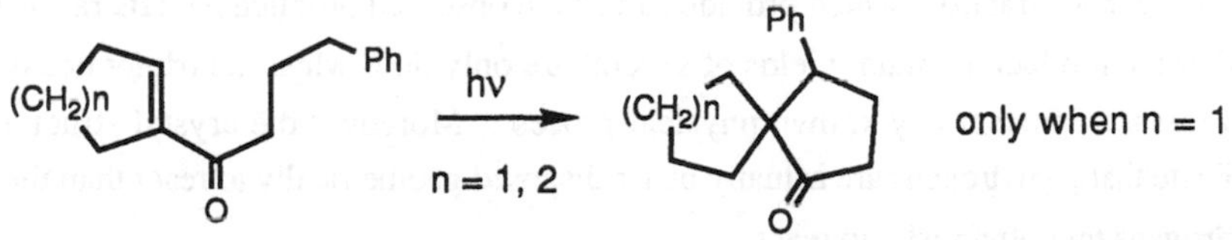

2. *1,4-Hydrogen transfers.* This geometry for hydrogen transfer is very rare, presumably because the transition state cannot normally be anywhere near linear. However, examples are known. Scheffer has carefully analyzed the photo-chemistry of various bicyclic enediones.[54] They undergo both singlet and triplet state hydrogen abstraction both in solution and in the solid. Those compounds that do both are state selective in that the singlet is responsible for β-hydrogen abstraction while the triplet undergoes only 1,5–hydrogen transfer.[319] The singlet β-hydrogen abstraction occurs from a nearly perfect geometry, with d = 2.5 Å, ω = 0°, and Δ = 85°. The triplet reaction involves a 1,5-transfer to the β-carbon of the enone with d = 2.8 Å. The lowest triplet is π,π*; so hydrogen abstraction by the carbonyl is retarded. In solution, the allylic biradicals close to several products ; but in the solid, only to those shown. These hydrogen transfers depend on the molecule adopting a twist conformation, whenever substitution disfavors this geometry, 2+2 cycloaddition of the double bonds occurs. Scheffer has also investigated various monoreduced reversions of these compounds, in which the enone carbons abstract allylic hydrogens from the cyclohexene unit.[54]

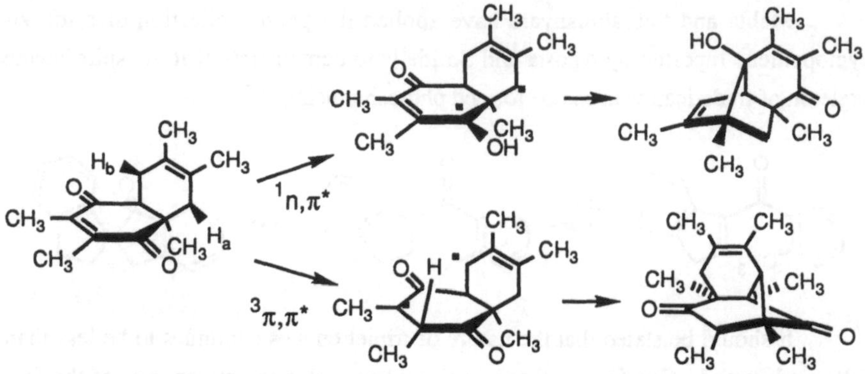

Scheffer has suggested that some of the α-cycloalkylacetophenones that he has studied may undergo totally revertible β-hydrogen abstraction in competition with γ-hydrogen abstraction, which provides all of the observed products.[54] His rationale is that the product quantum yields of several are only 50% when the triplet decay is too fast to represent any known physical process. Moreover the crystal structures indicate that β-hydrogens are actually better disposed geometrically to react than the γ-hydrogens that obviously do react.

α-Diketones without any highly reactive γ-hydrogens undergo a unique form of photoenolization in which 1-acylenols are formed.[320] The mechanism for this reaction is not clear; one possibility is a β-hydrogen abstraction that resembles a proton transfer followed by rapid 1,4-shift of hydrogen from one oxygen to the other.[48]

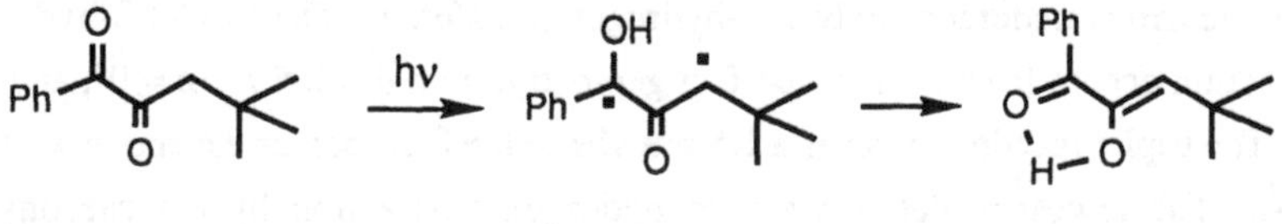

As we shall cover shortly, β-amino ketones undergo rapid internal CT that provides the products of β-hydrogen transfer. β-Carboxyamido ketones have been reported to form cyclopropanols[321]. Given the strong deactivating influence of a carbonyl group on direct hydrogen transfer as evidenced in Table 3, it is very unlikely that simple β-hydrogen abstraction explains this photoreaction. Slow CT from nitrogen to ketone may promote transfer of a proton α to the amide to the ketone oxygen. Another interesting facet of this report is the fact that the major product has the E rather than the Z configuration.

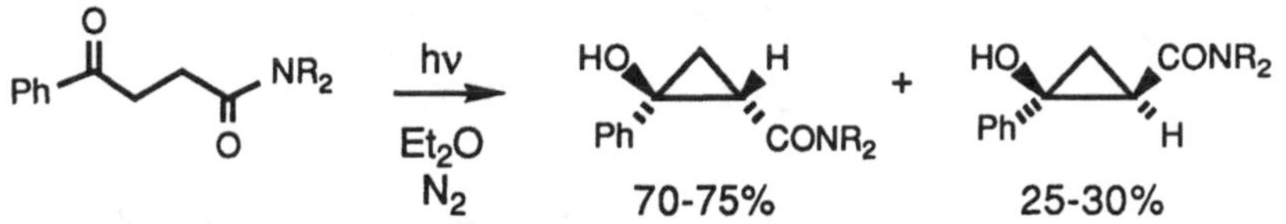

70-75%   25-30%

3. *N-Alkylimides.* Kanaoka first discovered that N-alkylimides undergo photoinduced intramolecular hydrogen abstractions when he observed the photocyclization of N-(o-tolyl)phthalimide.[322] He has conducted an extensive investigation of the photochemistry of such molecules, much of which is reviewed.[323] Unfortunately, not much mechanistic detail is available about these reactions. Phthalimides have n,π* lowest singlets and π,π* lowest triplets.[323] The photocyclization of the α-tolyl compound is quenchable by dienes, so it is triplet derived. The low quantum yields (< 1%) suggest slow reaction from an upper n,π* triplet or from an unreactive π,π* triplet.

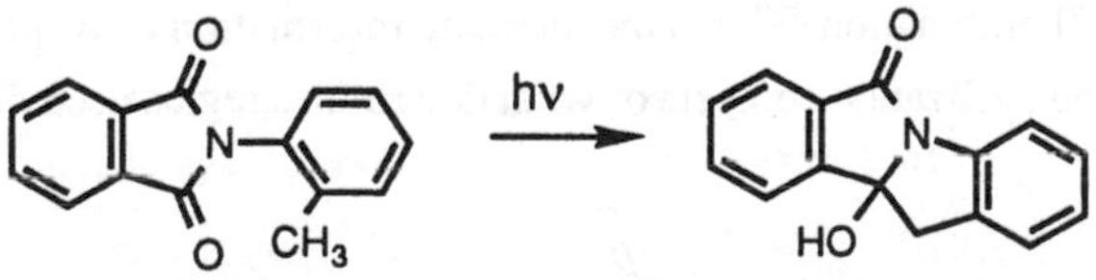

Straight n-alkylimides undergo both γ- and δ-hydrogen abstraction, with the latter preferred[324]. Thus the δ-methyl is half as reactive as the γ-methylene. A δ-methoxy group promotes 100% δ-hydrogen abstraction,[325] unlike the case for δ-methoxyvalerophenone, where the γ/δ ratio is 50:50.[52] Cyclization of the 1,4-biradicals gives a product that opens thermally to a benzazepinone lactam. Again quantum yields are very low (~0.01) but chemical yields are good. The reactions probably occur from excited singlets, since the lack of quenching by 0.08 M diene indicates a minimum triplet lifetime less than 1 nsec.[324] This value is too high for a π,π* triplet and would give higher quantum yields. N-cycloheptylphthalimide undergoes two consecutive γ-hydrogen abstractions, the second following opening of the hydroxylactam, to yield a novel tetracyclic product.[326]

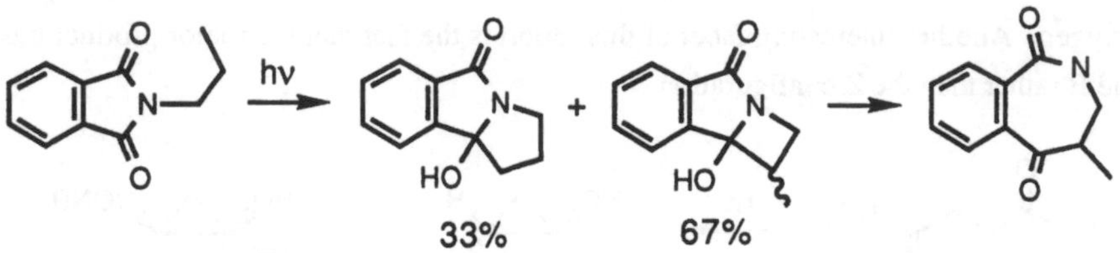

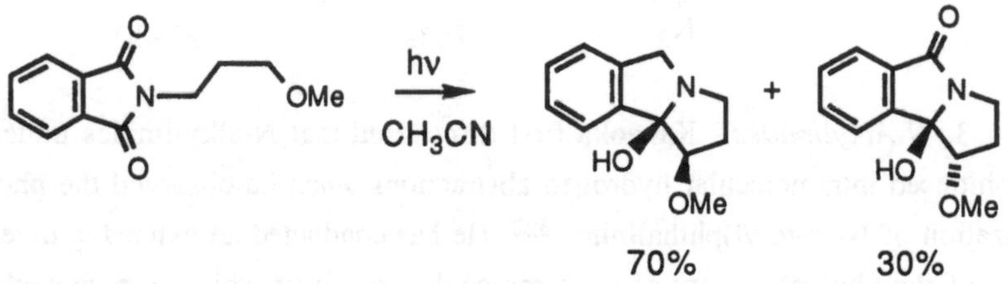

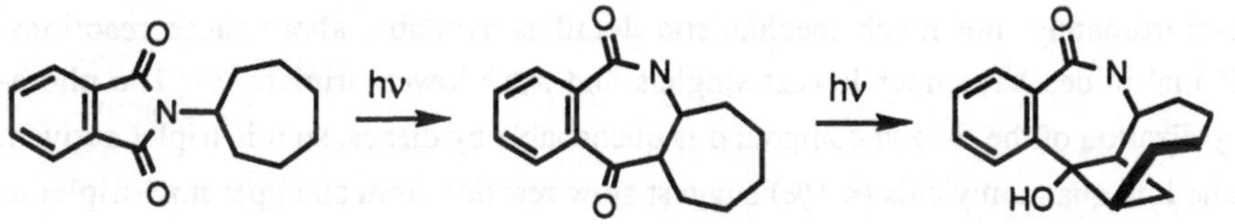

33%   67%

70%   30%

Aliphatic imides also undergo intramolecular hydrogen abstraction that leads to Norrish type II elimination.[327] No mechanistic information is available to explain how cleavage and cyclization compete or what determines regioselectivity.

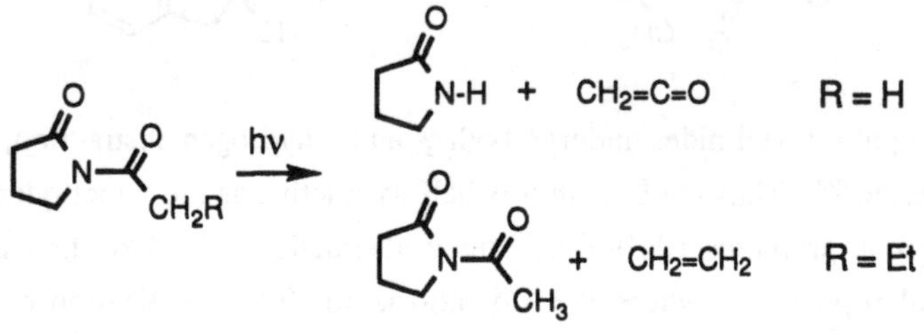

$N-H + CH_2=C=O$ R = H

$+ CH_2=CH_2$ R = Et

4. *Thiones*. de Mayo conducted a thorough investigation of the photochemistry of thioketones.[328] The main features are that reactions occur from two excited states, S_1 and S_2. The former is a n,π* state and behaves much like those of normal ketones except that quantum yields are very low, presumably because of the lowered reactivity caused by the low (44 kcal/mole) excitation energy. The upper state is some kind of π,π* state and shows much higher reactivity, which can compete with its >10^9 s[-1] decay. Intramolecular hydrogen abstraction occurs from β,

γ, δ, and ε positions, the latter two being preferred. The markedly different regio-selectivity from that shown by excited ketones presumably manifests the unpaired spin density being in the π system of the C=S bond rather than in an n orbital.

Among the more recent contributions in this area are Aoyama's demonstration that the thio derivatives of α-ketoamides[329] and N-acylimides[330] undergo efficient photocyclizations identical to those reported for the ketones.

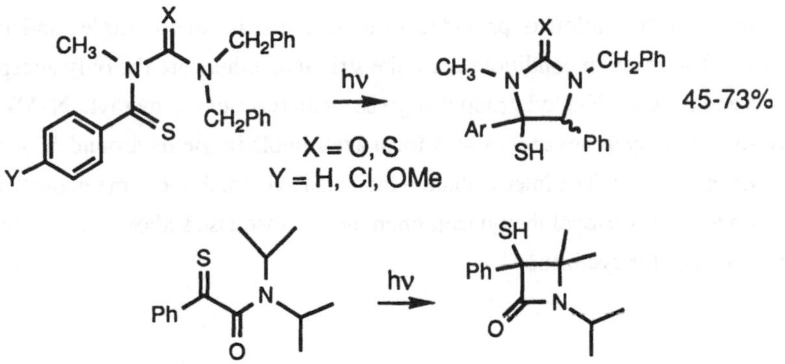

45-73%

X = O, S
Y = H, Cl, OMe

VI. HYDROGEN TRANSFER INITIATED BY CHARGE TRANSFER

A. α-Substituted Ketones

α-(N-alkylamino)ketones undergo facile γ-hydrogen transfer to produce either type II cleavage products or azetidinols, depending on substitution. Simple dimethylamino ketones undergo cleavage. Padwa first reported that ketones with π,π* lowest triplets undergo this reaction as well as those with n,π* triplets.[331] We later showed that intersystem crossing yields are <1% in benzene, presumably because of very rapid singlet state CT quenching, but are higher in methanol,[44] which retards CT by hydrogen bonding, as discussed earlier. We also showed that the reaction occurs from n,π* singlets in nonpolar media,and also from triplets in methanol. This triplet process represents another example of solvent protonation of an exciplex leading to overall hydrogen transfer. It is interesting that CT quenching is rather slow in the triplet and thus does not lead to much product.[109]

Padwa found that acylation of the amino group promotes triplet reaction, presumably by depressing singlet CT, although the of N-benzyl-N-benzoyl-α-aminoacetophenone has a triplet lifetime of only 0.1 nsec.[331] Such substitution also improves cyclization/cleavage ratios.[332] Hill has surveyed a variety of substituted α-amino ketones and found that N-aryl substitution enhances the production of azetidinol at the expense of type II cleavage, although total yields often are no better than 50%.[333] Naphthyl ketones and others with π,π* triplets react, although no photokinetics information is provided to distinguish between singlet and triplet reactivity. The reaction conditions are quite stringent; ethers are the only acceptable solvents and the only N-alkyl group that gives clean reactions is methyl. N-Alkyl-N-phenyl-α-amino cyclohexanones also form azetidinols in yields around 50% when irradiated in ether.[334] The intermediate biradicals have much the same geometries as those formed from α-cyclohexylacetophenone, as discussed above, and show the same preference for cyclization.

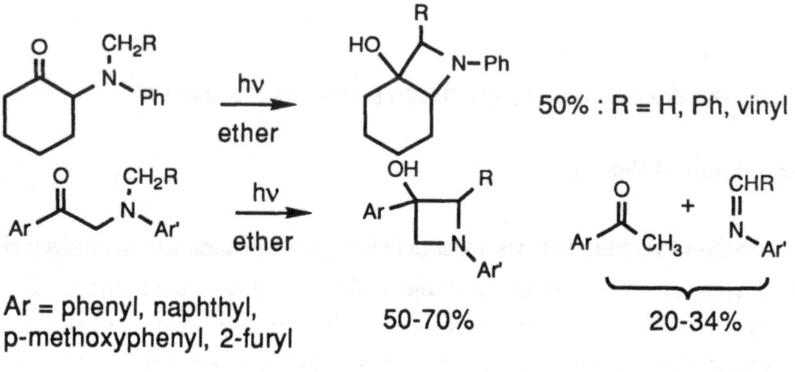

Ar = phenyl, naphthyl, p-methoxyphenyl, 2-furyl

50% : R = H, Ph, vinyl

50-70%

20-34%

Several N-phenacyl lactams also share the same basic structure and give good yields of bicyclic alcohols with no type II fragmentation observed.[335] The phenyl group and the lactam ring end up primarily *cis* (c/t = 2:1). Henning has studied N-benzyl-N-tosyl (or benzoyl) α-aminoacetophenones, which provide N-tosyl (25%) or N-benzoyl (65%) azetidinols in the yields noted[336] Again no photokinetics were performed to allow assignment of reactive states, but good diastereoselectivity was often observed.

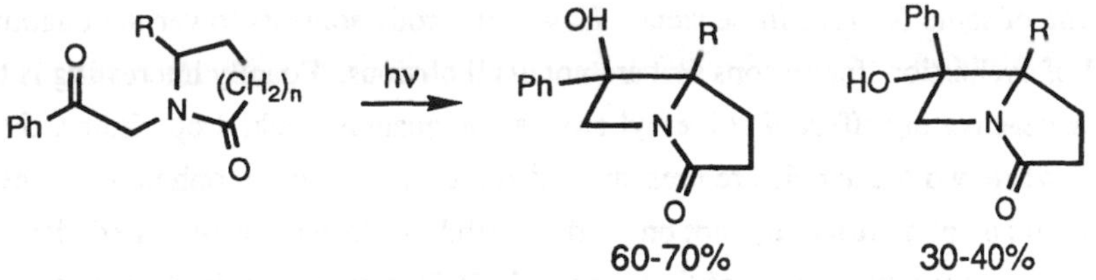

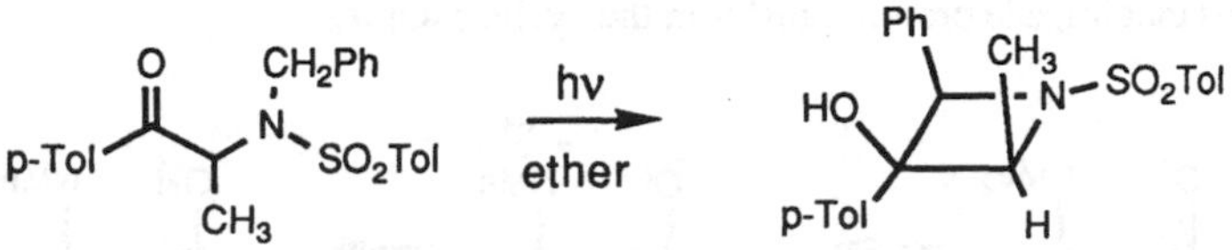

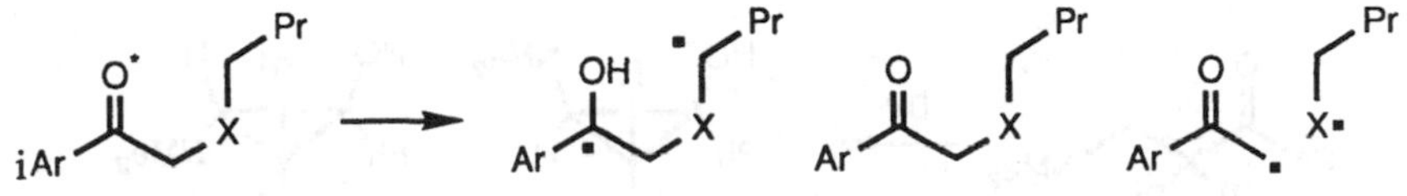

60-70%   30-40%

X = S	3.3 x 10^9	1.4 x 10^9	1.5 x 10^8
X = SO	< 1 x 10^8	6 x 10^8	5.4 x 10^9
X = SO$_2$	2 x 10^6	2 x 10^7	1 x 10^7

α-Ketosulfides undergo three competing triplet state reactions, two of which lead to cleavage; no thiacyclobutanols are formed.[337] The scheme above indicates how rate constants for γ-hydrogen abstraction, CT quenching, and radical cleavage compete. Hydrogen abstraction is slowed as the sulfur is oxidized. As with the α-amino ketones, it appears that none of the CT interaction leads to biradicals.

B. β-Substituted Ketones

Intramolecular CT quenching by β-aminopropiophenones is extremely rapid. We found triplet lifetimes <0.2 nsec by sensitization,[338] as did Scaiano.[339] Internal proton transfer now occurs quite readily to generate 2-aminocyclopropanols. DeVoe has recently clarified some confusing aspects of this reaction by showing how unstable these products are, especially to base.[340] Cyclopropanol formation is barely quenched by dienes; but it is quenched by diethyl fumarate, with $k_q\tau = 3$ M^{-1}. Since fumarates are known to quench internal aminonaphthalene exciplexes with diffusion controlled rate constants, DeVoe reasons that the β-aminoketone exciplex has a lifetime of some 0.2 nsec in benzene. However, protic solvents lower the quantum yield of cyclization, for reasons that are not at all obvious. Equally interesting is the strong deactivating effect of α-methyl groups; the quantum yield drops from 0.59 to <0.02 when two α-methyls are present on β-dimethylaminopropiophenone. These effects are on photoreactivity, not on product stability, since the *gem*-dimethylcyclo-propanols are the most stable of those studied. DeVoe ascribes this change to steric interactions that impede proton transfer in the cyclic exciplex.

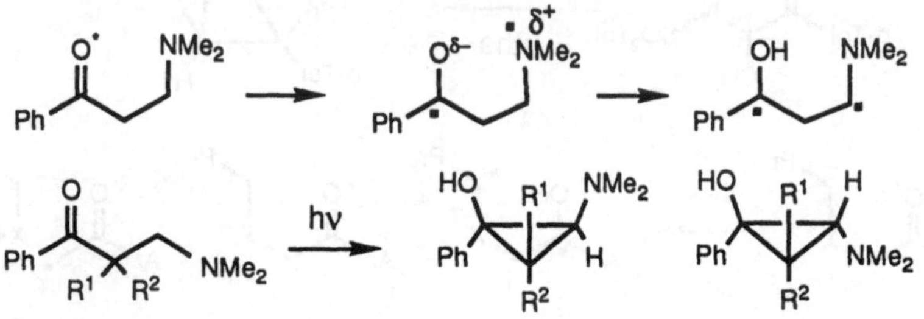

R¹ = R² = H	benzene	>95%	<5%
R¹ = R² = H	dioxane	50%	50%
R¹ = Me ; R² = H	either	<5%	>95%

Roth, who provided the first reports of aminocyclopropanol formation,[341] has shown that secondary amines also react.[342]

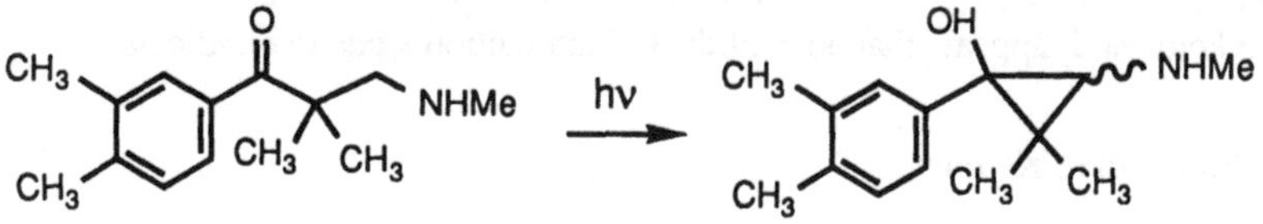

Henning has found that several β-amidopropiophenones undergo competing cyclopropanol and proline formation when the δ-carbon is substituted with a carboxy group.[343] The carbonyl obviously enhances the acidity of the δ C–H bonds, allowing the energetics of proton transfer to compensate for the entropic difference between 1,4 and 1,6 hydrogen transfers. There is an interesting solvent effect on the competition in that the proline, formed mainly as the E isomer, is the major product in hydrocarbon solvents; both processes occur about equally in ether.

Z = tosyl or benzoyl

Intramolecular CT quenching is also very rapid in β-ketosulfides, but no cyclopropanol products were found.[111] These compounds must be reinvestigated to determine product stability.

C. γ-Substituted Ketones.

We concluded years ago that there is no proton transfer leading to 1,4-biradicals following intramolecular CT quenching in triplet γ-amino ketones except in protic solvents.[109] Maximum type II quantum yields are only 5% in benzene but jump to 25% in methanol. Later we found more dramatic evidence that is similar to that described by Fessenden for intermolecular reactions.[93] The photochemistry of γ-dimethylamino-2-butyronaphthone is highly solvent dependent.[101] In benzene, a very low quantum yield, unquenchable type II elimination takes place that is typical for excited singlets of naphthyl ketones.[43] Sensitization experiments show that a long-lived but unreactive triplet is formed in high quantum yield. In protic solvents such as methanol and trifluoroethanol, however, this long-lived triplet does undergo type II elimination in moderate quantum efficiency. We postulated that in aprotic solvents, the π,π* triplet forms a normal intramolecular exciplex, with the low rate constant predicted by the triplet's low reduction potential. We blame the lack of γ-hydrogen transfer to an oxygen p-orbital on restricted rotational freedom in the exciplex whether the nitrogen lone pair overlaps with the carbonyl n-orbital or with an aromatic π orbital. A weakly acidic solvent catalyzes hydrogen transfer very efficiently, presumably by direct protonation of the oxygen in the exciplex followed by deprotonation of the resulting amine radical-cation. As an alternative explanation, solvation may allow full electron transfer and free rotation of a biradical zwitterion. This possibility seems less likely because of the poor thermodynamics for full electron transfer and the fact that a solvent molecule could get in the way of proton transfer. The drawing below indicates how one hydrogen α to the nitrogen is held in a 1,3-diaxial relationship to the oxygen p-orbital in the cyclic exciplex. An axial alkyl group on the nitrogen actually has a hydrogen much better aligned for transfer to the oxygen.

It is not clear why the β- and γ-dimethylamino ketones show such different solvent effects on the partitioning of their presumedly similar intramolecular triplet exciplexes. The energetics for proton transfer must be identical, whereas entropy would favor transfer in the β-amino systems. That internal proton transfer is efficient in the β-amino ketones and inefficient in the γ-amino ketones is puzzling enough, but the opposite solvent effects on that efficiency is more puzzling. We have concluded that γ-thiyl ketones behave similarly to the γ-amino ketones in that type II elimination proceeds only by γ-hydrogen transfer discrete from the competing CT interaction.[111] The rate constant derived for γ-hydrogen abstraction on the assumption that no 1,4-biradical is formed from the exciplex, contained in Table 3, is similar to that for γ-methoxybutyrophenone and would be even smaller if any biradical were formed following CT. Radical production from bimolecular sulfide quenching of triplet ketones is intrinsically inefficient.[344]

D. More remote CT

Our early study of δ-dialkylaminovalerophenones found that formation of any aminocyclopentanol product has a quantum efficiency <1% despite the fact that 94% of the triplets undergo rapid internal CT interaction. These compounds mimic the γ-amino ketones in undergoing little internal proton transfer following CT. There are very few examples of more remotely substituted ketones undergoing chemical reaction. Hasegawa has just reported that ω-dialkylaminoalkyl esters of benzoylacetic acid photocyclize in respectable yields.[345] α-Methylation increases the total cyclization quantum yield of the keto-ester with n = 2 from 0.07 to.45. Without such α-substitution, the enol is sufficiently populated to be an effective internal filter. The triplet decay rate was measured as 1.5×10^8 s^{-1} when n = 2, which compares to 1×10^9 for δ-pyrrolidinovalerophenone.[108] It is noteworthy that reaction at the external methyl or benzyl group is the only reaction when n = 2 and the predominant reaction when n = 3 or 4. This behavior is general, as the following examples will show.

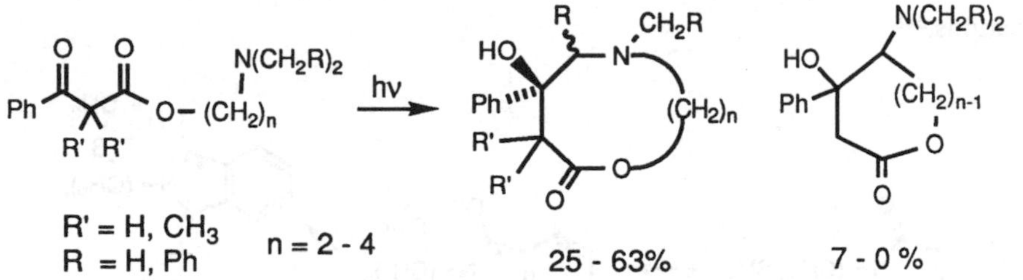

R' = H, CH$_3$
R = H, Ph   n = 2 - 4   25 - 63%   7 - 0 %

In contrast to the few examples for ketones, remote electron/proton transfer has been observed for a wide variety of imides. Kanaoka has reviewed his early work[323], which began with N-(ω-thioalkyl)phthalimides.[346] These undergo both ε and ζ-hydrogen abstractions at the expense of γ or δ. These reactions are thought to involve prior CT from sulfur to the readily reduced phthalimide chromophore, followed by proton transfer. This sort of picture is mandated by the regiospecificity,

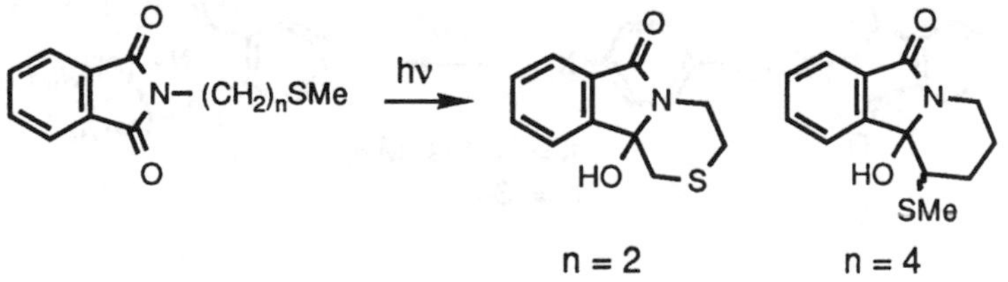

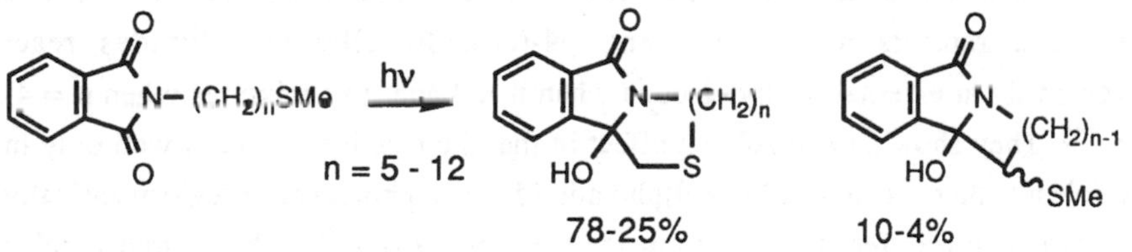

n = 2   n = 4

n = 5 - 12   78-25%   10-4%

since simple hydrogen abstraction from carbons α to sulfur is not that much faster than from unactivated methylenes (Table 3)

The preference for 1,7-transfer over 1,5-transfer when n = 2 is particularly noteworthy, since it relates to the facility of proton transfer within a cyclic exciplex, as discussed above. In fact, macrocyclic structures formed by attack at the terminal methyl group are the major products for n = 5-12[347]. This regioselectivity appears to be related to greater accessibility of the α-protons external to the cycle, since primary protons are not that much more reactive than secondary protons in such radical-cations.[348]

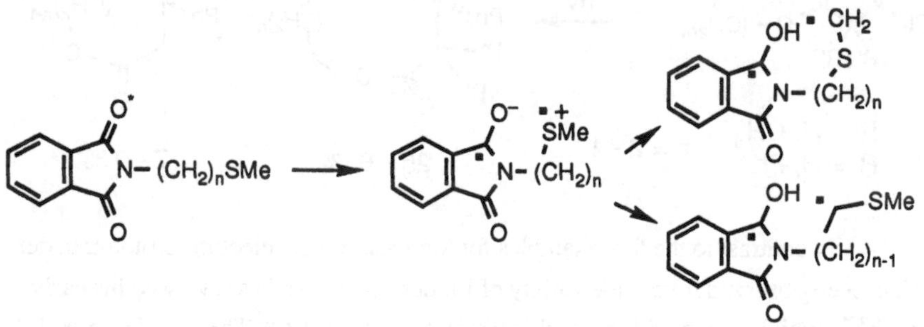

N-(aralkyl)phthalimides also undergo regiospecific transfer of a benzylic proton to oxygen when n = 3-6, with a maximum for n = 4.[349]

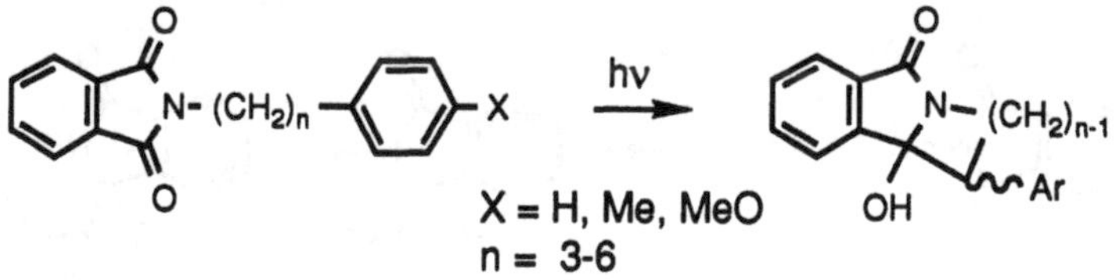

X = H, Me, MeO
n = 3-6

Kanaoka has looked more recently at aminophthalimides, in which remote hydrogen transfers are also efficient. N-(ω-anilinoalkyl)phthalimides react exclusively at external methyl groups when n < 4 and give mixtures when n = 4-18.[350] They show an odd solvent effect in that the reaction proceeds well only in petroleum ether. N-(ω-amidoalkyl)phthalimides also photocyclize regiospecifically so as to favor transfer of a proton outside the exciplex ring.[351] The chemical yields of these cyclizations vary from 10-60% for reasons that have not been fully explored.

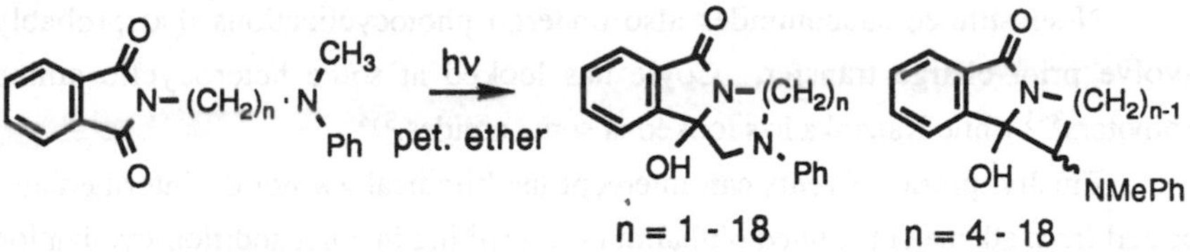

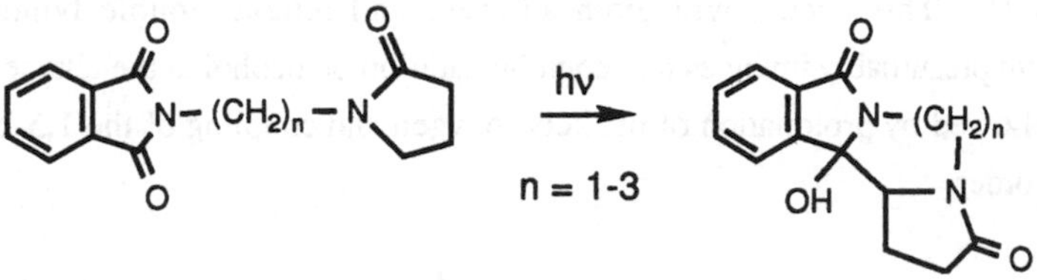

Coyle also has explored the synthetic potential in the photocyclization of N-(aminoalkyl)phthalimides [352] The example below points out the unresolved problem of controlling regioselectivity with unsymmetrically substituted imides.

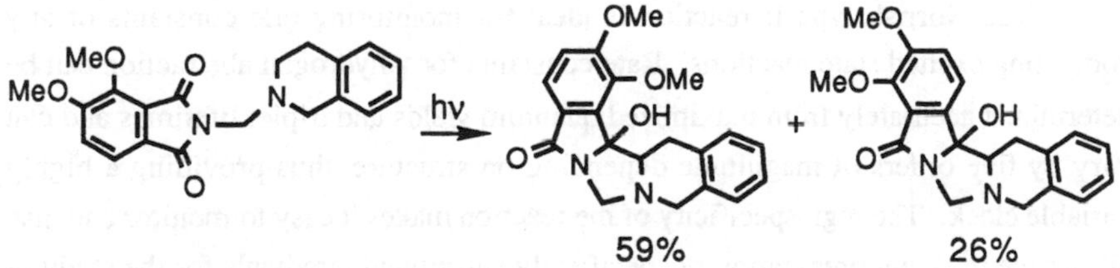

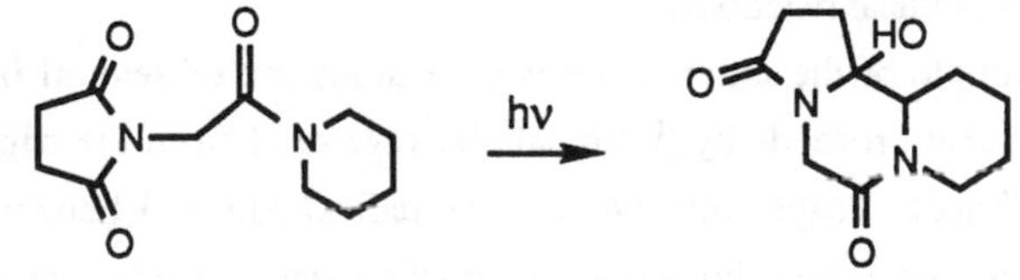

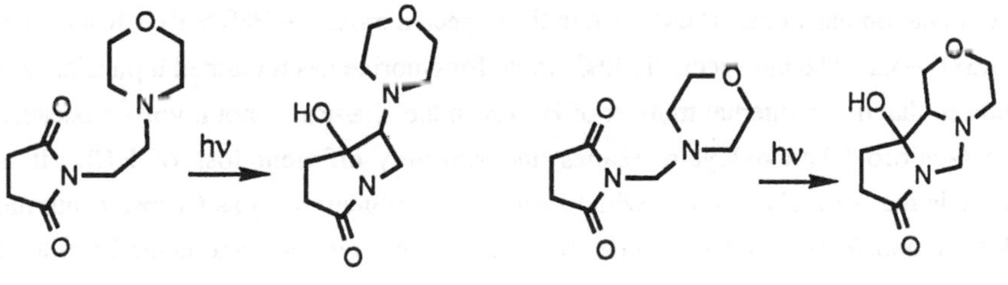

N-substituted succinimides also undergo photocyclizations that probably involve prior charge transfer. Coyle has looked at some heterocyclic amine promoters[353] while Kanaoka has looked at some amides.[354]

Finally, protic solvents can intercept the biradical-zwitterion intermediates formed from alkene-substituted phthalimides, resulting in a net addition-cyclization reaction.[355] This occurs with both adjacent and remote double bonds. The mechanism presumably involves nucleophilic addition of alcohol to the alkene radical-cation followed by protonation of the ketyl oxygen and coupling of the 1,5-biradical in either order.

VII. USE OF HYDROGEN ABSTRACTION TO MONITOR OTHER REACTIONS

The Norrish type II reaction is ideal for monitoring rate constants of any competing excited state reactions. Rate constants for γ-hydrogen abstraction can be determined accurately from maximized quantum yields and triplet lifetimes and can vary by five orders of magnitude depending on structure, thus providing a highly variable clock. The regiospecificity of the reaction makes it easy to monitor and also allows the transient preparation of specifically substituted biradicals for the study of both biradical and radical reactions.

One example of the latter is the photoelimination of several δ-substituted valerophenones that proceeds by β-radical cleavage of X from the triplet biradical followed by efficient in-cage conversion of the radical pair to 4-benzoylbutene and HX.[65] Simple product ratios allowed us to compile a list of rate constants for radical β-elimination reactions. These were in the expected order: I ~ PhS > Br ~ RSO > RS ~ RSO₂ ~ Cl. The unexpectedly high value for chlorine has remained a puzzle. We showed that direct internal transfer of H to X in the biradical is not involved because the 4-chloro-1-benzoylcyclohexanes undergo very efficient loss of HCl. It is possible that strongly electron-withdrawing δ-substituents such as Cl foster internal electron transfer within the biradical, followed by ionic cleavage and neutralization of the ion pair as drawn below. The original cleavage rate constants for the other less electron-demanding radicals are probably valid. We note that Bergmark's indanone formation from o-methylphenacyl chloride demands a similar ionization.[232] Likewise the low yield of type II products from γ-chlorobutyrophenone coupled with the formation of a dimeric product[3] suggests a dechlorination of the biradical.

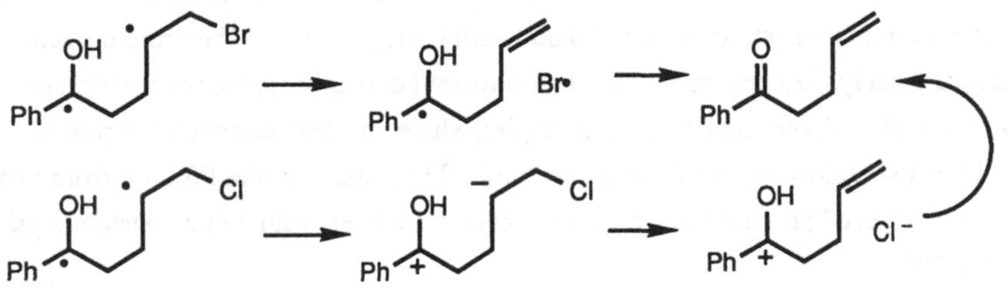

Hamer has looked at the photochemistry of several α-bromo-α-diketones and found that they retain the bromine through all of their reactions.[356] These all involve 2-keto 1,4- or 1,5-biradicals, and none undergo β-cleavage of bromine. There are several reasons why the conjugated keto group might prevent loss of bromine. It obviously lowers the spin density at the hydroxyradical site and would be expected to lower the rate constant for any radical reaction. It also forces spin density into the π system and the C–Br bond may not be properly aligned. It would also retard electron transfer.

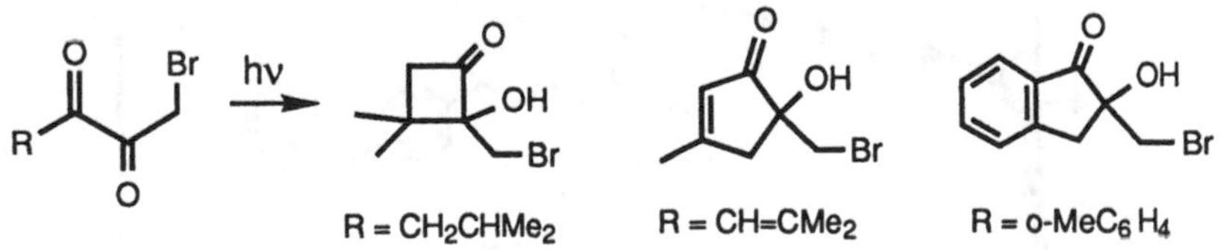

R = CH₂CHMe₂   R = CH=CMe₂   R = o-MeC₆H₄

Study of ω-amino and ω-thiyl ketones led to rate constants for internal CT quenching of the excited ketones[109,111]; study of ω-azidoketones provided rate constants for energy transfer to the azide group.[357] Study of diketones, especially unsymmetric ones, has provided measurements of reversible triplet energy transfer between nearly identical chromophores.[20,358,359] Careful manipulation of structure leads to some intriguing wavelength dependence, from which actual rate constants for energy transfer can be derived. For example, Figure 5 shows Stern-Volmer plots for quenching of acetophenone formation from δ-(p-benzoylphenyl)valerophenones.[360] At 365 nm., only the benzophenone chromophore is excited; its long-lived triplet transfers energy slowly uphill to the butyrophenone in competition with normal decay. Quenching reveals a single 1 μsec lifetime. With 313 nm. excitation, both chromophores are excited; at higher quencher concentrations, all of the benzophenone is quenched but reaction still proceeds by excitation of the butyrophenone fragment, which reacts in competition with rapid downhill energy transfer. Therefore the plot is curved. Finally, 313 nm irradiation of an o-methyl derivative produces reaction only from the short-lived butyrophenone triplet, since the benzophenone triplet now enolizes too rapidly to transfer energy uphill. The 2 nsec. triplet lifetime from this plot is composed primarily of irreversible energy transfer, with a rate constant equal to 4×10^8 s^{-1}.

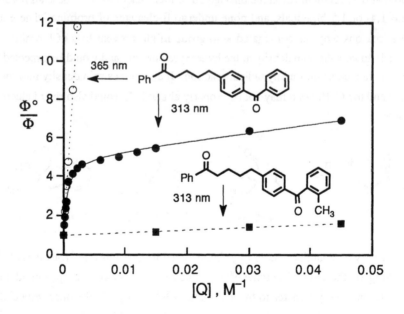

Figure 5. Wavelength effects on the quenching of acetophenone formation from δ-(p-benzoylphenyl)valerophenone.

The use of acylpyridines as ligands in ruthenium(II) complexes has allowed measurements of rate constants for photophysical processes in such compounds. In (p-acylpyridine)Ru(II)(NH$_3$)$_5$ compounds, the acylpyridine absorbs ~10% of the excitation at 313 nm and undergoes type II cleavage with that quantum efficiency independent of the actual rate constant for triplet state γ-hydrogen abstraction.[143] This result indicated that the rate constant for internal conversion from the locally excited ligand to the low-lying MLCT state is < 10^8 s^{-1}. In (p-acylpyridine)$_2$-Ru(II)(bpy)$_2$ complexes, the total rate of decay of the locally excited ligand triplet is ≤ 2 x 10^8 s^{-1} [361] This value represents the sum of internal conversion to the MLCT state and energy tranfer to the 2,2'-bipyridine ligands.

The first "slow" flash experiment on the type II reaction involved monitoring the formation and decay of enol by IR spectroscopy.[362] Kresge and Wirz have updated this idea by using modern flash spectroscopy to generate enols so that the kinetics of their acid and base catalyzed reketonizations can be measured.[363] These efforts and other photochemical methods of generating enols have revivified interest in one of the most important intermediates in organic chemistry and biochemistry.[364]

Finally, Scaiano showed that irradiation of valerophenone in methyl methacrylate induces its polymerization.[365] The 1,4-biradical adds to the vinyl monomer. This observation proves that 1,4-biradicals can live long enough to initiate polymerization and lends support to such a mechanism for the thermal polymerization of various monomers.

EPILOGUE : YANG PHOTOCYCLIZATION

The 1971 review of the Norrish type II reaction[8] lumped elimination and cyclization together mechanistically, as is appropriate. One totallly unanticipated and unintended consequence was the recent application by several investigators of this "name" reaction to all photocyclizations of ketones that proceed by intramolecular hydrogen abstraction. One cannot state the content of Norrish's wildest dreams, but it seems odd to associate his name with a reaction that he neither studied nor discovered. In fact it was N. C. Yang in 1958 who first reported that cyclobutanol formation accompanies Norrish type II elimination[366] and that more remote hydrogen abstraction could also compete.[136] He pointed out that these cyclizations strongly supported mechanisms involving biradical intermediates generated by intramolecular hydrogen atom abstractions. Therefore, if reactions following such a mechanism need a "name", then Yang's name certainly seems the most appropriate. We suggest that intramolecular hydrogen abstraction by photoexcited ketones followed by closure of the intermediate biradical to cyclic alcohols be known as the "Yang reaction".

Acknowledgement. We thank Dr. Robert DeVoe and Professors Tullio Corrona, Kendall Houk, John Scheffer, and Richard Fessenden for preprints of unpublished work and Professor Wolfgang Kirmse for describing his unpublished work. The writing of this review and all of the work at Michigan State U. herein reported were supported by the National Science Foundation.

References

1   J. C. Scaiano, *J. Photochem.*, **2**, 81 (1973).

2   (a) P. J. Wagner, in "Molecular Rearrangements," (P. de Mayo, Ed.), Academic Press, New York, Vol. 3, p. 381 (1980); (b) P. J. Wagner, *Top. Curr. Chem.*, **66**, 1 (1976).

3   P. J. Wagner and A. E. Kemppainen, *J. Am. Chem. Soc.*, **94**, 7495 (1972).

4   L. Giering, M. Berger and C. Steel, *J. Am. Chem. Soc.*, **96**, 953 (1974).

5   P. J. Wagner, R. J. Truman, A. E. Puchalski and R. Wake, *J. Am. Chem. Soc.*, **108**, 7727 (1986).

6   Y. M. A. Naguib, C. Steel, S. G. Cohen and M. A. Young, *J. Phys. Chem.*, **91**, 3033 (1987).

7   C. Walling and M. J. Gibian, *J. Am. Chem. Soc.*, **87**, 3361 (1965); A. Padwa, *Tetrahedron Lett.*, 3465 (1964).

8   P. J. Wagner, *Acc. Chem. Res.*, **4**, 168 (1971).

9   C. Walling, "Free Radicals in Solution", Wiley, New York (1957).

10   P. J. Wagner and E. J. Siebert, *J. Am. Chem. Soc.*, **103**, 7329 (1981).

11   D. R. Kearns and W. A. Case. *J. Am. Chem. Soc.*, **88**, 5087 (1966); S. Dym and R. M. Hochstrasser, *J. Chem. Phys.*, **51**, 2458 (1969).

12   P. J. Wagner, R. J. Truman and J. C. Scaiano, *J. Am. Chem. Soc.*, **107**, 7093 (1985).

13   P. J. Wagner, M. J. Thomas and A. E. Puchalski, *J. Am. Chem. Soc.*, **108**, 7739 (1986).

14   A. Beckett and G. Porter, *Trans. Faraday Soc.*, **59**, 2051 (1963); G. Porter and P. Supan, *ib id.*, **61**, 1664 (1965).

15   G. S. Hammond and P. A. Leermakers, *J. Am. Chem. Soc.*, **84**, 207 (1962).

16   N. C. Yang, D. S. McClure, S. L. Murov, J. J. Houser and R. Dusenbery, *J. Am. Chem. Soc.*, **89**, 5466 (1966).

17   N. C. Yang and R. L. Dusenbery, *J. Am. Chem. Soc.*, **90**, 5899 (1968).

18   P. J. Wagner, A. E. Kemppainen and H. N. Schott, *J. Am. Chem. Soc.*, **95**, 5604 (1973).

19   A. A. Lamola, *J. Am. Chem. Soc.*, **92**, 5045 (1970).

20   P. J. Wagner and T. Nakahira, *J. Am. Chem. Soc.*, **95**, 8474 (1973).

21   M. Berger, E. McAlpine and C. Steel, *J. Am. Chem. Soc.*, **100**, 5147 (1978).

22   M. V. Encina, E. A. Lissi, E. Lemp, A. Zanocco and J. C. Scaiano, *J. Am. Chem. Soc.*, **105**, 1856 (1983).

23   P. J. Kropp, *J. Am. Chem. Soc.*, **91**, 5783 (1969); J. M. Hornback, *iibid.*, **96**, 6773 (1974).

24   R. Reinfried, D. Bellus, and K. Shaffner, *Helv. Chim. Acta*, **54**, 1517 (1971); A. B. Smith and W. C. Agosta, *J. Am. Chem. Soc.*, **95**, 1961 (1973); J. R. Scheffer, B. M. Jennings, and J. P. Louwerens, *iibid.*, **98**, 7040 (1976).

25   F. Wilkinson and A. Farmilo, *J. Chem. Soc. Faraday 2*, **72**, 604 (1976).

26   P. J. Wagner, P. A. Kelso and R. G. Zepp, *J. Am. Chem. Soc.*, **94**, 7480 (1972).

27   J. C. Scaiano, *J. Am. Chem. Soc.*, **102**, 7747 (1980).

28   P. J. Wagner, *J. Am. Chem. Soc.*, **89**, 2503 (1967).

29   C. D. deBoer, W. G. Herkstroeter, A. P. Marchetti, A. P. Schultz, R. H. Schlessinger, *J. Am. Chem. Soc.*, **95**, 3963 (1973).

30   P. J. Wagner, B. P. Giri, J. C. Scaiano, D. L. Ward, E. Gabe and F. L. Lee, *J. Am. Chem. Soc.*, **107**, 5483 (1985).

31   H. E. Zimmerman, *Adv. Photochem.*, **1**, 183 (1963).

32   W. A. Pryor, "Free Radicals", McGraw-Hill, New York, Chap. 3 (1966).

33   G. Porter and P. Suppan, *Proc. Chem. Soc.*, 191 (1964).

34   S. G. Cohen, M. D. Saltzman, and J. B. Guttenplan, *Tetrahedron Lett.*, 4321 (1969).

35   R. E Connors and P. S. Walsh, *Chem. Phys. Lett.*, **52**, 436 (1977).

36   N. J. Turro, I. R. Gould, J. Liu, W. S. Jenks, H. Staab, and R. Alt, *J. Am. Chem. Soc.*
      111, 6378 (1989).

37   J. C. Scaiano, W. J. Leigh, M. A. Meador and P. J. Wagner, *J. Am. Chem. Soc.*, **107**,
      5806 (1985).

38   H. Suzuki, "Electronic Absorption Spectra and Geometry of Organic Molecules",
      Academic Press, New York, 1967, pp. 453-457.

39   M. J. S. Dewar and C. Doubleday, *J. Am. Chem. Soc.*, **100**, 4935 (1978).

40   P.J. Wagner and G.S. Hammond, *J. Am. Chem. Soc.*, **87**, 4009 (1965).

41   J. C. Dalton and N. J. Turro, *Mol. Photochem.*, **2**, 133 (1970).

42   P. J. Wagner in "Creation and Detection of the Excited State", (A.A. Lamola, Ed.)
      Marcel Dekker, New York, pp. 174-212 (1971).

43   N. C. Yang and A. Shani, *Chem. Commun.*, 815 (1971); J. C. Coyle, *J. Chem. Soc.*
      Perkin 2, 233 (1973).

44   P. J. Wagner and T. Jellinek, *J. Am. Chem. Soc.*, **93**, 7328 (1971).

45   C. M. Previtali and J. C. Scaiano, *J. Chem. Soc., Perkin Trans.*, **2** , 1667, 1672
      (1972).

46   J. E. Rudzki, J. L. Goodman, and K. S. Peters, *J. Am. Chem. Soc.*, **107**, 7849
      (1985).

47   N. J. Turro and R. Engel, *J. Am. Chem. Soc.*, **91**, 7113 (1969); N. J. Turro and T. Lee,
      J. Am. Chem. Soc., **91**, 5651 (1969).

48   P. J. Wagner, R. G. Zepp, K.-C. Liu, M. Thomas, T.-J. Lee and N. J. Turro, *J. Am.*
      Chem. Soc., **98**, 8125 (1976).

49   J. C. Scaiano, V. Wintgens, and J. C. Netto-Ferreira, *Photochem and Photobiol.*, **50**,
      707 (1989).

50   W. H. Urry and D. J. Trecker, *J. Am. Chem. Soc.*, **84**, 713 (1962).

51   N. J. Turro and D. S. Weiss, *J. Am. Chem. Soc.*, **90**, 2185 (1968).

52   P. J. Wagner, P. A. Kelso, A. E. Kemppainen and R. G. Zepp, *J. Am. Chem. Soc.*, **94**,
      7500 (1972).

53   S. Ariel, V. Ramamurthy, J. R. Scheffer, and J. Trotter, *J. Am. Chem. Soc.*, **105**, 6959
      (1983).

54   J. R. Scheffer, *Org. Photochem.*, **8**, 249 (1987).

55   E. S. Lewis in "Isotopes in Organic Chemistry", (E. Buncel and C. C. Lee, Eds.),
      Elsevier, Amsterdam, Vol. 2, p. 134 (1976).

56 W. D. Chandler and L. Goodman, *J. Mol. Spectr.*, **35**, 232 (1970).

57 C. T. Rettner and R. N. Zare, *J. Chem. Phys.*, **77**, 2416 (1982).

58 D. Severance, B. Pandey, and H. Morrison, *J. Am. Chem. Soc.*, **109**, 3231 (1987).

59 A. E. Dorigo and K. N. Houk, *J. Am. Chem. Soc.*, **109**, 2195 (1987); A. E. Dorigo, M. A. McCarrick, R. J. Loncharich, and K. N. Houk, *ibid.*, **112**, 7508 (1990)..

60 R. H. Hesse, *Advan. Free Rad. Chem.*, **1**, 83 (1969).

61 (a) N. Sugiyama, T. Nishio, K. Yamada, and H. Aoyama, *Bull. Chem. Soc. Japan*, **43**, 1879 (1970); (b) J. P. Colpa and D. Stehlik, *J. Chem. Phys.*, **81**, 163 (1983).

62 H. C. Chang, R. Popovitz-Biro, M. Lahav, and L. Leiserowitz, *J. Am. Chem. Soc.*, **109**, 3883 (1987).

63 (a) R. R. Sauers, A. Scimone, and H. Shams, *J. Org. Chem.* , **53**, 6084 (1988); (b) R. R. Sauers and K. Krogh-Jesperson, .*Tetrahedron Lett.*, **30**, 527 (1989).

64 H. L. Casal, W. G. McGimpsey, J. C. Scaiano, R. A. Bliss and R. R. Sauers, *J. Am. Chem. Soc.*, **108**, 8255 (1986).

65 P. J. Wagner, J. H. Sedon and M. J. Lindstrom, *J. Am. Chem. Soc.*, **100**, 2579 (1978).

66 P. J. Wagner and B. Zhou, *Tetrahedron Lett.*, **31**, 2251 (1990).

67 (a) S. G. Cohen and J. I. Cohen, *J. Am. Chem. Soc.*, **89**, 164 (1967); S. G. Cohen and H. M. Chao, *J. Am. Chem. Soc.*, **90**, 165 (1968); (b) R. S. Davidson and P. F. Lambeth, *Chem. Commun.*, 511 (1968).

68 P. J. Wagner and R. A. Leavitt, *J. Am. Chem. Soc.*, **95**, 3669 (1973).

69 (a) G. A. Davis, P. A. Carapellucci, K. Szoc, and J. D. Gresser, *J. Am. Chem. Soc.*, **91**, 2264 (1969); (b) S. G. Cohen, G. A. Davis, and W. D. K. Clark, *ibid.*., **94**, 869 (1972); (c) R. S. Davidson and M. Santhanam, *J. Chem. Soc. Perkin 2*, 2355 (1972).

70 I. Kochevar and P. J. Wagner, *J. Am. Chem. Soc.*, **94**, 3859 (1972); J. B. Guttenplan and S. G. Cohen, *ibid.*, **94**, 4040 (1972).

71 R. D. Small Jr. and J. C. Scaiano, *J. Am. Chem. Soc.*, **100**, 296 (1978).

72 W. M. Wolf, R. E. Brown, and L. A. Singer, *J. Am. Chem. Soc.*, **99**, 526 (1977).

73 R. A. Caldwell and D. Creed, *Acc. Chem. Res.*, **13**, 45 (1980).

74 L. J. Johnston, J. C. Scaiano and T Wilson, *J. Am. Chem. Soc.*, **109**, 1291 (1987).

75 M. J. Thomas and P. J. Wagner, *J. Am. Chem. Soc.*, **99**, 3845 (1977).

76 M. L. Manion-Schilling, R. S. Hutton, and H. D.Roth, *J. Am. Chem. Soc.*, **99**, 7792 (1977).

77   W. M. Moore and C. Baylor, *J. Am. Chem. Soc.*, **88**, 5677 (1966).

78   C. H. DePuy, H. L. Jones, and W. M. Moore, *J. Am. Chem. Soc.*, **95**, 477 (1973).

79   P. J. Wagner and A. E. Puchalski, *J. Am. Chem. Soc.*, **102**, 7138 (1980).

80   D. R. Arnold and L. J. Lamont, *Can. J. Chem.*, **67**, 2119 (1989).

81   (a) H. D. Becker, *J. Org. Chem.* , **32**, 2115, 2124, 2140 (1967); (b) S. M. Rosenfeld, R. G. Lawler and H. R. WArd, *J. Am. Chem. Soc.*, **95**, 946 (1973).

82   N. J. Turro and R. Engel, *Mol. Photochem.*, **1**, 143 (1969).

83   P. K. Das, M. V. Encinas, S. Steenken and J. C. Scaiano, *J. Am. Chem. Soc.*, **103**, 4162 (1981).

84   P. K. Das, M. V. Encinas and J. C. Scaiano, *J. Am. Chem. Soc.*, **103**, 4154 (1981).

85   J. C. Scaiano, W. G. McGimpsey, W. J. Leigh and S. Jakobs, *J. Org. Chem.*, **52**, 4540 (1987).

86   H. Shizuka, H. Hagiwara, and M. Fukushima, *J. Am. Chem. Soc.*, **107**, 7816 (1985).

87   S. G. Cohen, A. Parola, and G. H. Parsons, *Chem. Rev.*, **73**, 141 (1973).

88   D. Griller, J. A. Howard, P. R. Marriott and J. C. Scaiano, *J. Am. Chem. Soc.*, **103**, 619 (1981).

89   P. J. Wagner and A. Kemppainen, *J. Am. Chem. Soc.*, **91**, 3085 (1969).

90   S. Inbar, H. Linschitz and S. G. Cohen, *J. Am. Chem. Soc.*, **103**, 1048 (1981).

91   D. Rehm and A. Weller, *Ber. Bunsenges. Physik. Chem.*, **73**, 834 (1969).

92   S. Inbar, H. Linschitz and S. G. Cohen, *J. Am. Chem. Soc.*, **102**, 1419 (1980).

93   C. Devadoss and R. W. Fessenden, *J. Phys. Chem.*, **94**, 4540 (1990)..

94   H. Masuhara, Y. Maeda, H. Nakajo, N. Mataga, K. Tomita, H. Tatemitsu, Y. Sakata, and S. Misumi, *J. Am. Chem. Soc.*, **103**, 634 (1981).

95   P. W. Atkins, K. A. McLauchlen, and P. W. Percival, *J. Chem. Soc. Chem. Commun.*, 121 (1973); H. D. Roth and A. A. Lamola, *J. Am. Chem. Soc.*, **96**, 6270 (1974).

96   S. Arimitsu, H. Masuhara, N. Mataga, and H. Subomura, *J. Phys. Chem.*, **79**, 1255 (1975).

97   K. S. Peters, S. C. Freilich and C. G. Schaeffer, *J. Am. Chem.Soc.*, **102**, 5701 (1980).

98   (a) C. G. Schaefer and K. S. Peters, *J. Am. Chem.Soc.*, **102**, 7566 (1980); (b) H. Ohtani, T. Kobayoshi, K. Suzuki, S. Nagakura, *Nippon Kagaku Kaishi*, **10**, 1479 (1984); M. Hoshino and H. Shizuka, *J. Phys. Chem.*, **91**, 714 (1987); N. A. Borisevich, N. A. Lysak, S. V. Mel'nichuk, S. A. Tikhominov, and G. B. Tolstorozlllhev,

J. Appl. Spectrosc., **49**, 1077 (1988); C. Devadoss and R. W. Fessenden, J. Phys. Chem., in press.

99   S. G. Cohen and J. I. Cohen, Tetrahedron Lett., 4823 (1968).

100   A. H. Parola and S. G. Cohen, J. Photochem., **12**, 41 (1980).

101   P. J. Wagner and D. A. Ersfeld, J. Am. Chem.Soc., **98**, 4515 (1976).

102   J. D. Simon and K. S. Peters, J. Am. Chem.Soc., **104**, 6542 (1982).

103   H. Miyasaka and N. Mataga, Bull. Chem. Soc. Japan, **63**, 131 (1990).

104   H. Fischer, R. I. Walter, and B. M. P. Hendricks, J. Am. Chem.Soc.,**101**, 2378 (1979).

105   J. K. S. Wan and S.-K. Wong, Acc. Chem. Res., **7**, 58 (1974).

106   M. J. Thomas, P. J. Wagner, M. L. Manion Schilling and H. D. Roth, J. Am. Chem. Soc., **99**, 3842 (1977).; P. J. Wagner and M. J. Thomas, J. Am. Chem. Soc., **102**, 4173 (1980).

107   M. A. Souto, P. J. Wagner, and M. A. El-Sayed, Chem. Phys., **6**, 193 (1974) and references therein.

108   P. J. Wagner and B. J. Scheve, J. Am. Chem. Soc., **99**, 1858 (1977).

109   P. J. Wagner, A. E. Kemppainen and T. Jellinek, J. Am. Chem. Soc., **94**, 7512 (1972).

110   P. J. Wagner, Acc. Chem. Res., **16**, 461 (1983).

111   P. J. Wagner and M. J. Lindstrom, J. Am. Chem. Soc., **109**, 3057 (1987).

112   G. L. Closs, M. D. Johnson, J. R. Miler, and P. Piotrowiak, J. Am. Chem. Soc., **111**, 3755 (1989).

113   D. Griller and K. U. Ingold, Acc. Chem. Res., **9**, 13 (1976).

114   H. Fischer, J. Am. Chem. Soc., **108**, 3925 (1986).

115   Y. M. A. Naguib, S. G. Cohen, and C. Steel, J. Am. Chem. Soc., **108**, 128 (1985); J. C. Scaiano and P. J. Wagner, unpublished results.

116   J. N. Pitts Jr., R. L. Letsinger, R. P. Taylor, J. M Patterson, G. Recktenwald, and R. B. Martin, J. Am. Chem. Soc., **81**, 1068 (1969).

117   D. I. Schuster and P. B. Karp, J. Photochem., **12**, 332 (1980).

118   M. B. Rubin, Tetrahedron Lett., **23**, 4615 (1982).

119   (a) S. F. Nelsen and P. D. Bartlett, J. Am. Chem. Soc., **88**, 137 (1966); (b) H. Langals and H. Fischer, Chem. Ber., **111**, 543 (1978).

120   J. Chilton, L. Giering and C. Steel, J. Am. Chem. Soc., **98**, 1865 (1976).

121   S. A. Weiner, J. Am. Chem. Soc., **93**, 425 (1971).

122   Y. M. A. Naguib, C. Steel and S. G. Cohen, *J. Phys. Chem.*, **92**, 6574 (1988).

123   A. E. Puchalski, Ph. D. Thesis, Michigan State University, 1980; Y. Zhang, M. S. Thesis, Michigan State University, 1989.

124   E. S. Huyser and D. C. Neckers, *J. Am. Chem. Soc.*, **85**, 3641 (1963).

125   P. G. Stone and S. G. Cohen, *J. Am. Chem. Soc.*, **102**, 5685 (1980).

126   R. Breslow, S. Kitabatake, and J. Rothbard, *J. Am. Chem. Soc.*, **100**, 8156 (1978).

127   V. P. Rao and N. J. Turro, *Tetrahedron Lett.*, **30**, 4641 (1989).

128   J. Cossy, D. Belotti, and J. P. Pete, *Tetrahedron Lett.*, **28**, 4545, 4547 (1987).

129   M. A. Winnik, *Chem. Rev.*, **81**, 491 (1981).

130   F. D. Lewis, R. W. Johnson, and D. E. Johnson, *J. Am. Chem. Soc.*, **96**, 6090 (1974).

131   P. J. Wagner and C.-P. Chen, *J. Am. Chem. Soc.*, **98**, 239 (1976).

132   S. Winstein and N. J. Holness, *J. Am. Chem. Soc.*, **77**, 5562 (1955).

133   M. A. Winnik, C. K. Lee, S. Basu, and D. S. Saunders, *J. Am. Chem. Soc.*, **96**, 6182 (1974).

134   F. D. Lewis, R. W. Johnson and D. R. Kory, *J. Am. Chem. Soc.*, **96**, 6100 (1974).

135   E. C. Alexander and J. A. Uliana, *J. Am. Chem. Soc.*, **96**, 5644 (1974).

136   (a) M. Barnard and N. C. Yang, *Proc. Chem. Soc.*, London, 302 (1958); (b) R. R. Sauers and S.-Y. Huang, *Tetrahedron Lett.*, **31**, 5709 (1990).

137   F. M. Menger, *Acc. Chem. Res.*, **18**, 128 (1985).

138   D. R. Coulson and N. C. Yang, *J. Am. Chem. Soc.*, **88**, 4511 (1966).

139   P. J. Wagner and A. E. Kemppainen, *J. Am. Chem. Soc.*, **90**, 5896 (1968).

140   D. DeKeukeleire, G. Blondeel, and A. deBruyn, *Tetrahedron Lett.*, **25**, 2055 (1984).

141   P. J. Wagner and J. H. Sedon, *Tetrahedron Lett.*, 1927 (1978); P. J. Wagner, M. J. Lindstrom, J. H. Sedon and D. R. Ward, *J. Am. Chem. Soc.*, **103**, 3842 (1981).

142   P. S. Skell and K. J. Shea in "Free Radicals", Vol. 2, (J. K. Kochi, Ed.), Wiley-Interscience, New York, p. 809 (1973).

143   P. J. Wagner and R. Bartoszek-Loza, *J. Am. Chem. Soc.*, **103**, 5587 (1981).

144   F. S. Heirtzler, M. S. Thesis, Michigan State University, (1989.)

145   P. J. Wagner and M. J. Thomas, *J. Org. Chem.*, **46**, 5431 (1981).

146   (a) R. W. Kilb, C. C. Lin, and E. B. Wilson, *J. Chem. Phys.*, **26**, 1695 (1957); (b) G. J. Karabatsos and D. J. Fenoglio, *Top. Stereochem.*, **5**, 167.(1970); (c) K. B. Wiberg and E. J. Martin, *J. Am. Chem. Soc.*, **107**, 5035 (1985).

147 S. Ariel, S. Evans, N. Omkaram, J. R. Scheffer, and J. Trotter, *J. Chem. Soc. Chem. Commun.*, 372 (1986).

148 P. J. Wagner and R. G. Zepp, *J. Am. Chem. Soc.*, **94**, 287 (1972).

149 J. C. Scaiano, E. A. Lissi and M. V. Encina, *Rev. Chem. Intermediates*, **2**, 139 (1978).

150 R. B. Gagosian, J. C. Dalton, and N. J. Turro, *J. Am. Chem. Soc.*, **92**, 4752 (1970).

151 F. D. Lewis and T. A. Hilliard, *J. Am. Chem. Soc.*, **94**, 3852 (1972).

152 P. J. Wagner and M. J. Thomas, *J. Am. Chem. Soc.*, **98**, 241 (1976).

153 F. D. Lewis, R. W. Johnson, and R. A. Ruden, *J. Am. Chem. Soc.*, **94**, 4292 (1972).

154 T. Hasegawa, Y. Arata, and A. Kageyama, *Tetrahedron Lett.*, **24**, 1995 (1983).

155 P. J. Wagner and K.-C Liu, *J. Am. Chem. Soc.*, **96**, 5952 (1974,).

156 R. D. Small Jr. and J. C. Scaiano, *Chem. Phys. Lett.*, **50**, 431 (1977).

157 R. D. Small and J. C. Scaiano, *J. Phys. Chem.*, **81**, 828 (1977); R. D. Small and J. C. Scaiano, *J. Photochem.*, **6**, 453 (1976/77).

158 J. C. Scaiano, *Acc. Chem. Res.*, **15**, 252 (1982).

159 R. D. Small Jr. and J. C. Scaiano, *Chem. Phys. Lett.*, **59**, 246 (1978).

160 R. D. Small Jr. and J. C. Scaiano, *J. Phys. Chem.*, **81**, 2126 (1977).

161 (a) R. D. Small Jr. and J. C. Scaiano, *J. Am. Chem. Soc.*, **100**, 4512 (1978); (b) W. Adam, S. Grabowski, and R. M. Wilson, *Chem. Ber.*, **122**, 561 (1989).

162 M. V. Encinas and J. C. Scaiano, *J. Photochem.*, **11**, 241 (1979).

163 R. A. Caldwell, T. Majima and C. Pac, *J. Am. Chem. Soc.*, **104**, 629 (1982).

164 P. J. Wagner, K.-C. Liu and Y. Noguchi, *J. Am. Chem. Soc.*, **103**, 3837 (1981).

165 M. V. Encinas, P. J. Wagner and J. C. Scaiano, *J. Am. Chem. Soc.*, **102**, 1357 (1980).

166 R. D. Small Jr. and J. C. Scaiano, *J. Phys. Chem.*, **82**, 2662 (1978).

167 R. Bernardi, T. Caronna, D. Coggiola, and S. Morrocchi, unpublished results.

168 P. Maruthamuthu and J. C. Scaiano, *J. Phys. Chem.*, **82**, l588 (1978).

169 J. C. Scaiano, *Tetrahedron*, **38**, 819 (1982).

170 L. Salem and C. Rowland, *Angew. Chem., Int. Ed. Engl.*, **11**, 92 (1972).

171 D. H. R. Barton, B. Charpiot, K. U. Ingold, L. J. Johnston, W. B. Motherwell, J. C. Scaiano and S. Stanforth, *J. Am. Chem. Soc.*, **107**, 3607 (1985).

172 M. B. Zimmt, C. Doubleday Jr., I. R. Gould and N. J. Turro, *J. Am. Chem. Soc.*, **107**, 6724 (1985); M. B. Zimmt, C. Doubleday Jr. and N. J. Turro, *ibid.*,**107**, 6726 (1985).

173   R. A. Caldwell, S. N. Dhawan and T. Majima, *J. Am. Chem. Soc.*, **106**, 6454 (1984).

174   W. T. Dixon and R. O. C. Norman, *J. Chem. Soc.*, 4857 (1964); A. Carrington and I. C. P. Smith, *Mol. Phys,*, **9**, 137 (1965).

175   L. J. Johnston, J. C. Scaiano, J. W. Sheppard and J. P. Bays, *Chem. Phys. Lett.*, **124**, 493 (1986).

176   R. A. Caldwell and S. C. Gupta, *J. Am. Chem. Soc.*, **111**, 740 (1989).

177   N. J. Turro and F. D. Lewis, *J. Am. Chem. Soc.*, **92**, 311 (1970).

178   P. J. Wagner, I. E. Kochevar and A. E. Kemppainen, *J. Am. Chem. Soc.*, **94**, 7489 (1972).

179   R. A. Caldwell in "Kinetics and Spectroscopy of Carbenes and Biradicals" (M. S. Platz, Ed.) Plenum, New York, p. 77 (1990).

180   P. J. Wagner, *Acc. Chem. Res.*, **22**, 83 (1989).

181   R. C. Neuman and C. T. Berge, *Tetrahedron Lett.*, 1709 (1978).

182   N. J. Turro, K.-C. Liu and M.-F. Chow, *Photochem. Photobiol.*, **26**, 413 (1977).

183   N. J. Turro and G. C. Weed, *J. Am. Chem. Soc.*, **105**, 1861 (1983).

184   S. P. Pappas and A. Chattopadhyay, *J. Am. Chem. Soc.*, **95**, 6484 (1973).

185   F. D. Lewis, R. T. Lauterbach, H.-G. Heine, W. Hartmann, and H. Rudolph, *J. Am. Chem. Soc.*, **97**, 1519 (1975).

186   P. J. Wagner and M. J. Lindstrom, unpublished work.

187   P. deMayo, A. Nakamura, P. W. K. Tsang, and S. K. Wong *J. Am. Chem. Soc.*, **104**, 6824 (1982)

188   S. Devanathan and V. Ramamurthy, *J. Phys. Org. Chem.*, **1**, 91 (1988).

189   G. Dasarathu Reddy, K. V. Ramanathan, and V. Ramamurthy, *J. Org. Chem.*, **51**, 3085 (1986).

190   G. Dasarathu Reddy and V. Ramamurthy, *J. Org. Chem.*, **52**, 5521 (1987)

191   N. J. Turro and P. Wan, *Tetrahedron Lett.*, **25**, 3655 (1984).

192   P. J. Wagner and J. M. McGrath, *J. Am. Chem. Soc.*, **94**, 3849 (1972).

193   Z. He and R. G. Weiss, *J. Am. Chem. Soc.*, **112**, 5535 (1990).

194   W. J. Leigh in *Photochemistry on Solid Surfaces* (M. Anpo and T. Matsuura, Eds.) Eldevier, Amsterdam, p. 481 (1989).

195   P. R. Worsham; D. W. Eaker; D. G. Whitten, *J. Am. Chem. Soc.*, **100**, 7091 (1978).

196   J. E. Guillet, *Polymer Photophysics and Photochemistry*, Cambridge University Press, Cambridge, Ch. 10 (1985); J. E. Guillet, *Adv. Photochem.*, **14**, 91 (1988).

197   G. H. Hartley and J. E. Guillet, *Macromolecules*, **1**, 165 (1968).

198   F. J. Golemba and J. E. Guillet, *Macromolecules*, **5**, 212 (1973); I. Lukac and P. Hrdlovic, *Polym. Photochem.*, **2**, 277 (1982).

199   P. Hrdlovic and J. E. Guillet, *Polym. Photochem.*, **7**, 423 (1986).

200   (a) M. V. Encinas, K. Funabashi and J. C. Scaiano, *Macromolecules*, **12**, 1167 (1979); (b) T. Kilp and J. E. Guillet, *ibid.*, **14**, 1680 (1981).

201   P. Hrdlovic, J. C. Scaiano, I. Lukac, and J. E. Guillet, *Macromolecules*, **19**, 1637 (1986).

202   S. V. Evans, M. Garcia-Garibay, N. Omkaram, J. R. Scheffer, and J. Trotter, *J. Am. Chem. Soc.*, **108**, 5648 (1986).

203   S. V. Evans, N. Omkaram, J. R. Scheffer, and J. Trotter, *Tetrahedron Lett.*, **26**, 5903 (1985); *ibid.*, **27**, 1419 (1986).

204   T. J. Lewis, S. J. Rettig, J. R. Scheffer, J. Trotter, and F. Wireko, *J. Am. Chem. Soc.*, **112**, 3679 (1990).

205   S. Ariel, S. V. Evans, M. Garcia-Garibay, B. R. Harkness, N. Omkaram, J. R. Scheffer, and J. Trotter, *J. Am. Chem. Soc.*, **111**, 5591 (1989).

206   C. C. Wamser and W. R. Wagner, *J. Am. Chem. Soc.*, **103**, 7232 (1981).

207   S. Mohr, *Tetrahedron. Lett.*, **21**, 593 (1980).

208   N. C. Yang and M. J. Jorgensen, *Tetrahedron Lett.*, 1203 (1964).

209   (a) I. A. Skinner and A. C. Weedon, *Tetrahedron Lett.*, **24**, 4299 (1983); (b) J.-P. Pete, F. Henin, R. Montezaei, J. Muzart, and O. Piva, *Pure Appl. Chem.*, **58**, 1257 (1986).

210   F. Henin, R. Montezaei, J. Muzart, J.-P. Pete, and O. Riva, *Tetrahedron*, **45**, 6171 (1989).

211   P. G. Sammes, *Tetrahedron*, **32**, 405 (1976).

212   G. Porter and M. F. Tchir, *J. Chem. Soc. A*, 3772 (1971); D. M. Findley and M. F. Tchir, *J. Chem. Soc. Faraday 1*, **72**, 1096 (1976).

213   P. J. Wagner, *Pure Appl. Chem.*, **49**, 259 (1977).

214   P. K. Das, M. V. Encinas, R. D. Small Jr. and J. C. Scaiano, *J. Am. Chem. Soc.*, **101**, 6965 (1979).

215   R. Haag, J. Wirz and P. J. Wagner, *Helv. Chim. Acta*, **60**, 2595 (1977).

216   K. H. Grellmann, H. Weller, E. Tauer, *Chem. Phys. Lett.*, **95**, 195 (1980); U. Baron, G. Bartelt, A. Eychmueller, K. H. Grellmann, U. Schmitt, E. Tauer, and H. Weller, *J. Photochem.*, **28**, 187 (1985).

217   T. Matsuura and Y. Kitaura, *Tetrahedron*, **25**, 4487 (1969).

218   Y. Kitaura and T. Matsuura, *Tetrahedron*, **27**, 1597 (1971).

219   B. J. Arnold, P. G. Sammes, T. W. Wallace, *J. Chem. Soc.Perkin 1*, 415 (1974).

220   M. C. Carre, M.-L. Viriot-Vilaume, and P. Caubere *J. Chem. Soc.Perkin 1*, 1395, 2542 (1979).

221   R. M. Wilson and K. Hannemann, *J. Am. Chem. Soc.*, **109**, 4741 (1987).

222   R. M. Wilson, K. Hannemann, W. R. Heineman and J. R. Kirchhoff, *J. Am. Chem. Soc.*, **109**, 4743 (1987).

223   B. J. Arnold, S. M. Mellow, and P. G. Sammes, *J. Chem. Soc.Perkin 1*, 401 (1974).

224   D. Subrahmanyam and B.-S. Park, unpublished results.

225   N. K. Hamer, *J. Chem. Soc. (C)* 1193 (1970).

226   Y. Ogata and K. Takagi, *J. Org. Chem.*, **39**, 1385 (1974).

227   N. K. Hamer, *J. Chem. Soc.Perkin 1*, 508 (1979).

228   Y. Ito, H. Nishimura, Y. Umehara, Y. Yamada, M. Tone, and T. Matsuura, *J. Am. Chem. Soc.*, **105**, 1590 (1983).

229   Y. Ito, T. Matsuura, and K. Fukuyama, *Tetrahedron Lett.*, **29**, 3087 (1988).

230   Y. Ito, N. Inada, and T. Matsuura, *J. Chem. Soc.Perkin 2*, 1857 (1983).

231   Y. Ito, Y. Umehara, Y. Yamada, T. Matsuura, and F. Imashiro, *J. Org. Chem.* , **46**, 4359 (1981).

232   W. R. Bergmark, C. Barnes, J. Clark, S. Paparoan, and S. Marynowski, *J. Org. Chem.*, **50**, 5612 (1985).

233   V. B. Rao, S. Wolff, and W. C. Agosta, *J. Am. Chem. Soc.*, **107**, 521 (1985).

234   W. C. Agosta, R. A. Caldwell, J. Jay, L. J. Johnston, B. R. Venepall, J. C. Scaiano, M. Singh, and S. Wolff, *J. Am. Chem. Soc.*, **109**, 3030 (1987).

235   (a) A. Amaro, M. C. Carreno, and F. Farina, *Tetrahedron Lett.*, 3983 (1979); (b) G. Quinkert *et al.*, *Angew. Chem. Int. Ed. Engl.*, **22**, 637 (1983)..

236   J. L. Charlton and T. Hurst, *Tetrahedron Lett.*, **25**, 2663 (1984).

237   L. A. Paquette and T. Sugimura, *J. Am. Chem. Soc.*, **108**, 3841 (1986).

238   H. Aoyama, T. Hasegawa, and Y. Omote, *J. Am. Chem. Soc.*, **101**, 5343 (1979).

239   H. Aoyama, M. Sakamoto, and Y. Omote, *Chem. Lett.*, 1211 (1982); *Tetrahedron Lett.*, **24**, 1169 (1983); *J. Chem. Soc.Perkin 1*, 1357 (1981); H. Aoyama, S. Suzuki, T. Hasegawa, and Y. Omote, *ibid.*, 247 (1982).

240   S. C. Shim and D. W. Kim, *Heterocycles*, **20**, 575 (1983).

241   F. Toda, M. Yagi, and S. Soda, *J. Chem. Soc. Chem. Commun.*, 1413 (1987).

242   (a) P. A. Leermakers, P. C. Warren, and G. F. Vesley, *J. Am. Chem. Soc.*, **86**, 1768 (1964); E. S. Huyser and D. C. Neckers, *J. Org. Chem.*, **29**, 276 (1964).

243   H. Aoyama, M. Sakamoto, K. Yoshida, and Y. Omote, *J. Heterocycl. Chem.*, **20**, 1099 (1983).

244   H. A. J. Carless and G. K. Fekarabobo, *Tetrahedron Lett.*, **24**, 107 (1983)

245   (a) F. D. Lewis and R. H. Hirsch, *Mol. Photochem.*, **2**, 259 (1970); J. C. Arnould, A. Enger, A. Feigenbaum, and J. P. Pete, *Tetrahedron*, **35**, 2501 (1979).

246   G. Descotes, L. Cottier, and C. Bernasconi, *Nouv. J. chim.*, **2**, 79 (1978); G. Descotes *et al.*, *Bull. Soc. chim. Fr.*, 332 (1979).

247   R. W. Binkley and H. F. Jarrell, *J. Carbohydr. Nucleotides*, **7**, 347 (1980).

248   G. Descotes *et al*, *Synthesis*, 711 (1979).

249   D. D. K. Manh, J. Ecoto, M. Fetizon, H. Colin, and J.-C. Diez-Masa, *J. Chem. Soc. Chem. Commun.*, 953 (1981).

250   G. Hilgeis and H.-D. Scharf, *Liebigs Ann. Chem.*, 1498 (1985).

251   J. L. Goodman and J. A. Berson, *J. Am. Chem. Soc.*, **106**, 1867 (1984).

252   J. P. Bays, M. V. Encinas and J. C. Scaiano, *Macromolecules*, **12**, 348 (1979).

253   G. R. Lappin and J. S. Zannucci, *J. Org. Chem.* , **36**, 1808 (1971).

254   F. R. Sullivan and L. B. Jones, *J. Chem. Soc. Chem. Commun.*, 312 (1974).

255   S. P. Pappas, B. C. Pappas, and J. E. Blackwell, Jr., *J. Org. Chem.* , **32**, 3066 (1967); S. P. Pappas, and R. D. Zehr, *J. Am. Chem. Soc.* , **93**, 7112 (1971); S. P. Pappas, J. E. Alexander, Jr., and R. D. Zehr, *J. Am. Chem. Soc.*, **96**, 6928 (1974).

256   E. J. O'Connell, *J. Am. Chem. Soc.*, **90**, 6550 (1968).

257   (a)D. J. Coyle, R. V.Peterson, and J. Heicklen, *J. Am. Chem. Soc.*, **86**, 3850 (1964); (b) P. Yates and J. M. Pal, *J. Chem. Soc. D* , 553 (1970); (c) L. M. Stephenson, J. L. and Parlett, *J. Org. Chem.* , **36**, 1093 (1971).

258   M. J. Jorgensen and N. C. Yang, *J. Am. Chem. Soc.*, **85**, 1698 (1963).

259   P. J. Wagner and R. G. Zepp, *J. Am. Chem. Soc.*, **93**, 4958 (1971).

260   C. Walling and A. Padwa, *J. Am. Chem. Soc.*, **85**, 1597 (1963).

261   L. A. Paquette and D. W. Balogh, *J. Am. Chem. Soc.*, **104**, 774 (1982).

262   P. Beak, A. Basha, B. Kokko, and D. Loo, *J. Am. Chem. Soc.*, **108**, 6016 (1986).

263   J. Ounsworth and J. R. Scheffer *J. Chem. Soc. Chem. Commun.* 232 (1986).

264   P. J. Wagner and C. Chiu, *J. Am. Chem. Soc.*, **101**, 7134 (1979).

265   G. Descotes, *Bull. Soc. Chim. Belg.*, **91**, 973 (1982); *Top. Curr. Chem.*, **154**, 41 (1990).

266   T. Koziuk, L. Cottier, and G. Descotes, *Tetrahedron*, **37**, 1875 (1981).

267   G. A. Kraus and L. Chen, *J. Am. Chem. Soc.*, **112**, 3464 (1990).

268   H. G. Henning and K. Walther, *Pharmazie*, **37**, 810 (1982).

269   T. Hasegawa *et al.*, *J. Chem. Soc. Perkin 1*, 963 (1979); 541 (1986).

270   P. J. Wagner and R. Pabon, paper in preparation.

271   W. R. Bergmark and G. D. Kennedy, *Tetrahedron Lett.*, 1485 (1979).

272   S. R. Ditto, S. J. Card, P. D. Davis, and D. C. Neckers, *J. Org. Chem.*, **44**, 894 (1979).

273   P. J. Wagner, B. P. Giri, R. Pabon and S. B. Singh, *J. Am. Chem. Soc.*, **109**, 8104 (1987).

274   P. J. Wagner, M. J. Thomas and E. Harris, *J. Am. Chem. Soc.*, **98**, 7675 (1976).

275   S. Miki, K. Matsuo, M. Yoshida, and Z.-I. Yoshida, *Tetrahedron Lett.*, **29**, 2211 (1988).

276   P. J. Wagner, M. A. Meador, and B-S. Park, *J. Am. Chem. Soc.*, **112**, 5199 (1990).

277   P. J. Wagner and G. Laidig, unpublished results.

278   P. J. Wagner, M. A. Meador and J. C. Scaiano, *J. Am. Chem. Sco.*, **106**, 7988 (1984).

279   J. H. van der Westhuizen, D. Ferreira, and D. G. Roux, *J. Chem. Soc. Perkin 1*, 1540 (1980).

280   A. L. J. Beckwith and G. Moad, *J. Chem. Soc. Perkin 2*, 1473 (1980).

281   H. A. J. Carless and D. J. Haywood, *J. Chem. Soc. Chem. Commun.*, 657 (1980).

282   A. Padwa and D. Eastman, *J. Am. Chem. Soc.*, **91**, 462 (1969).

283   ref. 2, p. 395.

284   P. J. Wagner, P. A. Kelso, A. E. Kemppainen, J. M. McGrath, H. N. Schott and R. G. Zepp, *J. Am. Chem. Soc.*, **94**, 7506 (1972).

285   W. Kirmse, private communication.

286   J. P. Praly and G. Descotes, *Carbohydr. Res.*, **95**, C1 (1981); L. Cottier, G. Descotes, M. F. Grenier, and F. Metras, *Tetrahedron*, **37**, 2515 (1981).

287 G. A. Kraus, P. J. Thomas, and M. D. Schwinden, *Tetrahedron Lett.*, **31**, 1819 (1990).

288 H. Aoyama, T. Nishio, Y. Hirabayashi, T. Hasegawa, H. Noda, and N. Sugiyama, *J. Chem. Soc. Perkin 1* , 298 (1975).

289 P. J. Wagner and Q. Cao, unpublished results.

290 M. A. Meador and P. J. Wagner, *J. Am. Chem. Soc.*, **105**, 4484 (1983).

291 P. J. Wagner and B. Zhou, *J. Am. Chem. Soc.*, **110**, 611 (1988).

292 P. J. Wagner and B. Zhou, *Tetrahedron Lett.*, **30**, 5389 (1989).

293 P. J. Wagner, B. Zhou, and T. Hasegawa, *J. Am. Chem. Soc.,* in press.

294 J. E. Guillet, W. K. MacInnis, and A. E. Redpath, *Can. J. Chem.*, **63**, 1333 (1985).

295 P. J. Wagner and B.-S. Park, unpublished results.

296 P. J. Wagner and M. A. Meador, *J. Am. Chem. Soc.*, **106**, 3684 (1984).

297 M.B. Zimmt, C. Doubleday Jr., and N. J. Turro, *Chem. Phys. Lett.*, **134**, 549 (1987).

298 R. Kaptein, F. J. J. deKanter, and G. M. Rist, *J. Chem. Soc. Chem. Commun.*, 499 (1981).

299 G. Closs and O. D. Redwine, *J. Am. Chem. Soc.* , **107**, 4543 (1985).

300 M. A. Meador and P. J. Wagner, *J. Org. Chem.*, **50**, 419 (1985).

301 B. Zhou and P. J. Wagner, *J. Am. Chem. Soc.*, **111**, 6796 (1989).

302 D. H. R. Barton, P. D. Magnus, and J. I. Okugun, *J. Chem. Soc. Perkin 1*, 1103 (1972).

303 J. Wirz, *Helv. Chim. Acta*, **57**, 1283 (1974).

304 H. A. J. Carless and G. K. Fekarurhobo, *Tetrahedron Lett.*, **25**, 5946 (1984).

305 H. A. J. Carless and S. Mwesigye-Kibende, *J. Chem. Soc. Chem. Commun.*, 1673 (1987).

306 G. Adam, A. Preiss, P. D. Hung, and L. Kutschabsky, *Tetrahedron*, **43**, 5815 (1987).

307 R. Breslow, S. Baldwin, T. Fletchner, P. Kalicky, S. Liu, and W. Washburn, *J. Am. Chem. Soc.*, **95**, 3251 (1973); R. Breslow, *Acc. Chem. Res.*, **13**, 170 (1980)..

308 R. Breslow, J. Rothbard, F. Herman, and M. L. Rodriguez, *J. Am. Chem. Soc.*, **100**, 1213 (1978).

309 B. Dors, H. Luftmann, and H. J. Schafer, *Chem. Ber.*, **116**, 761 (1983).

310 for a review, oce D. 3. Schuster in ref. 2, Vol. 3, Ch. 17, pp. 199-206.

311 S. Wolff, W. L. Schreiber, A. B. Smith, and W. C. Agosta, *J. Am. Chem. Soc.*, **94**, 7797 (1972).

312 S. Ayral-Kaloustian, S. Wolff, and W. C. Agosta, *J. Am. Chem. Soc.*, **99**, 5984 (1977).

313 B. Byrne, C. A. Wilson, S. Wolff, and W. C. Agosta, *J. Chem. Soc. Perkin 1* , 1550 (1979).

314 T. Kobayashi, M. Kuromo, H. Sata, and K. Nakanishi, *J. Am. Chem. Soc.*, **94**, 2863 (1972).

315 W. Herz, V. S. Iyer, M. G. Nair, and J. Saltiel, *J. Am. Chem. Soc.*, **99**, 2704 (1977).

316 A. B. Smith and W. C. Agosta, *J. Am. Chem. Soc.*, **95**, 1961 (1973).

317 G. Mehta and D. Subrahmanyam, *Tetrahedron Lett.* **28**, 479 (1987); **30**, 2709 (1989).

318 M. Tada, T. Maeda, and H. Saiki, *Bull. Chem. Soc. Japan*, **51**, 1516 (1978).

319 J. R. Scheffer, B. M. Jennings, and J. P. Louwerens, *J. Am. Chem. Soc.*, **98**, 7040 (1976).

320 R. G. Zepp and P. J. Wagner, *J. Am. Chem. Soc.*, **92**, 7466 (1970); N. J. Turro and T. Lee, *ibid.*, **92**, 7467 (1970).

321 H. G. Henning, R. Beilinghoff, A. Mahlow, H. Koeppel, and K. D. Schleinitz, *J. Prakt. Chem.*, **323**, 914 (1981).

322 Y. Kanaoka and K. Koyama, *Tetrahedron Lett.*, 4517 (1972).

323 Y. Kanaoka, *Acc. Chem. Res.*, **11**, 407 (1978).

324 M. Wada, H. Nakai, Y. Sato, and Y. Kanaoka, *Chem. Pharm. Bull.*, **30**, 3414 (1982).

325 M. Wada, H. Nakai, K. Kotera, Y. Sato, and Y. Kanaoka, *Chem. Pharm. Bull.*, **30**, 2275 (1982).

326 Y. Kanaoka, K. Koyama, J. L. Flippen, I. L. Karle, and B. Witkop, *J. Am. Chem. Soc.*, **96**, 4719 (1974).

327 M. Machida, H. Takechi, A. Sukushima, and Y. Kanaoka, *Heterocycles*, **15**, 479 (1981).

328 P. de Mayo, *Acc. Chem. Res.*, **9**, 52 (1976).

329 H. Aoyama, S. Suzuki, T. Hasegawa, and Y. Omote, *J. Chem. Soc. Perkin 1* , 247 (1982).

330 H. Aoyama, M. Sakamoto, and Y. Omote, *Chem. Lett.*, 1397 (1983).

331 A. Padwa, W. Eisenhardt, R. Gruber, and D. Pashayan, *J. Am. Chem. Soc.*, **93**, 6998 (1971).

332 E. H. Gold, *J. Am. Chem. Soc.*, **93**, 2793 (1971).

333   K. L. Ailworth, A. A. El-Hamamy, M. M. Hesab, and J. Hill, *J. Chem. Soc. Perkin 1*, 1671 (1980); M. M. Hesab, J. Hill and A. A. El-Hamamy, *ibid.*, 2371 (1980).

334   A. A. El-Hamamy, J. Hill, and J. Townsend, *J. Chem. Soc. Perkin 1*, 573 (1983).

335   J. C. Gramain, L. Ouzzani-Chahd, and Y. Troin, *Tetrahedron Lett.*, **22**, 3185 (1981).

336   J. Fuhrmann, M. Haupt, and H. G. Henning, *J. Prakt. Chem.*, **326**(2), 177 (1984).

337   P. J. Wagner and M. J. Lindstrom, *J. Am. Chem. Soc.*, **109**, 3062 (1987).

338   P. J. Wagner and W. Mueller, unpublished results.

339   M. V. Encinas and J. C. Scaiano, *J. Am. Chem. Soc.*, **101**, 2146 (1979).

340   R. J. DeVoe, A. E. Osborn, and S. V. Pathre, *J. Org. Chem.*, in press.

341   H. J. Roth and M. H. El Raie, *Tetrahedron Lett.*, 2445 (1970); H. J. Roth and M. H. El Raie, *Arch. Pharm.*, **305**, 209 (1972); H. J. Roth, M. H. El Raie, and T. Schrauth, *ibid.*, **307**, 584 (1974).

342   A. Abdul-Baki, F. Rotter, T. Schrauth, and H. J. Roth, *Arch. Pharm.*, **311**, 341 (1978).

343   H. G. Henning, R. Sukale, H. Buchholz, and H. Haber, *J. Prakt. Chem.*, **327**, 51 (1985).

344   S. Inbar, H. Linschitz and S. G. Cohen, *J. Am. Chem. Soc.*, **104**, 1679 (1982).

345   T. Hasegawa *et al.*, *J. Chem. Soc. Perkin 1*, 901 (1990).

346   Y. Sato, H. Nakai, T. Ogiwata, Y. Mizoguchi, Y. Mitiga, and Y. Kanaoka, *Tetrahedron Lett.*, 4565 (1973).

347   Y. Sato, H. Nakai, T. Mizoguchi, Y. Hatanaka, and Y. Kanaoka, *J. Am. Chem. Soc.*, **98**, 2349 (1976).

348   F. D. Lewis and T.-I. Ho, *J. Am. Chem. Soc.*, **102**, 1751 (1980).

349   Y. Kanaoka and Y. Migita, *Tetrahedron Lett.*, 3693 (1974).

350   M. Machida, H. Takechi, and Y. Kanaoka, *Chem. Pharm. Bull.*, **30**, 1579 (1982).

351   Y. Kanaoka, *Synthesis*, 1078 (1982).

352   L. R. B. Bryant, J. D. Coyle, J. F. Chaliner, and E. J. Haws, *Tetrahedron Lett.*, **25**, 1087 (1984).

353   J. D. Coyle and L. R. B. Bryant, *J. Chem. Res. (S)*, 164 (1982).

354   M. Machida, S. Oyadamari, H. Takechi, K. Ohno, and Y. Kanaoka, *Heterocycles*, **19**, 2057 (1982).

355   (a) K. Maruyama and Y. Kubo, *J. Am. Chem. Soc.*, **100**, 7772 (1978); (b) K. Maruyama, Y. Kubo, M. Machida, K. Oda, Y. Kanaoka, and K. Fukuyama, *J. Org.*

Chem., **43**, 2303 (1978); (c) M. Michida, K. Oda, and Y. Kanaoka, *Heterocycles*, **18**, 211 (1982)..

356 N. K. Hamer, *J. Chem. Soc. Perkin 1*, 61, (1983).

357 P. J. Wagner and B. J. Scheve, *J. Am. Chem. Soc.*, **101**, 378 (1979).

358 P.J. Wagner and T. Nakahira, *J. Am. Chem. Soc.*, **96**, 3668 (1974); J. P. Bays, M. V. Encinas, R. D. Small Jr. and J. C. Scaiano, *ibid.*, **102**, 727 (1980).

359 H. J. Frerking, Ph. D. Thesis, Michigan State University, 1978.

360 B. P. Giri and P. J. Wagner, unpublished results.

361 P. J. Wagner and N. Leventis, *J. Am. Chem. Soc.*, **109**, 2188 (1987).

362 G. R. McMillan, J. G. Calvert, and J. N. Pitts, Jr., *J. Am. Chem. Soc.*, **86**, 3602 (1964).

363 P. Haspra, A. Sutter, J. Wirz, *Angew. Chem., Int. Ed. Engl.*, **18**, 617 (1979); Y. Chiang, A. J. Kresge, Y. S. Tang, and J. Wirz, *J. Am. Chem. Soc.*, **106**, 460 (1984).

364 J. R. Keeffe and A. J. Kresge in *The Chemistry of Enols* (Z. Rappoport, Ed.) Wiley, Chichester, Ch. 7, p. 399 (1990); H. Hart, Z. Rappoport, and S. E. Biali, *ibid.*, Ch. 8, p. 481; A. C. Weedon, *ibid.*, Ch. 9, p. 591.

365 M. Hamity and J. C. Scaiano, *J. Photochem.*, **4**, 229 (1975).

366 N. C. Yang and D.-H. Yang, *J. Am. Chem. Soc.*, **80**, 2913 (1958).

5

Selected Topics in the Matrix Photochemistry of Nitrenes, Carbenes, and Excited Triplet States

MATTHEW S. PLATZ

ELISA LEYVA

Ohio State University
Columbus, Ohio

KARL HAIDER

Yale University
New Haven, Connecticut

I   Introduction

Mechanistic chemists have long appreciated that short lived, high energy species such as nitrenes, carbenes, and excited triplet states play a crucial role in the photochemistry of azides[1], diazo compounds[2], diazirines[3] and various π electronic systems.[4] One of the goals of physical organic chemistry is to characterize transient species using the same spectroscopic tools available to chemists studying kinetically stable materials. In general this can be accomplished in either of two ways; by using nano or picosecond time resolved spectroscopy, or by generating the species of interest at cryogenic temperatures in a rigid inert medium. In the past decade there has been a great deal of progress in time resolved spectroscopy and it is now possible to obtain UV-VIS absorption[5], fluorescence[6], infrared[7] and magnetic resonance[8] spectra of species with very short lifetimes in solution.

In the low temperature approach a precursor molecule such as a diazo-methane derivative may be condensed with an inert gas (e.g. Argon) on a cold finger cooled to 10K.[9] In a matrix isolation experiment of this kind the ratio of Argon to the diazo-methane will typically be of the order of several thousand to one, hence, each diazo molecule is completely surrounded by a rigid Argon matrix.

$$\underset{R}{\overset{R}{>}}=N_2 \quad \xrightarrow[\text{10K, h}\nu]{\text{Argon}} \quad \underset{R}{\overset{R}{>}}C: \ + N_2$$

Photolysis of the matrix isolated diazomethane derivative leads to fragmentation to form a carbene and molecular nitrogen. Whereas the lifetime of the carbene may be less than 1 nanosecond (ns) in an organic solvent at ambient temperature[10], its lifetime in a cold rigid matrix may be as long as minutes, hours or even days.

Time resolved spectroscopy and matrix isolation spectroscopy of carbenes, nitrenes, biradicals and radicals are now standard tools used in the study of photogenerated species and will not be reviewed here. However, the same methodology which enables the success of both time resolved spectroscopy (intense pulsed laser radiation) and matrix isolation spectroscopy (long lifetimes of the species of interest) may inadvertently or deliberately lead to secondary photolysis of the initially produced nitrene, carbene or excited triplet state.

The photochemistry of reactive intermediates in fluid solution has recently been reviewed and will not be covered herein.[11] Instead the secondary photolysis of certain matrix isolated nitrenes, carbenes, and triplet states will be considered. This chapter is not meant to be encyclopedic, instead the case histories of 4 diverse systems; phenyl nitrene, diphenylcarbene, triplet mesitylene and chlorinated benzyl radicals will be considered in some detail, in an attempt to demonstrate the complications and opportunities present in this area of research.

II The Low Temperature Photochemistry of Phenyl Nitrene

The photochemistry of phenyl azide **1** has been described as "wonderfully complex."[12] The complexity is due in part to the large number of candidate species which may in principle be formed and trapped upon photolysis of **1**. These include the singlet (**1S***) and triplet (**1T***) photochemical excited states of phenyl azide,

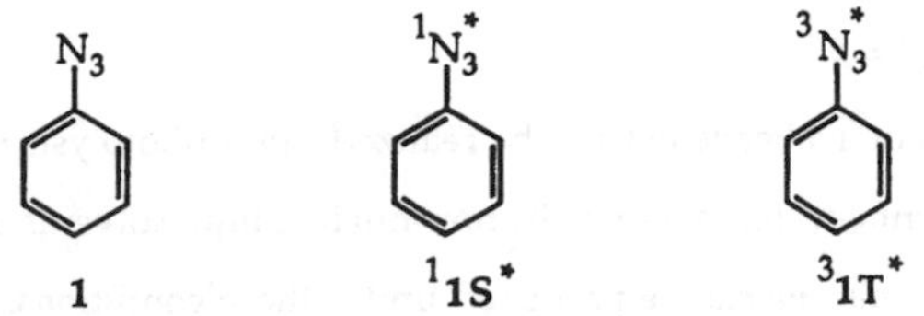

1   **^{1}1S***   **^{3}1T***

and reactive intermediates **2-4**.

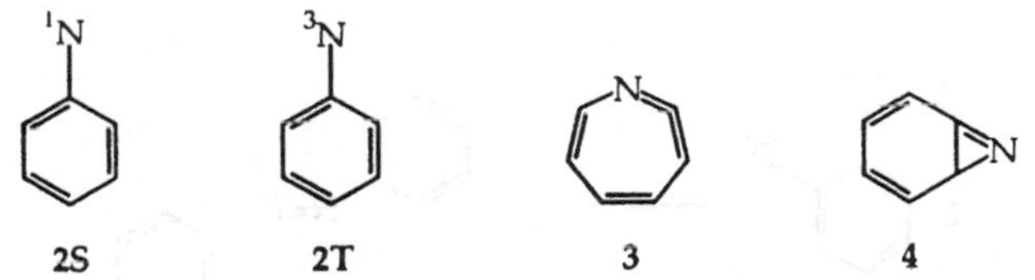

2S   **2T**   **3**   **4**

Early studies of the photochemistry of aryl azides produced evidence that supported the intermediacy of most of these types of species under certain conditions.[13] In 1964 Doering and Odum discovered that upon photolysis of phenyl azide in diethylamine (DEA) azepine **5** can be isolated in good yield. Simple mechanisms can be written to rationalize the formation of azepine **5** by trapping of either ketenimine **3** or bicyclic azirine **4**.[14]

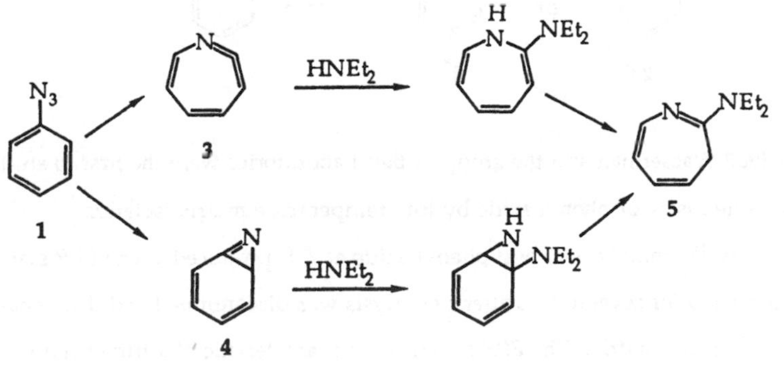

It is now clear that azepine **5** is formed by interception of ketenimine **3** rather than bicyclic azirine **4**,[12,15] however as we will discuss, benzo derivatives of **4** are indeed produced in the photochemistry of the naphthyl azides.

Polymeric tar is the major product produced upon photolysis of **1** in a non nucleophilic solvent such as benzene. The polymer, which has recently been characterized by Meijer, is most likely formed in a chain reaction involving both **3** and phenyl azide **1**.[15,16]

Modest yields of azobenzene **6** can be realized upon photolysis of highly dilute solutions of **1** at ambient temperature in non-nucleophilic solvents implying that some triplet phenyl nitrene can be produced under these conditions.[12] Larger yields of azobenzene and aniline are realized upon triplet sensitized photolysis of phenyl azide.[17]

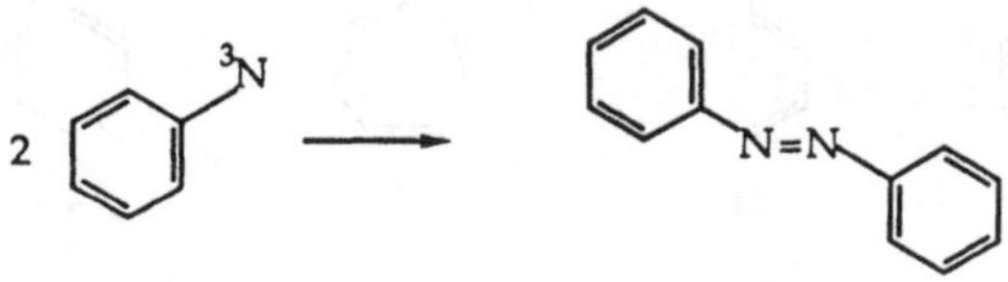

The formation of aniline among the products of photolysis of phenyl azide may be due to hydrogen atom abstraction reactions of triplet phenyl nitrene **2T** or due to a reaction of the excited triplet state of phenyl azide (³**1***)

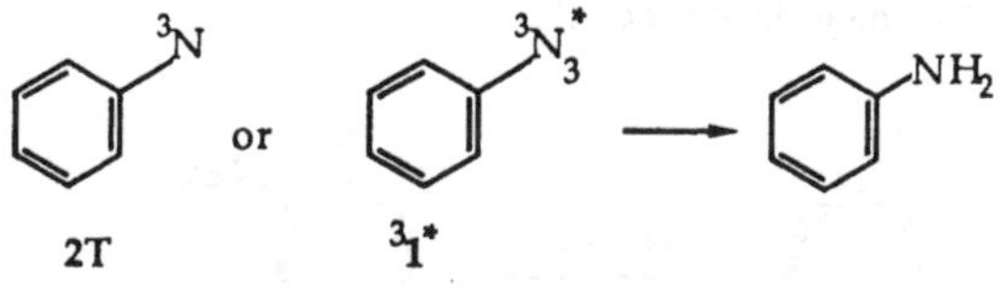

In 1962 Wasserman and the group at Bell Laboratories were the first to study the photochemistry of phenyl azide by low temperature matrix isolation spectroscopy.[18] Photolysis of neat phenyl azide at 77K produced a new EPR signal which persisted for several days after photolysis was discontinued and disappeared upon thawing the matrix. The EPR spectrum is characteristic of a triplet state species ($|D/hc| = 0.9978$ cm^{-1}) and quite distinct from that of a doublet state (free

radical) molecule. Wasserman and co-workers associated the carrier of the EPR

spectrum with triplet phenyl nitrene, an interpretation which has not been

challenged in the ensuing 27 years. Subsequent to this seminal study, the EPR

spectra of many aryl, alkyl, sulfonyl and carboalkoxy substituted nitrenes have been

reported.[19]

Shortly after Wasserman's study was described, Reiser and coworkers reported

their analysis of the low temperature photochemistry of phenyl azide.[20] A dilute

solution of phenyl azide in an ether-pentane-alcohol (EPA) mixture was cooled to

77K to form a clear rigid glass. Upon exposure of the glass to light, the UV-VIS

spectrum of phenyl azide disappeared and was replaced by a series of new

absorption bands which extended to longer wavelengths than the absorption of the

precursor. These photo induced absorption bands disappeared upon melting and

refreezing the glass. Reiser and coworkers followed Wasserman's lead and

associated the photo induced UV-VIS bands with triplet phenyl nitrene. This

assignment was supported by the fact that the UV-VIS spectrum attributed to 2T

resembled that of closely related open shell systems such as benzyl and anilino

radical.

In 1978 Chapman deposited phenyl azide in an Argon matrix at 10K. Upon

brief photolysis of the matrix the intense IR band of the azide group near 2200

cm^{-1} disappeared and was replaced by a new absorption band at 1895 cm^{-1}.[21] The

same band could be observed in solution at ambient temperature by Schuster and

Poliakof et al. utilizing time resolved IR spectroscopy, and is most reasonably

attributed to ketenimine 3, of all of the possible candidate structures.[15]

However, Chapman found that the same matrix isolated sample which gave
an intense IR band of 3 also produced an intense EPR spectrum of the triplet
nitrene 2T. However, due to the inherent sensitivity differences in these
techniques, the ratio of [2T]/[3] formed in the matrix could not be determined. In
fact, following exhaustive photolysis of the mixture of the matrix isolated aryl
nitrene and ketenimine, the EPR spectra of an aryl carbene was also observed. The
carbene was identified as 2-pyridylcarbene (7T) based upon a comparison of the EPR
spectra produced from prolonged photolysis of a matrix containing phenyl azide
with the authentic spectra of 7T prepared by photolysis of 2-pyridyl diazomethane 9.

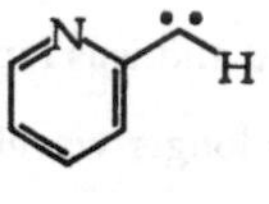

7T

Upon photolysis of 7T in Argon at 10K it was possible to detect the IR spectrum
of 3. Chapman and co-workers interpreted these observations in the manner
described in Scheme 1.

In view of Chapman's results one can not help but wonder if Reiser's UV-VIS
spectrum is due to triplet phenyl nitrene 2T, ketenimine 3, pyridylcarbene 7T or to
some mixture of the above species. One can also wonder which compound is the
initial species formed upon photolysis of phenyl azide at low temperature.

Scheme 1

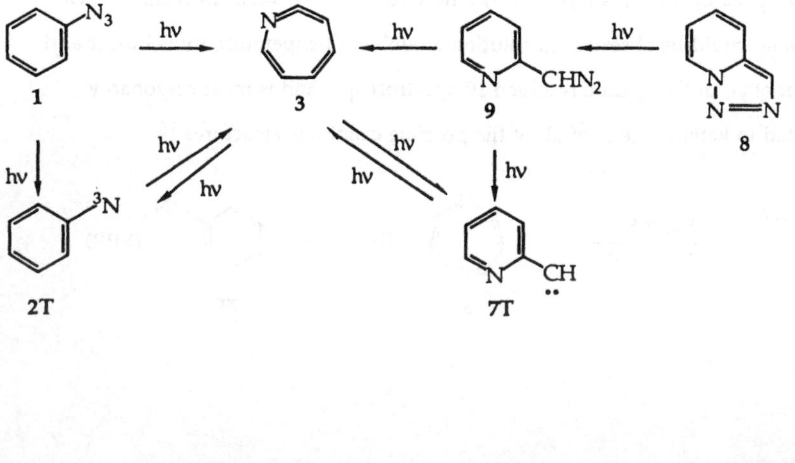

Reiser's low temperature UV-VIS spectrum has been reproduced by several research groups.[22] Leyva et al. were able to detect a weak fluorescence from a sample of phenyl azide which had been briefly photolyzed at 77K in an ether-pentane-alcohol (EPA) glass.[23] This emission had a mirror image symmetry relationship with the long wavelength tail of the absorption band produced at low temperature (Figure 1). Interestingly, the excitation spectrum of the fluorescence was in excellent agreement with the low temperature UV-VIS absorption spectrum. This demonstrated that only a single UV-VIS active species was produced by brief photolysis of phenyl azide in EPA at 77K.[24]

The same pattern of results were observed in the low temperature photochemistry of 2,6 dimethylphenyl azide and pentafluorophenyl azide as well.[23]

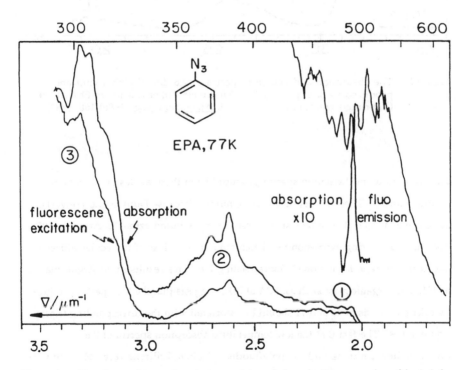

Figure 1. The absorption and emission spectra of phenyl nitrene produced by brief photolysis of phenyl azide in EPA at 77K. (Reprinted with permission from J. Am. Chem. Soc. (1986) 108, 3783)

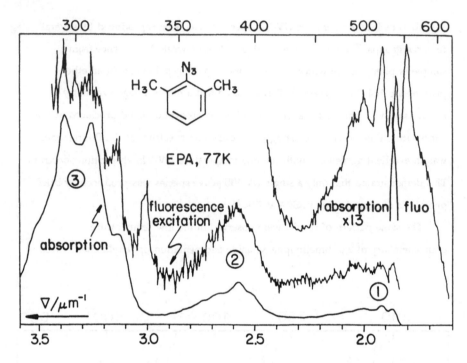

Figure 2.   The absorption and emission spectra of 2,6-dimethylphenylnitrene
produced by brief photolysis of 2,6-dimethylphenyl azide in EPA at 77K.
(Reprinted with permission from J. Am. Chem. Soc. (1986) **108**, 3783)

The absorption and excitation spectra produced from these azides are shown in
Figures 2 and 3. These observations are significant because Dunkin and co-workers
have demonstrated that photolysis of these two substituted azides in Argon at 10K
does not lead to ring expansion to a ketenimine, but, to loss of molecular nitrogen
and triplet nitrene formation.[25] Extrapolating Dunkin's results in an Argon matrix
at 10K to an organic glass such as, EPA at 77K, suggests that the UV spectra produced
by brief photolysis of **10** and **11** should be associated with the corresponding triplet
aryl nitrenes. The fact that the low temperature absorption spectra of 2,6-
dimethylphenyl nitrene and the polyfluorinated phenyl nitrene resembles the UV-
VIS spectrum of the product of photolysis of phenyl azide in EPA at 77K, implies
that in all three cases a triplet phenyl nitrene has been produced.

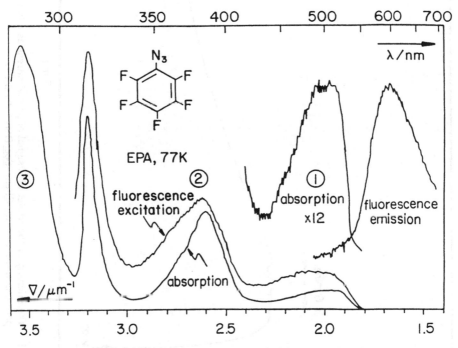

12   **10**   **13T**

15   **11**   **14T**

It is important to note that the spectra shown in Figures 1, 2 and 3 are produced by very brief, monochromatic (254 nm), photolysis of the azide

Figure 3. The absorption and emission spectra of pentafluorophenyl nitrene produced by brief photolysis of pentafluorophenyl azide in EPA at 77K. (Reprinted with permission from J. Am. Chem. Soc. (1986) 108, 3783)

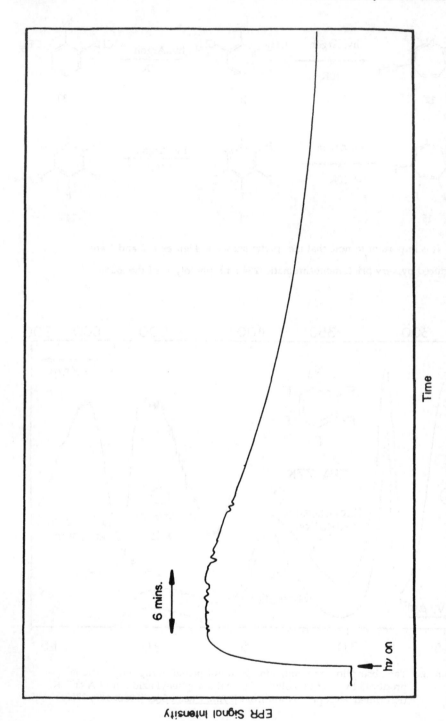

Figure 4.   The intensity of the triplet EPR signal of phenyl nitrene as a function of photolysis time at 77K.   (Reprinted with permission from J. Am. Chem. Soc. (1986) 108, 3783)

precursors. Extended photolysis of a glass containing phenyl azide leads to rapid destruction of the spectrum shown in Figure 1. The sharp bands initially present at 308 and 370 nm are replaced by a broad featureless absorption band around 330 nm. The latter absorption band is presumably due to ketenimine 3, formed by extended photolysis of the nitrene.

Upon photolysis of phenyl azide in EPA at 77K a strong EPR spectrum of triplet phenyl nitrene is produced. It is interesting to note that this spectrum is produced by brief photolysis of the glass, however, extended photolysis of the sample rapidly bleaches the EPR spectrum of the triplet nitrene (Figure 4).[23] Thus the growth of Reiser's UV-VIS absorption bands and their subsequent

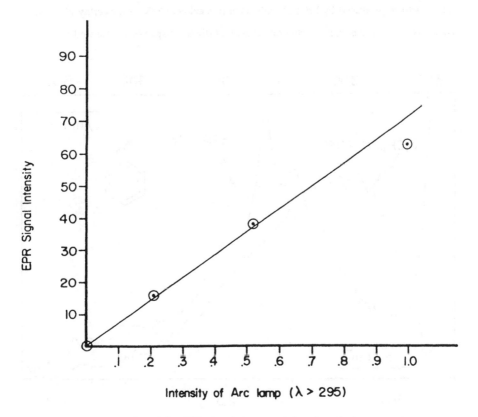

Figure 5.   The intensity of the EPR signal due to triplet phenyl nitrene as a function of light intensity.   (Reprinted with permission from J. Am. Chem. Soc. (1986) 108, 3783)

disappearance is mirrored by the behavior of the triplet nitrene EPR signal. Leyva et al. have also measured the yield of triplet nitrene as determined by EPR spectroscopy, as a function of light intensity. At very low light levels the intensity of the triplet EPR signal produced from phenyl azide is found to be a linear function of the light intensity (λ>295 nm). This demonstrates that triplet phenyl nitrene is produced from phenyl azide in a monophotonic process (Figure 5). Thus, these two observations further imply that Reiser's UV-VIS spectrum is indeed due to triplet phenyl nitrene.

It is interesting to compare the UV-VIS spectrum of the transients produced by laser flash photolysis of phenyl azide at room temperature with the persistent UV-VIS spectrum produced by brief photolysis of this azide at 77K. An overlay of these two spectra (Figure 6) rapidly demonstrates that different species are formed at

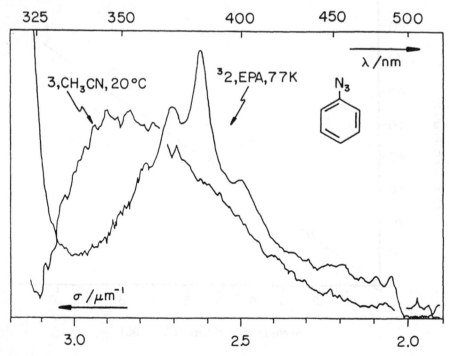

Figure 6.   An overlay of the transient spectrum of ketenimine 3 produced by flash photolysis of phenyl azide at ambient temperature with that of triplet phenyl nitrene 2T produced at 77K.   (Reprinted with permission from J. Am. Chem. Soc. (1986) 108, 3783)

ambient and at low temperature. Spectroscopic differences are even more pronounced upon comparing the room temperature flash spectrum and the low temperature matrix spectrum generated from 2,6-dimethylphenyl azide (Figure 7). The broad spectra produced by flash photolysis of **1** and **10** at room temperature are due to the ketenimines **3** and **12** whereas the more structured absorptions observed at low temperature are due to the triplet nitrenes **2T** and **13T**. Thus, the photochemistry of phenyl azide is a sensitive function of temperature. This is further demonstrated by monitoring the products formed upon photolysis of phenyl azide in the presence of diethylamine (Table 1). Photolysis of phenyl azide in 2-methyltetrahydrofuran (2-MTHF) containing 1.0 M diethylamine at ambient temperature produces the azepine product **5** reported by Doering and Odum.[14] The

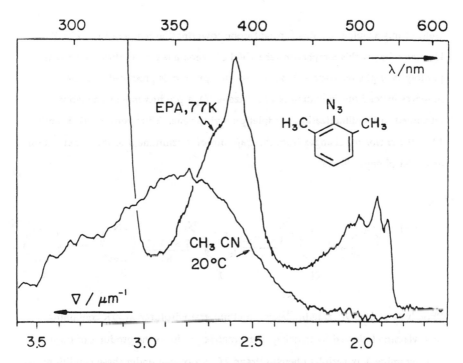

Figure 7.   An overlay of the transient spectrum of ketenimine **12** produced by flash photolysis of 2,6-dimethylphenyl azide at ambient temperature with that of 2,6-dimethylphenylnitrene **13T** produced at 77K. (Reprinted with permission from J. Am. Chem. Soc. (1986) **108**, 3783)

Table 1. Distribution of Products Formed on Photolysis of **1** in 2-Methyltetra-
hydrofuran-diethylamine as a Function of Temperature[23]

T/K	5	6
293	59	0
273	59	0
253	58	0
233	58	0
213	43	0
193	25	0
173	0	11
153	0	24
77	16	7

yield of this azepine remains constant upon lowering the temperature to 233K.
However, below this temperature the yield of azepine begins to drop. Finally upon
photolysis of phenyl azide at 173 K no azepine product is produced and one
observes instead the formation of azobenzene (**6**), a product most reasonably
attributed to the dimerization of triplet phenyl nitrene. Thus between 173K and
193K the chemistry changes from the trapping of ketenimine **3** to the dimerization
of triplet phenyl nitrene.

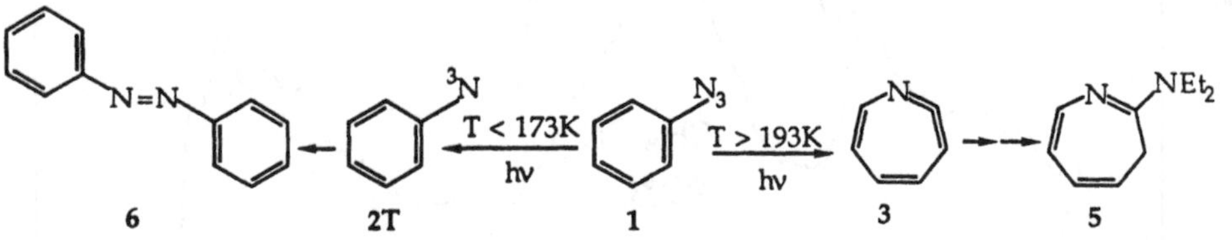

Photolysis of phenyl azide in glassy 2-methyltetrahydrofuran at 77K containing
diethylamine followed by thawing of the matrix produces the products of trapping
of ketenimine **3** and triplet phenyl nitrene 2T. Of course under these conditions we
are not trapping reactive intermediates as per the work done in fluid solutions. At
77K the sample is immobilized in a rigid glass and the primary photo products are

not reactive intermediates but are persistent species which can undergo their own characteristic photochemistry.

The nature of the photoreaction which consumes matrix isolated triplet phenyl nitrene has been elucidated by studying the products obtained upon thawing the matrix as a function of photolysis time (Table 2). Upon brief photolysis of phenyl azide in a 2-MTHF glass containing diethylamine at 77K the only product detected upon thawing the matrix is azobenzene, a characteristic reaction product formed by dimerization of triplet phenyl nitrene. If the glass is photolyzed for a longer period of time at 77K before it is thawed, both the triplet nitrene derived product (azobenzene) and the ketenimine derived product (azepine) are formed in nearly equal amounts. If the matrix is subjected to a rather long photolysis at 77K, the yield of azobenzene produced upon thawing the matrix remains constant, while the azepine is now the major product formed. The data clearly indicate that azobenzene is derived from a primary photo product; triplet phenyl nitrene, while the precursor to azepine 5 is a secondary photo product. The simplest interpretation of the data is that photolysis of matrix isolated triplet phenyl nitrene in 2-MTHF at 77K leads to its isomerization to the ketenimine . Thus the photochemistry of phenyl azide shows a rather unusual temperature dependence. Photolysis of 1 at ambient temperature produces a ketenimine, whereas photolysis

Table 2. Distribution of Products Formed on Photolysis of 1 in 2-Methyltetra-hydrofuran-diethylamine at 77K as a Function of Photolysis Time[23]

350 Rayonet Lamp				366 nm 1000 W 11g Xe Arc Lamp			
time/min	1	5	6	time/min	1	5	6
5	89	0	8	1	95	0	1.0
10	63	7	11	2	93	0	1
15	58	7	12	5	93	1	2
30	56	10	11	15	90	1	2
60	48	21	13	30	83	2.5	1
				63	83	3.8	1

in an organic glass at 77K initially forms triplet phenyl nitrene. Ironically,
prolonged photolysis at 77K of an organic glass originally containing phenyl azide
isomerizes the initially formed triplet nitrene to the species that is the primary
photoproduct in solution at ambient temperature.

Singlet phenyl nitrene 2S is the branching point in the favored mechanism
and is responsible for the temperature dependence of the photochemistry of phenyl
azide. We speculate that the ring expansion reaction of 2S involves migration of
the electron pair of a C-C sigma bond into an empty orbital of the singlet nitrene
which is in the plane of the benzene ring.

This unimolecular process should have a very large Arrhenius pre-exponential
factor but a non zero activation enthalpy as bonds are broken and reformed in the
transition state. The competing reaction is intersystem crossing, which may have
zero activation energy as no bonds are broken in this process. However, the
intersystem crossing process is most likely associated with a relatively low pre-
exponential Arrhenius parameter.

High temperature favors the process with the larger pre-exponential term (ring
expansion) but low temperature favors the process with the lower activation
enthalpy (ISC).

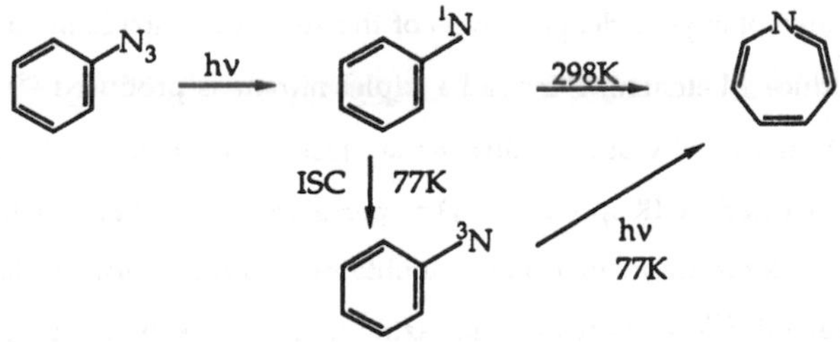

The data and conclusions of Leyva et al.[23] are consistent with the transformations discovered by Chapman and co-workers[21] in Argon matrices. However, as we will see it remains to be demonstrated whether the primary photoproduct in Argon and in organic glasses is exactly the same.

Substituted Phenyl Nitrenes

We have briefly alluded to the low temperature photochemistry of pentafluorophenyl azide 11, and 2,6 dimethylphenyl azide 10. It is clear that photolysis of the 2,6 dimethylphenyl azide produces different reactive intermediates at room temperature and at low temperature (Figure 7). The photochemistry of 2,6 dimethylphenyl azide is quite similar to that of phenyl azide. Photolysis of the dimethylated azide in solution at ambient temperature releases a ketenimine whereas photolysis of this azide in an organic glass at 77K produces a triplet nitrene.[23]

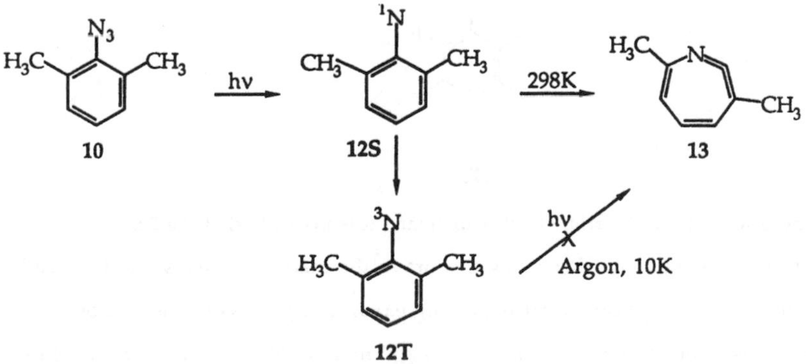

However, unlike phenyl azide, photolysis of the substituted azide in Argon at 10K
does not produce a ketenimine, instead a triplet nitrene is produced.[25a] Exhaustive
photolysis of the dimethylphenyl nitrene may lead to its isomerization to the
ketenimine (detected by IR spectroscopy) as per parent phenyl nitrene but only in
trace amounts. More dramatic results are obtained with pentafluorophenyl azide
11 in Argon at 10K.[25b] Photolysis of this azide under these conditions again leads to
the formation of a nitrene. In this case there is no spectroscopic evidence for the
formation of the ketenimine upon prolonged photolysis of this nitrene. Photolysis
of 11 in organic solvents at ambient temperature indicates that singlet nitrene 14S
does not isomerize to ketenimine 15 under these conditions, although evidence
exists that gas phase pyrolysis of 11 leads to loss of nitrogen and ketenimine
formation.[26] It appears that in general 2,6 disubstitution of phenyl azide and the
corresponding reactive intermediates produced photochemically suppresses the
ring expansion reaction in Argon at 10K, although it does not necessarily suppress
this reaction in solution at ambient temperature,[23] or in the gas phase at several
hundred degrees centigrade.

The origin of the di-ortho effect is unclear. It is possible that the 2,6
substituents on the phenyl ring act as anchors in the matrix. Perhaps the steric bulk
of the substituents physically retards the ring expansion process by the friction
which may accompany ring expansion in a rigid matrix. This is usually referred to

as the "Topological Principle."[27] This explanation seems reasonable in the case of the dimethyl substituents but it seems less likely in the case of fluorine which is a rather small substituent. If a phenyl nitrene is substituted with only a single ortho substituent, photolysis of this nitrene at 10K leads to isomerization and ring expansion to the ketenimine away from the ortho subsitutent.[28]

There is reason to suspect that the photochemistry observed in Argon at 10K is not necessarily the same as the photochemistry that prevails in organic glasses at 77K. Photolysis of phenyl azide in Argon at 10K will produce molecular nitrogen and a singlet nitrene with excess vibrational energy. Because Argon is a rather poor heat sink the vibrationally hot singlet nitrene formed in this matrix may isomerize to form the ketenimine before it can "cool off." Photolysis of phenyl azide in an organic glass at 77K produces a singlet phenyl nitrene which can rapidly transfer its excess heat to the organic glass and undergo intersystem crossing to the ground state triplet nitrene. It is ironic to note that the photochemistry of phenyl azide in Argon at 10K may in fact more closely resemble the gas phase pyrolysis of the azide at several hundred degrees centigrade, than its photolysis in an organic glass at 77K. In fact Wentrup has shown that gas phase pyrolysis of aryl azides leads to ring expansion chemistry.[29] The 2,6 di-ortho substituent effect may have its origin in vibrational deactivation as the 2,6 substituents may be able to absorb some of the excess vibrational energy produced in the nascent singlet nitrene, and thus facilitate its deactivation.

There is evidence for this type of an effect in the recent work of LeBlanc and Sheridan.[30] Photolysis of 16 in toluene-d_8 at 77K gives predominantly 7-norbornadienone 17.

However, photolysis of 16 in Argon at 10K produces mainly benzene and carbon monoxide. Higher yields of 17 are realized upon photolysis of 16 in a nitrogen

matrix at 10K and considerably higher yields of **17** are realized upon photolysis of **16**
in 3-methylpentane at this temperature. The relative ratio of compound **17** to
benzene produced in Argon, nitrogen and 3-methylpentane matrices are 1.0, 2.9,
and 6.5 respectively. Presumably benzene is formed by decomposition of
vibrationally hot **17** before it is deactivated. Apparently the ability of a low
temperature host to quench a vibrationally excited molecule is matrix dependent.
An analogous "hot molecule" explanation could resolve the disparity between
Chapman's results in Argon at 10K with phenyl azide, and the work carried out in
organic glasses at higher temperatures. Work of this type with phenyl azide is
indicated.

Light Induced Hydrogen Atom Transfer Reactions of Substituted Phenyl Nitrenes

Photolysis of 2,6-difluorophenyl azide **14** and pentafluorophenyl azide **12** in an
organic glass at 77K produces the triplet EPR spectra of nitrenes **15T** and **18T**.[31]
Upon continued photolysis of the nitrenes in an optically clear glass such as
2MTHF, the triplet nitrene EPR signals are rapidly bleached and replaced by EPR
signals that are characteristic of radical pairs. Attribution of the EPR signals to
radical pairs **19T** and **20T** stems from three observations; one that perdeuteration of
the matrix substantially retards the formation of the EPR signal of the radical pair
and two, that in a perfluorinated matrix where hydrogen atom transfer is
impossible, nitrenes **15T** and **18T** are remarkably stable towards prolonged
photolysis. Furthermore, radical pair EPR spectra are not generated upon
photolysis of **12** or **14** in a perfluorinated matrix.[31] Finally, if one analyzes the

stable products formed upon exhaustive photolysis of 2,6 difluorophenyl azide and pentafluorophenyl azide in polycrystalline toluene one finds near quantitative yields of the corresponding products of formal CH insertion. These products are easily visualized as resulting from the collapse of radical pairs **19T** and **20T**. (Scheme 2)

In these two cases hydrogen atom abstraction reactions of the nitrene with the matrix are favored because the 2,6-di-ortho substitution pattern suppresses the competing ring expansion reaction of the phenyl nitrene. Similar results are realized for the same reason in the low temperature photochemistry of 2,6 dimethylphenyl azide in organic glasses.[32]

Despite repeated attempts, several investigations of the photochemistry of parent triplet phenyl nitrene in rigid organic media failed to turn up evidence of hydrogen transfer reactions of **1T**.[32] This is consistent with efficient ring expansion to ketenimine **3**.

The photochemically induced hydrogen atom abstraction reactions of triplet nitrenes occur to some extent even when there is only a single substituent on the

Scheme 2

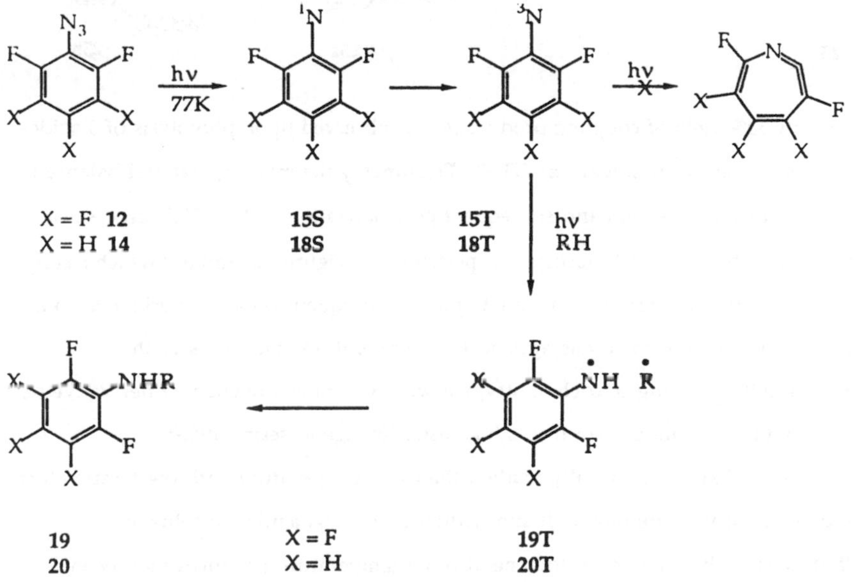

| X = F 12 | 15S | 15T | hv |
| X = H 14 | 18S | 18T | RH |

| 19 | X = F | 19T |
| 20 | X = H | 20T |

benzene ring, and that substituent need not in fact be placed at the ortho position.
The first example of this reaction was provided by meta-nitrophenyl azide.[33]
Prolonged photolysis of a polycrystalline toluene matrix (77K) containing 21 ,
followed by thawing of the matrix produces 22, the product of coupling in 30%
yield. This product (22) is not formed upon photolysis of the azide in toluene
solution at ambient temperature.

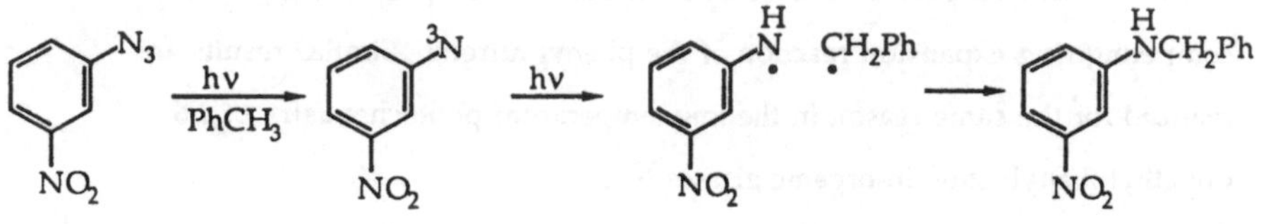

A nearly 30% yield of coupling product 24 was observed upon photolysis of 3 azido-
methylbenzoate 23 in toluene at 77K.[34] Presumably the missing material balance is
a result of ring expansion and subsequent polymerization of 25a. This process can
be followed by matrix UV-absorption spectroscopy (Figures 8a and 8b) which clearly
show the initial formation of a triplet and its subsequent photoisomerization to a
ketenimine. Kanakarajan has applied low temperature conditions to the
photoaffinity labelling of α-chymotrypsin with a meta azidobenzoate derivative of
23 bound to α-chymotrypsin through an ester linkage to serine 195.[34]

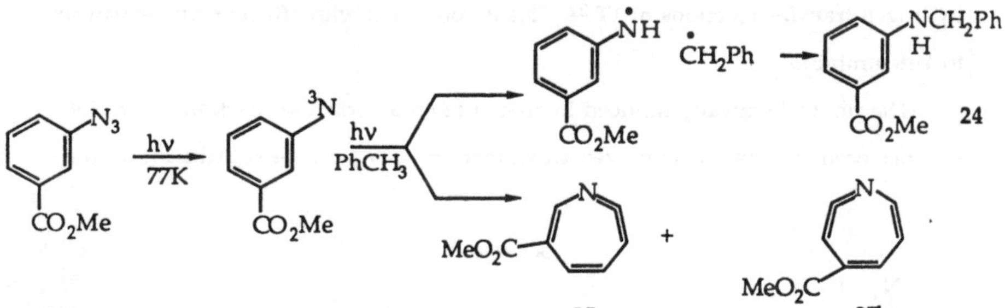

Munoz has systematically studied the room temperature and low temperature
photochemistry of the mono, di and trifluorinated aryl azides in toluene.[35]
Photolysis of these azides in toluene at room temperature produces mainly the

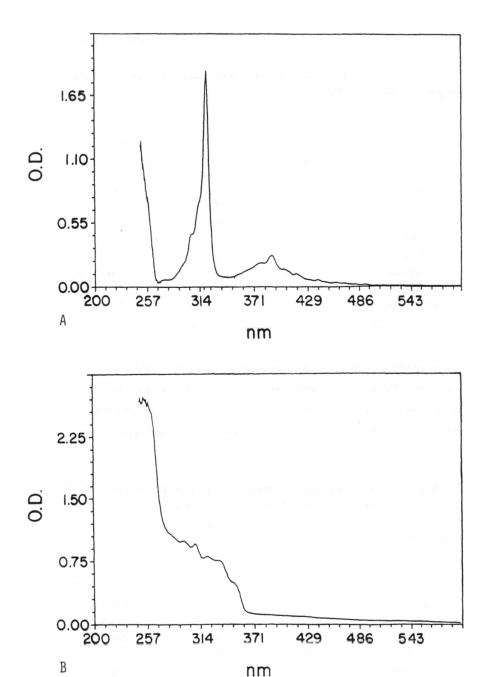

Figure 8. The UV-VIS spectrum of meta-carbomethoxyphenylnitrene produced by brief photolysis of the azide in EPA at 77K, and that of the isomeric ketenimine produced by extended photolysis of the triplet nitrene. (Reprinted with permission from J. Am. Chem. Soc. (1988) 110, 6536)

Table 3. Distribution of Products Formed on Photolysis of Fluorinated Phenyl
Azides in Toluene at 298K.[35]

Subsituted Phenyl Azide	Aniline	Azo	Toluene Adduct
2 - fluoro	9	24	4
3 - fluoro	3	8	traces
4 - fluoro	traces	7	-
2,3 - difluoro	4	16	2
2,4 - difluoro	5	24	11
2,5 - difluoro	9	16	4
3,4 - difluoro	traces	12	traces
2,6 - fifluoro	17	3	13
2,3,4,5,6 - pentafluoro	12	-	52

corresponding anilines and azo compounds and only trivial yields of CH insertion
adducts (Table 3). The situation is not improved by triplet photosensitized
photolysis of these azides in toluene solution (Table 4) or upon photolysis at 195K
(Table 5). However, photolysis of these azides in frozen polycrystalline toluene at

Table 4. Distribution of Products Formed on Photolysis of Fluorinated Phenyl
Azides in Toluene at 298K under Sensitized Conditions.[35]

Subsituted Phenyl Azide	Aniline	Azo	Toluene Adduct
2 - fluoro	10	34	-
3 - fluoro	4	21	traces
4 - fluoro	traces	22	traces
2,3 - difluoro	6	28	traces
2,4 - difluoro	6	48	5
2,5 - difluoro	traces	33	traces
3,4 - difluoro	8	16	3
2,6 - difluoro	54	10	5
2,3,4,5,6 - pentafluoro	17	-	8

Table 5. Distribution of Products Formed on Photolysis of Fluorinated Phenyl Azides in Toluene at 198K.[35]

Subsituted Phenyl Azide	Aniline	Azo	Toluene Adduct
2 - fluoro	5	30	-
3 - fluoro	3	13	traces
4 - fluoro	3	16	traces
2,3 - difluoro	5	35	traces
2,4 - difluoro	traces	30	-
2,5 - difluoro	3	22	traces
3,4 - difluoro	traces	10	traces
2,6 - difluoro	16	24	traces
2,3,4,5,6 - pentafluoro	10	17	33

77K, followed by thawing of the matrix, produces fair to modest yields of insertion products (Table 6). The yields of insertion products were largest in cases in which at least one fluorine was ortho to the azide group. Monitoring of the nitrene by EPR spectroscopy indicates that the C-H insertion products arise via a photochemical reaction of the matrix isolated triplet nitrene rather than through a thermal

Table 6. Distribution of Products Formed on Photolysis of Fluorinated Phenyl Azides in Toluene at 77K.[35]

Subsituted Phenyl Azide	Aniline	Azo	Toluene Adduct
2 - fluoro	4	14	35
3 - fluoro	4	8	8
4 - fluoro	traces	16	6
2,3 - difluoro	traces	16	14
2,4 - difluoro	5	18	39
2,5 - difluoro	traces	7	25
3,4 - difluoro	traces	25	8
2,6 - difluoro	traces	6	94
2,3,4,5,6 - pentafluoro	6	8	77

reaction of the nitrene. Neither triplet sensitization nor photolysis of the azide in solution phase at low temperature succeeds in producing insertion products. These products are only found when the host medium is rigid. Under these circumstances the photogenerated triplet nitrene is long lived and can undergo its own characteristic secondary photochemistry. A similar pattern of results has been observed by Soundararajan who studied the ambient and low temperature photochemistry of a series of fluorine substituted azido methyl benzoates.[36]

Perhaps one of the most interesting systems is 2,4,6 tribromophenyl azide.[23] In this system the same photogenerated species is produced in a glass at low temperature and in solution at ambient temperature (Figure 9). By comparison to Figures 1-3 the spectrum observed at high and low temperature clearly indicates

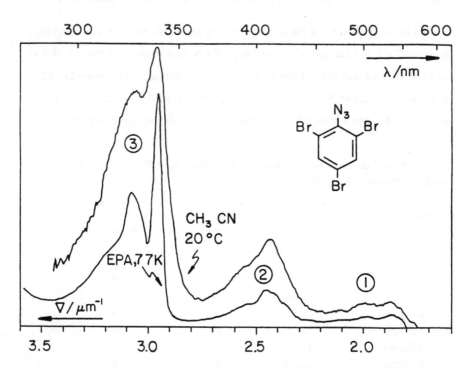

Figure 9.   The UV-VIS spectrum of 2,4,6 tribromophenylnitrene produced by flash photolysis of 2,4,6 tribromophenylazide at ambient temperature, and by photolysis of the azide at 77K.   (Reprinted with permission from J. Am. Chem. Soc. (1986) 108, 3783)

Figures 1-3 the spectrum observed at high and low temperature clearly indicates that the transient species is the triplet nitrene. In this case photolysis of the azide again leads to fragmentation to form a singlet nitrene and molecular nitrogen. However, due to the presence of the three heavy atom substituents, intersystem crossing is accelerated[37] and is now the kinetically favored process even at room temperature. It is interesting to note that heavy atom effects are not generally observed in the photochemsitry of diazo compounds and their associated carbenes.[38] This would seem to indicate that intersystem crossing in aryl nitrenes may be slower than that of aryl carbenes.

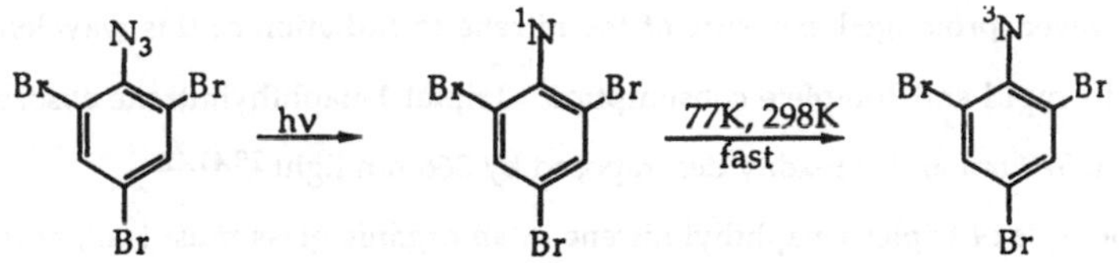

1 and 2 Naphthyl Azide

The low temperature photochemistry of 1 and 2-naphthyl azides differs from that observed with phenyl azide. Upon photolysis of 1-naphthyl azide in Argon at 10K the infrared bands of the azide disappear and are replaced by new bands between 1708 and 1730 cm[-1]. These bands are attributed to the two isomeric azirines 26 and 27.[39]

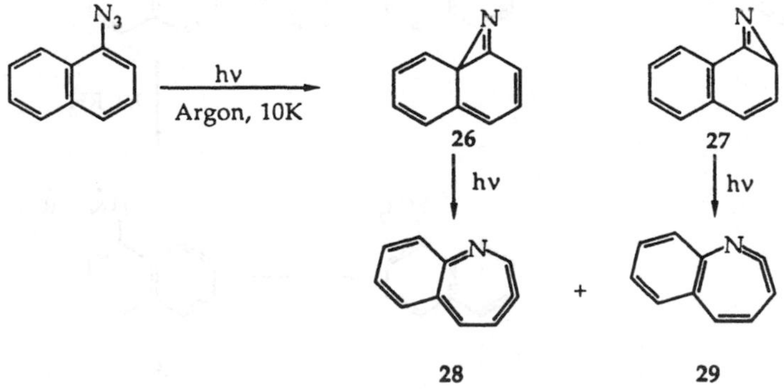

Upon exhaustive photolysis of azirines **26** and **27** in Argon at 10K the bands at 1730 cm[-1] disappear and are replaced by new bands between 1911 and 1926 cm[-1]. This indicates that the two isomeric azirines can be photochemically converted into two isomeric ketenimines **28** and **29**. Similar results were obtained with 2-naphthyl azide.

Photolysis of 1-napthyl azide in an organic glass at 77K produces the persistent EPR[19] and UV-VIS[20] spectrum of triplet 1-naphthylnitrene. It can be demonstrated that triplet 1-naphthylnitrene is a primary photo product in an organic host matrix as the intensity of the triplet EPR signal is a linear function of light intensity.[40] Triplet 1-naphthylnitrene is formed rapidly upon exposure of the azide to 366 nm light, however, prolonged exposure of the nitrene to radiation of this wavelength leads to its rapid and complete consumption. Triplet 1-naphthylnitrene absorbs strongly at 370 nm and is readily decomposed by 366 nm light.[20,41,42]

Photolysis of triplet 1-naphthyl nitrene in an organic glass must lead, at least in part, to hydrogen atom abstraction from the matrix.[39,43] Indeed the EPR

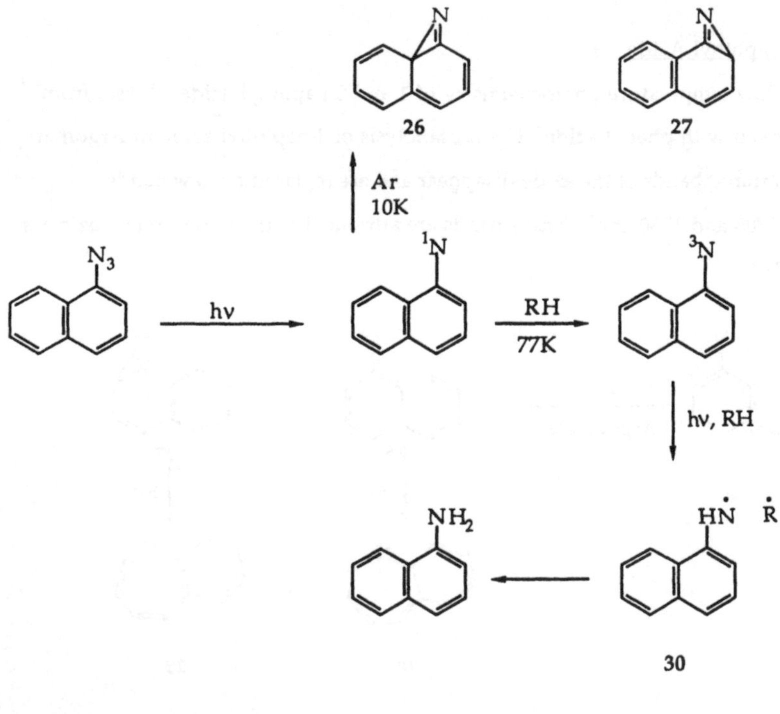

30

spectrum of a radical pair (30) can be observed upon low temperature photolysis of the azide in an organic host. Furthermore one of the stable products (1-naphthylamine) formed upon thawing an exhaustively photolyzed matrix containing 1.0 M diethylamine are consistent with this interpretation. In this glass adducts formed by coupling of radical pair 30 were not observed. Thus the primary product(s) obtained upon photolysis of 1-napthyl azide in Argon at 10K, and in an organic glass at 77K are probably the azirines (26 and 27) and triplet 1-naphthylnitrene respectively.

III The Low Temperature Photochemistry of Triplet Aryl Carbenes

Diphenylcarbene (DPC) was among the first reactive intermediates to be extensively studied by low temperature matrix isolation methods.[44] Hutchison and Closs were the first to obtain the low temperature EPR spectrum of DPC upon photolysis of diphenyldiazomethane (DPDM).[45] This work was confirmed and continued by Wasserman and Trozzolo[46] and co-workers at Bell Laboratories. Hutchison and Closs also obtained the first low temperature UV-VIS spectrum of the carbene.[47] In later work Hutchison and Doetschman[48] prepared diphenylcarbene in a single crystal of 1,1 diphenylethylene and measured its absolute rate of reaction with the olefinic host at low temperature. Trozzolo and Gibbons[49] were the first to observe the low temperature fluorescence spectrum of diphenylcarbene. In this experiment, the carbene was produced upon photolysis of diphenyldiazomethane in a clear glass of 2-methyltetrahydrofuran at 77K. Subsequent excitation of the glass produced the fluorescence emission spectrum of triplet diphenylcarbene. Upon exhaustive photolysis of the glass the emission spectrum of the carbene disappeared and was replaced by the emission spectrum of the benzhydryl radical. This was the first report of a photochemically initiated hydrogen atom abstraction reaction of a carbene in a low temperature matrix.[49]

DPDM   DPC

Photolysis of triplet DPC at low temperature initiates a reaction which is well known at room temperature,[45] but is thermally retarded in a low temperature matrix. Other examples of this phenomenon have been reported by Chapman and McMahon.[50] These workers prepared ortho methyl phenylcarbene 31 and phenylmethylcarbene 32 in Argon and other rare gas matrices at 10K. At this temperature carbenes 31 and 32 are relatively long lived because the thermally activated hydrogen shift reactions of these carbenes to form ortho-xylene and styrene respectively, are suppressed.

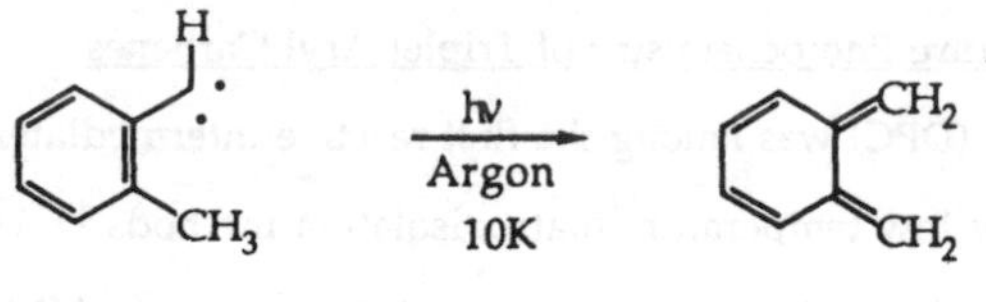

$$\text{31} \quad \xrightarrow[\text{10K}]{\substack{h\nu \\ \text{Argon}}}$$

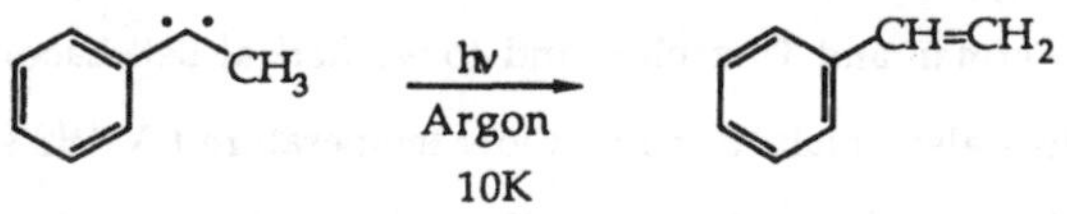

$$\text{32} \quad \xrightarrow[\text{10K}]{\substack{h\nu \\ \text{Argon}}}$$

$$\xrightarrow[\text{10K}]{\substack{h\nu \\ \text{Argon}}}$$

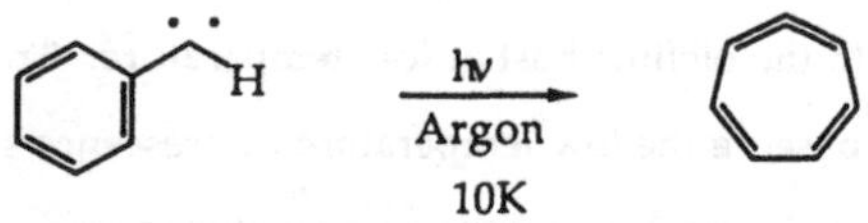

As carbenes 31 and 32 accumulate in the matrix they can absorb light and undergo photochemical reactions. Photolysis of 31 and 32 at low temperature leads to their rearrangement. The reaction products were identified by their characteristic UV-VIS and infrared spectra. Matrix photolysis of phenylcarbene leads to ring expansion, a reaction that occurs thermally in the gas phase at elevated temperatures.[51]

Kesselmayer and Sheridan have reported many photoreactions of singlet carbenes immobilized in rare gas matrices.[9,52] Usually the carbenes rearrange to known compounds upon photolysis. These observations can support the assignment of a matrix IR spectrum (produced by photolysis of a diazirine or diazo compound) to a particular carbene structure. The action spectrum for the light induced disappearance of the IR spectrum of the carbene yields the UV-VIS spectrum of the carbene. Several examples of this type are shown below.[52]

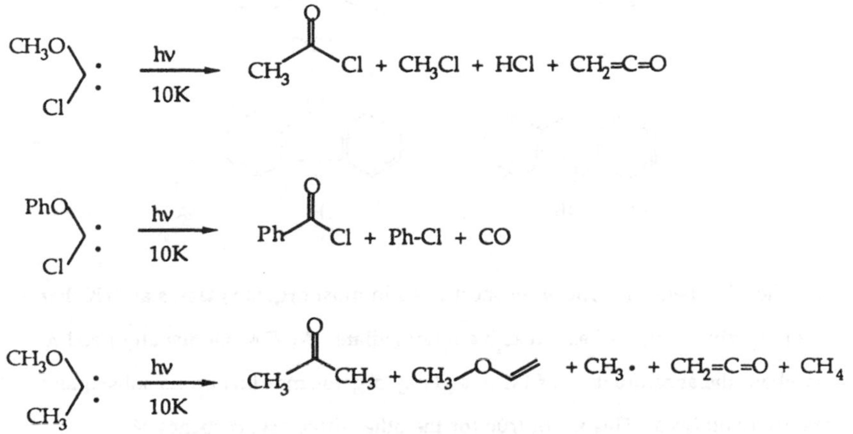

$$CH_3O\text{—}C(\text{:})(\text{:})\text{—}Cl \xrightarrow[10K]{h\nu} CH_3\text{—}C(=O)\text{—}Cl + CH_3Cl + HCl + CH_2=C=O$$

$$PhO\text{—}C(\text{:})(\text{:})\text{—}Cl \xrightarrow[10K]{h\nu} Ph\text{—}C(=O)\text{—}Cl + Ph\text{-}Cl + CO$$

$$CH_3O\text{—}C(\text{:})(\text{:})\text{—}CH_3 \xrightarrow[10K]{h\nu} CH_3\text{—}C(=O)\text{—}CH_3 + CH_3\text{—}O\diagdown\diagup + CH_3\cdot + CH_2=C=O + CH_4$$

Contrary to widespread belief, aryl carbenes are not necessarily long lived species in organic glasses at 77K. In fact, triplet carbenes experience a wide variation in lifetime at this temperature as a function of the structure of the carbene, the concentration of the diazo precursor, and the nature of the organic matrix itself.[53] Carbenes are sufficiently reactive species that they can undergo thermally activated processes even at 77K, albeit at rates that are orders of magnitude smaller than their rates of reaction in solution. Hydrogen atom abstraction processes are common reaction pathways of triplet carbenes and can proceed at a reasonable rate at 77K by a quantum mechanical tunneling mechanism. One of the challenges facing the carbene chemist is to distinguish thermal from photochemical reactions of a carbene at 77K.

Good examples of this dichotomy are provided by phenylcarbene (PC), diphenylcarbene (DPC), dibenzocycloheptadienylidene (DBC) and fluorenylidine (Fl). Of these carbenes phenylcarbene is by far the most thermally reactive at 77K. The reactivity of fluorenylidine in organic glasses is modest, the lifetime of this carbene is usually on the order of several minutes at 77K. Diphenylcarbene and

dibenzocycloheptadienylidine are relatively long lived (hours) under these
conditions.

PC   DPC

DBC   Fl

In fact phenylcarbene is sufficiently short lived in most organic glasses at 77K that it
can be properly thought of as a reactive intermediate. At 77K one usually need not
worry about the accumulation of PC in a glassy organic matrix, and its subsequent
secondary photolysis. This is not true for the other three aryl carbenes.[53]

Moss and later Tomioka and co-workers discovered that many reactions of
triplet ground state aryl carbenes are drastically different in solution phase and in
low temperature polycrystalline matrices.[54] For example both direct and

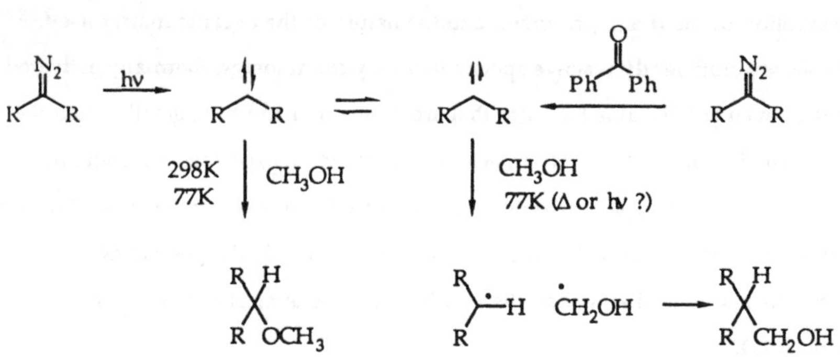

33, 34   37   35, 36

33, 35, 37 R=R=C_6H_5

34, 36 R, R=Fluorenylidene

photosensitized decomposition of diphenyldiazomethane (DPDM) or diazofluorene (DAF) in fluid methanol produces only a single product, the corresponding methyl ethers 33 and 34.[53-57] This product is widely believed to derive from an insertion reaction of the singlet state of the carbene into the OH bond of methanol. The fact that these products are formed in solution upon triplet sensitized photolysis of the diazo precursors is consistent with a very small singlet triplet energy separation in the aryl carbenes, and a mobile equilibrium. However, photolysis of diphenyldiazomethane in frozen polycrystalline methanol at 77K produces both OH and CH insertion products 33 and 35 respectively. Tomioka[55] concluded that the OH insertion product 33 was a product of reaction of the singlet carbene with the methanol matrix as per the reaction in solution phase. The CH insertion product was attributed to a two step, thermal reaction of triplet diphenylcarbene with the matrix host. The initial event was thought to involve thermally driven hydrogen atom transfer from the CH bond of methanol to produce a radical pair (37) which subsequently collapses to produce the observed product. Apparently the low temperature solid state conditions have enhanced the yield of triplet state chemistry relative to singlet state chemistry. Similar results were obtained with phenylcarbene as well. In both of these carbenes the triplet is known to be the ground state by EPR spectroscopy.[45] It is reasonable to expect that low temperature conditions will suppress a significant population of the higher energy singlet state of the carbene, relative to that present at ambient temperature, and thereby enhance the yield of triplet derived products. However, Leyva et al.[58] have demonstrated that much of the alcohol product derived from formal CH insertion reaction of diphenylcarbene and methanol (35), is due to secondary photolysis of the carbene. In fact the thermal reaction of the carbene with methanol at 77K may proceed completely through the low lying singlet state to form ether 33 cleanly. The effect of the matrix is to extend the lifetime of diphenylcarbene and allow the carbene to undergo secondary photochemical reactions. This was demonstrated by measuring the yield of OH and CH insertion products formed upon photolysis of DPDM in polycrystalline methanol, and 2-propanol as a function of both light intensity and length of photolysis of the matrix.[58]

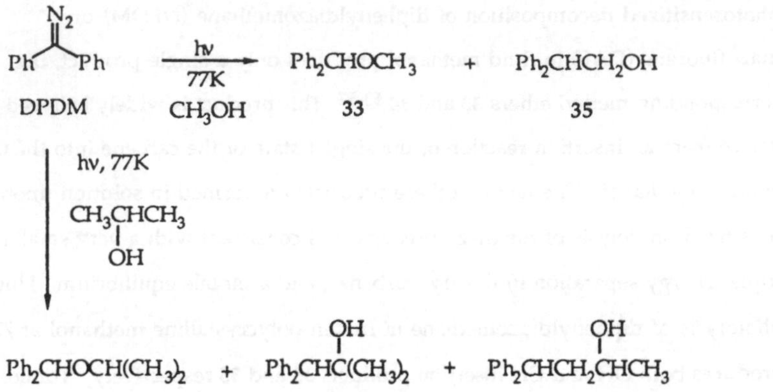

$$\underset{\text{DPDM}}{\overset{N_2}{\underset{\text{Ph}}{\bigvee}}_{\text{Ph}}} \xrightarrow[\underset{\text{CH}_3\text{OH}}{77K}]{\text{hv}} \underset{33}{\text{Ph}_2\text{CHOCH}_3} + \underset{35}{\text{Ph}_2\text{CHCH}_2\text{OH}}$$

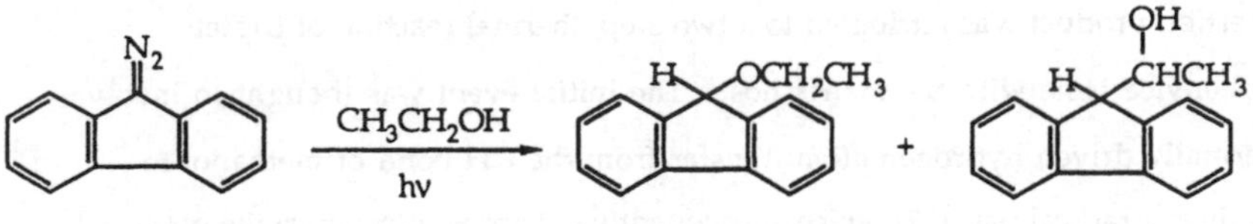

hv, 77K

CH₃CHCH₃
|
OH

$$\underset{36}{\text{Ph}_2\text{CHOCH(CH}_3)_2} + \underset{37}{\text{Ph}_2\text{CHC(CH}_3)_2} + \underset{38}{\text{Ph}_2\text{CHCH}_2\text{CHCH}_3}$$

with OH groups shown above 37 and 38.

As seen in Tables 7 and 8 the major products derived from DPC generated with low light intensities or with short exposure of the matrix to light, is that of OH insertion in both of the alcohols studied. The diphenylcarbene CH insertion product is formed appreciably only when more intense light levels are used, or following long exposure of the matrix to the photolysis source. In fact, at high light intensity and at long exposure times the CH insertion product is the major product of the matrix photolysis.

This is not the case, however, with fluorenylidine (Table 9). In the case of fluorenylidene and ethanol the major product detected at all light levels and at all light exposure times is that of CH insertion (Table 9).

Thermally driven reactions of fluorenylidene in ethanol at 77K lead to the formation of products of both CH and OH insertion. As mentioned previously it is generally believed that the product of OH insertion of the carbene and alcohol proceeds through the low lying singlet state of fluorenylidene. However, the

Table 7. Distribution of Products Formed on Photolysis of 0.01 M DPDM in
Methanol at 77K for 30 min as a Function of Light Intensity.[55]

light intensity	33	35	alcohol/ether
1.0	9.5	13.7	1.44
0.53	6.3	7.2	1.14
0.39	5.4	7.0	1.29
0.30	3.7	2.3	0.62

Distribution of Products Formed on Photolysis of 0.01 M DPDM in Methanol at 77K
as a Function of Photolysis Time.[55]

time, min	33	35	alcohol/ether
60	22	29	1.3
30	15	15	1
15	14	5	0.36
5	9	2	0.22

enhanced yields of CH insertion product realized at low temperature may be due to
reaction of either the singlet or the triplet state of this carbene under these
conditions.

Diphenylcarbene clearly undergoes secondary photolysis in alcohol matrices
whereas the extent of this reaction in the case of fluorenylidene is quite modest.
Thus Tomioka[55] was correct: the increased yields of CH insertion product formed
by DPC in alcohols are triplet carbene derived, but for this carbene they are a
consequence of photochemical rather than thermal reactions.

The different behavior of diphenylcarbene and fluorenylidene may be a result
of different excited state energies of these carbenes. The major electronic absorption
maxima of diphenylcarbene is near 300 nm with a weaker band near 500 nm.[45,56,57]
Diphenylcarbene fluoresces with a maxima near 505 nm in solution[11] and at 480

Table 8. Distribution of Products Formed on Photolysis of 0.01 M DPDM in 1-
Propanol at 77K as a Function of Light Intensity[55]

rel. light intensity	36	37	38	alcohol/ether
1.0	3.2	3.2	1.2	1.38
0.52	2.2	1.4	1.2	1.18
0.39	2.1	1.4	0.6	0.95
0.30	2.0	1.8	0.4	1.10

Distribution of Products Formed on Photolysis of 0.01 M DPDM in 2-Propanol as a
Function of Photolysis Time[55]

time, min	36	37	38	alcohol/ether
30	2.9	2.3	0.9	1.1
15	2.4	1.1	0.4	0.63
5	2.3	0.4	0.2	0.26
2	1.9	0.0	0.0	0.0

nm in a rigid matrix.[49] Fluorenylidene is not known to fluoresce, perhaps because
the excited state energy of this carbene is much lower than that of diphenylcarbene.
This in turn may lead to its lower reactivity in the matrix. It is interesting to note
that photolysis of phenyldiazomethane in alcohol matrices also forms a mixture of
OH and CH insertion products,[54] but, as mentioned previously phenylcarbene is
indeed a reactive intermediate at 77K and thus can not undergo secondary
photolysis. In the case of phenylcarbene and fluorenylidene Tomioka's original
interpretation may be appropriate; low temperature conditions do indeed direct the
thermal chemistry of phenylcarbene towards enhanced yields of CH insertion
reactions. It is not clear, however, whether the enhanced yield of CH insertion

Table 9. Distribution of Products Formed on Photolysis of 0.01 M DAF in Ethanol at 77K as a Function of Photolysis Time[55]

time, min	39	40	alcohol/ether
75	4.9	11.4	2.3
60	5.1	11.2	2.1
45	4.3	9.2	2.1
30	2.9	4.8	1.7
15	1.9	3.2	1.7

Distribution of Products Formed on Photolysis of 0.01 M DAF in Ethanol for 30 min at 77K as a Function of Light Intensity[55]

light intensity	39	40	alcohol/ether
1.00	4.9	16.7	3.41
0.52	4.6	14.9	3.24
0.39	3.2	8.0	2.50
0.30	2.5	7.7	3.08

product formed in the matrix reactions of phenylcarbene originate from the singlet state or the triplet state of this carbene.

It is interesting to note that the thermal reactions of the low lying singlet states of diphenylcarbene and fluorenylidene with the OH bond of an alcohol can proceed within minutes at 77K.[53,54] This is perhaps not too surprising in the case of fluorenylidene where the singlet-triplet energy separation (ΔH_{ST}) has been deduced to be only 1 kcal/mol.[57] However, this seems to be inconsistent with the calculated singlet-triplet gap of diphenylcarbene which has been deduced to be somewhat larger, 3-4 kcal/mol.[56] If indeed the energy gap in DPC is as high as 3 kcal/mol then the equilibrium ratio of triplet to singlet DPC at 77K is approximately 10^9. It is hard to imagine a reaction involving a very minute equilibrium quantity of singlet carbene proceeding at a rapid rate at very low temperature in a rigid matrix.[54]

These results suggest that the singlet-triplet energy gap in diphenylcarbene is considerably smaller than originally estimated or that the OH insertion of DPC may proceed by a surface crossing type of mechanism.[59]

Barcus et al.[60] have attempted to distinguish between thermal and photochemical reactions of DBC and fluorenylidene in various glasses at 77K utilizing EPR spectroscopy. Barcus found that upon photolysis of diazo compound 41 in various matrices the known EPR spectrum of the DBC was readily observed along with a radical pair type of EPR spectrum. The carbene disappears upon standing in the dark over a period of many minutes at 77K. The disappearance of the EPR spectrum of the carbene is accompanied by an increase in the intensity of the radical pair spectrum (Figure 10, 11).

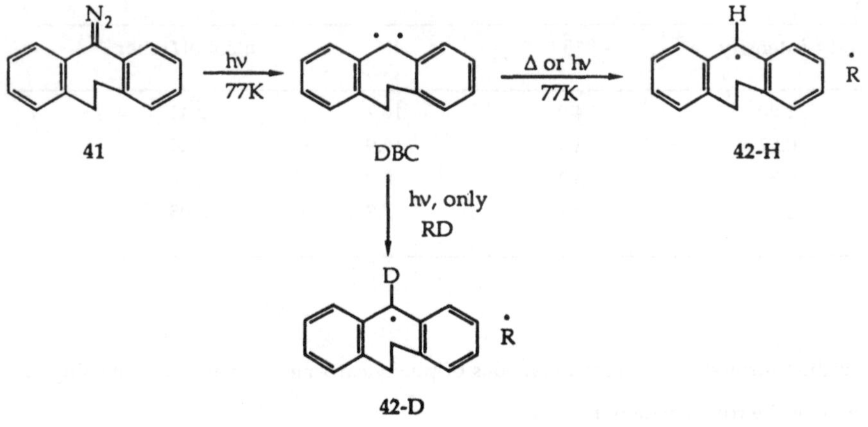

In the hydrocarbon matrix the radical pair can be formed by a thermally driven reaction of the carbene. If the experiment is repeated in a perdeuterated matrix, photolysis of the diazo compound again produces the EPR spectrum of both a carbene and a radical pair (Figure 12). However, once photolysis is discontinued the triplet carbene disappears once again (albeit more slowly than in the hydrocarbon matrix) but the radical pair does not increase in intensity as it does in a hydrogen atom containing matrix (Figure 13). These results demonstrate that although the carbene is thermally labile, thermally driven deuterium atom transfer reactions of ^{3}DBC do not occur readily at 77K. Therefore, the radical pair spectra detected in a

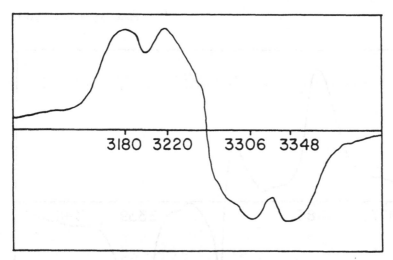

Figure 10. The DBC-toluene derived radical pair. (Reprinted with permission from J. Phys. Chem. (1987) 91, 6677)

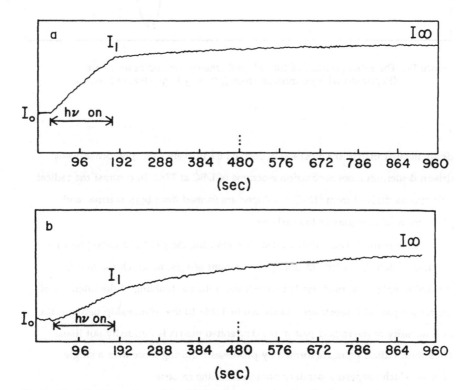

Figure 11. The formation of the DBC-toluene derived radical pair with time measured at (a) 3180G and at (b) 3220G. (Reprinted with permission from J. Phys. Chem. (1987) 91, 6677)

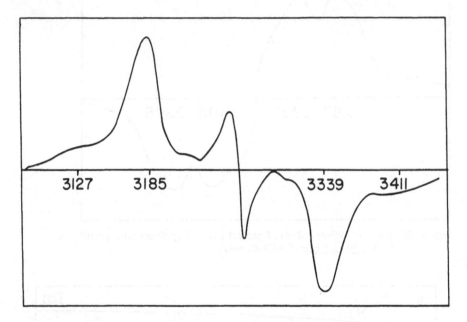

Figure 12. The EPR spectrum of the DBC-toluene-d8 derived radical pair.
            (Reprinted with permission from J. Phys. Chem. (1987) 91, 6677)

perdeuterated matrix are most likely derived exclusively from photochemically
driven deuterium atom abstraction reactions of DBC at 77K. In contrast the radical
pair spectra derived from ^{3}DBC in toluene are formed from both thermal and
photochemical reactions of the carbene.

   This hypothesis was substantiated by measuring the yield of radical pair as a
function of light intensity. Doubling the amount of light to which the matrix
isolated sample is exposed, leads to much more than a doubling of the intensity of
the radical pair EPR spectrum. As shown in Table 10 the relationship between the
EPR intensity of the radical pair in the deuterated matrix is decidedly not linear.
Thus these radical pairs are formed by processes requiring more than a single
photon, which suggests secondary photolysis of the carbene.

   Similar results were obtained with diazofluorene and fluorenylidene. Long
wavelength photolysis of a matrix containing 9-diazofluorene produced the EPR

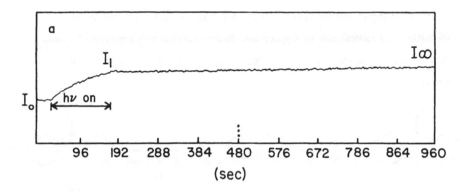

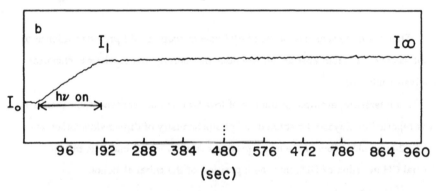

Figure 13. The formation of the DBC-toluene-d$_8$ derived radical pair with time measured at (a) 3118G and at (b) 3180G. (Reprinted with permission from J. Phys. Chem. (1987) **91**, 6677)

Table 10. EPR Signal Intensity of Radical Pairs as a Function of Light Intensity[60]

carbene	matrix	rel light transmission, %	RP EPR intensity
DBC	ether-d$_{10}$	100	79
DBC	ether-d$_{10}$	50	10
DBC	2-MTHF	100	59
DBC	2-MTHF	50	16
Fl	ether	100	47
Fl	ether	50	11

spectrum of triplet fluorenylidene but these conditions did not result in the formation of a radical pair EPR spectrum. Short wavelength photolysis of matrix isolated fluorenylidene led to the production of radical pair like spectra in this system.

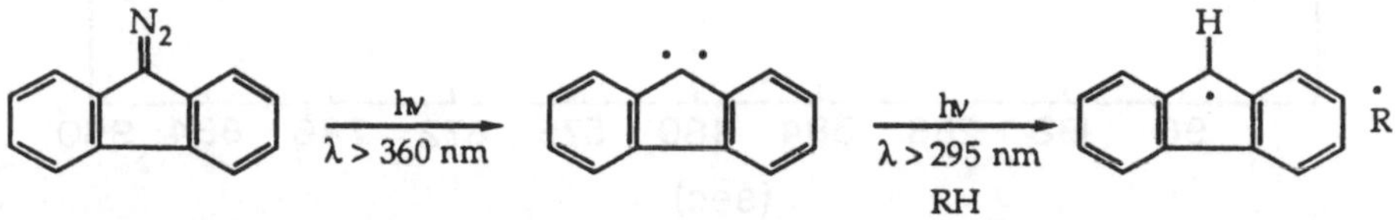

43

Thus the more sensitive probe of EPR spectroscopy could provide evidence of photochemical H atom abstraction reactions of fluorenylidene, whereas chemical analysis could not.

An interesting example of the use of low temperature carbene photochemistry was reported by Zayas who studied the photochemistry of diphenyldiazomethane in (S)-2-butanol.[61] One of the major products formed in this matrix is that of formal CH insertion of DPC into the 2 position of the chiral alchohol.

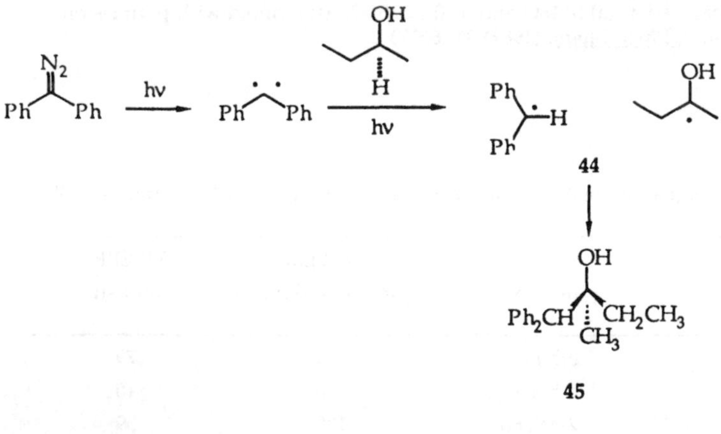

44

45

Zayas found that the alcohol product produced in the (S)-2-butanol matrix was formed with complete retention of enantiomeric purity. Thus, even though the reaction proceeds via a radical pair intermediate and each component of the radical pair is itself achiral, the rigid matrix forces the collapse of the radical pair in only one sense to produce a chiral product.

Zayas and Platz reasoned that a hydrogen atom transfer reaction between diphenylcarbene and (S)-2-butanol formed radical pair **44**, which subsequently collapsed to form the product of formal C-H insertion (**45**). These workers assumed that the formation of radical pair **44** was thermally driven. With the benefit of hindsight it is now clear that the hydrogen atom abstraction reaction of diphenylcarbene with polycrystalline (S)-2-butanol at 77K is a photochemical process.

IV Low Temperature Photochemistry of Mesitylene Triplet

It is well understood that photolysis of toluene, isolated in a rigid organic matrix at 77K can lead to benzylic CH bond cleavage to produce benzyl radical and a hydrogen atom.[62] Presumably the nascent hydrogen atom can rapidly extract another hydrogen atom from a matrix CH bond to produce hydrogen molecules. Albrecht has demonstrated that fragmentation of toluene is a 2-photon process at 275 nm. The first photon promotes toluene to an excited singlet state (S_1) which rapidly relaxes to its phosphoresent triplet state (T_1). The phosphorescent triplet state can absorb a second photon of light to produce a highly excited triplet state of toluene (T_N) which can undergo the bond cleavage reaction. This reaction does not occur to a measurable extent in organic solution at ambient temperature with conventional

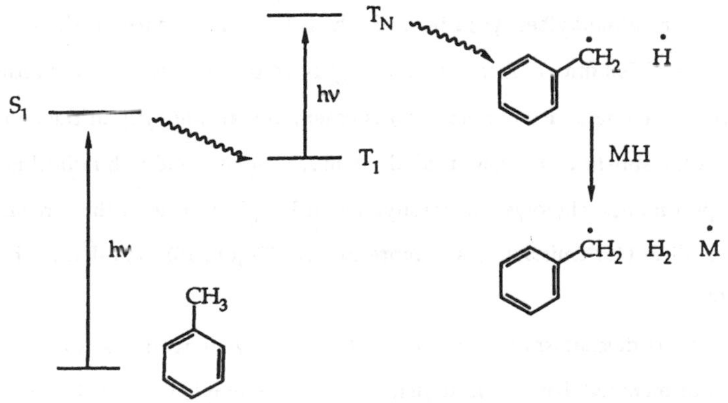

light sources, because the phosphorescent triplet state of toluene (T_1) has a half life
of only a few microseconds in solution whereas it may persist for several seconds in
a matrix.

It may be possible to achieve photochemical fragmentation of toluene in solution
phase using one or two high energy pulsed lasers in combination.[11]

Many years ago Migirdicyan discovered that 254 nm photolysis of meta-xylene
and ortho-xylene isolated in rigid matrices produced the emission of the methyl
substituted benzyl radicals along with new, weak emission bands at shorter
wavelength.[63] High resolution spectroscopy of the emission bands was obtained in
Shpolskii type matrices and led to their assignment as meta and ortho

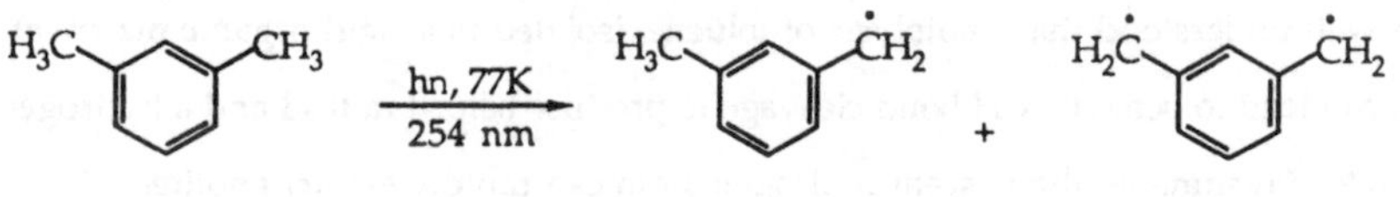

46   **47**

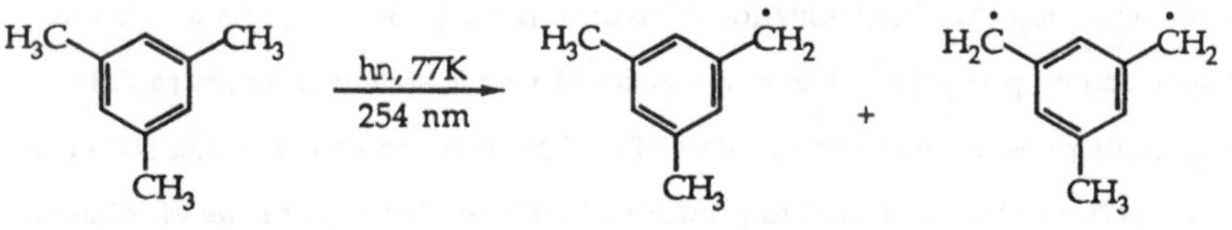

48   **49**

xylylene respectively. It is important to note that the major emitting species in the
matrix is the meta-methylbenzyl radical **46** which has fluorescence maxima
between 480 and 540 nm. In fact as seen in Figure 14 the emission of the biradical
(fluorescence maxima at 440 and 465 nm) is essentially an impurity in the emission
spectrum of the substituted benzyl radical. A much higher yield of biradical is
realized upon matrix photolysis of mesitylene, although even with this precursor
the mono radical (**48**) emission is still more intense (Figure 15) than that of the
biradical (**49**).

An obvious mechanism for formation of meta-xylyene would be the
conversion of meta-xylylene or mesitylene to a meta-methylbenzyl radical as

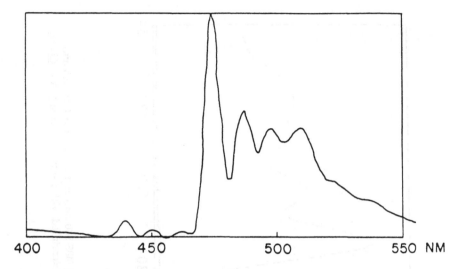

Figure 14. The emission spectra of the meta-methylbenzyl radical 46 (475-550 nm) and meta-xylylene biradical 47 (440-470 nm) produced by 254 nm photolysis of meta-xylene at 77K. (Reprinted with permission from J. Am. Chem. Soc. (1990) 112, 733)

described by Albrecht, followed by secondary photolysis of the mono radical to produce the biradical. (Scheme 3)

However, this simple mechanism could be easily discarded. The fluorescence intensities of monoradical 48 and biradical 49 derived from mesitylene were studied as a function of lamp source, light intensity and photolysis time. The same ratio of monoradical (48) to biradical (49) emission was obtained using a low pressure mercury arc (254 nm) continuous light source or by using the pulses of a krypton fluoride excimer laser which operates at 249 nm. Furthermore, the same

Scheme 3

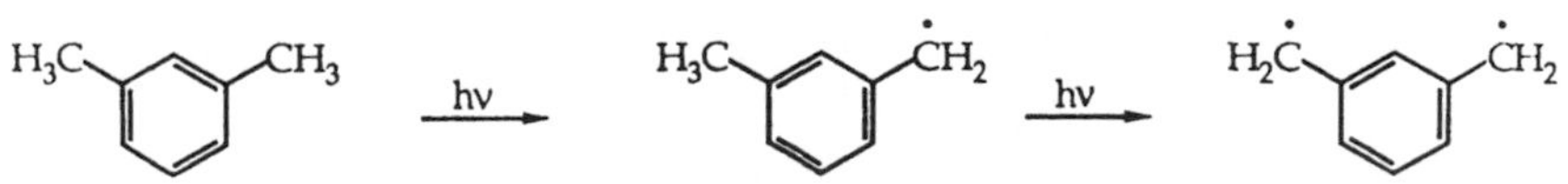

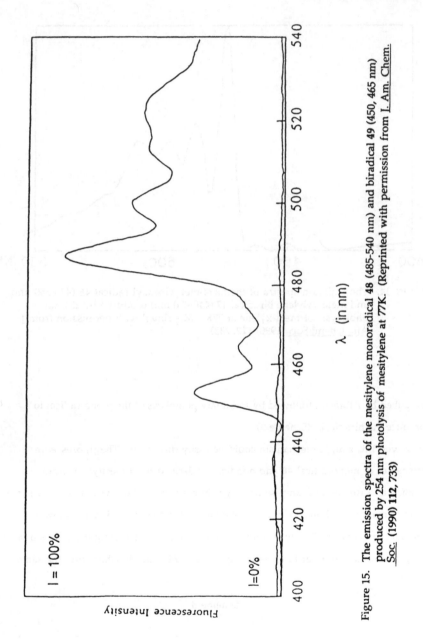

Figure 15. The emission spectra of the mesitylene monoradical **48** (485-540 nm) produced by 254 nm photolysis of mesitylene at 77K. (Reprinted with permission from J. Am. Chem. Soc. (1990) **112**, 733)

ratio of mono to biradical emission was produced upon modulation of the laser
light intensity with neutral density filters or by varying the number of laser pulses
administered. This demonstrates that the monoradical and biradical are both
formed by processes involving the same number of photons, a result that is
inconsistent with a sequential mechanism as shown in Scheme 3.[64]

The second piece of evidence against the sequential mechanism of Scheme 3
was provided by studying peroxide 50. Upon photolysis of 50 in an optically clear
glass at 77K, the emission spectrum of the monoradical was easily observed.
However, exhaustive photolysis of the monoradical prepared from the peroxide
precursor did not produce the emission spectrum of the biradical, even though the
emission spectrum of biradical 49 was easily detected upon exposure of mesitylene
to the same photolysis conditions. These experiments clearly rule out a
mechanism in which mesitylene is photochemically converted to a monoradical
that is subsequently photolyzed to form a biradical.

A second mechanistic alternative was suggested by Berson.[65] In this
mechanism photolysis of meta-xylene or mesitylene leads to an alkyl substituted
benzyl radical and a hydrogen atom. In a thermal reaction the nascent hydrogen
atom can either react with the matrix to produce molecular hydrogen and a matrix
derived radical, or, the hydrogen atom can react with a methyl group of the
substituted benzyl radical to form molecular hydrogen and a biradical. As these
reactions involve a small light atom and are likely to be very exothermic they may
proceed rapidly by quantum mechanical tunneling.[66]

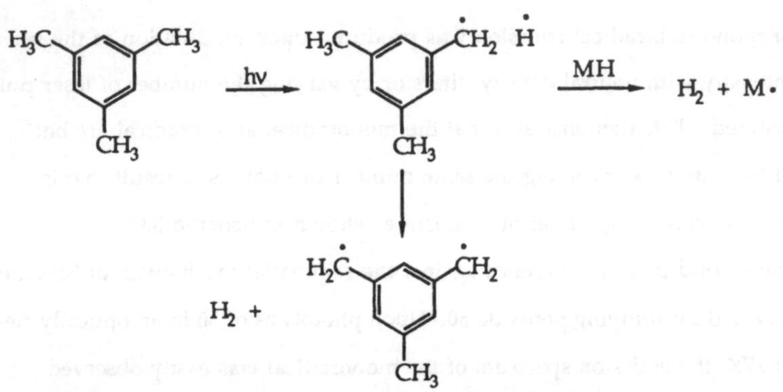

This mechanism was tested by measuring the ratio of monoradical to biradical emission as a function of matrix. It was reasoned that in a good hydrogen atom donating matrix the yield of biradical would be relatively small. Conversely the use of a poor hydrogen atom donating matrix would enhance the yield of biradical. The data are given in Table 11. The ratio of monoradical/biradical is indeed larger in the better hydrogen atom donating matrices. In fact perdeuteration of the matrix leads to a large enhancement in the yield of biradical relative to monoradical. We believe it is difficult to envision an alternative mechanism which will produce a matrix deuterium isotope effect.

Table 11. The relative ratios of biradical/monoradical (49/48) fluorescence intensities produced from photolysis of matrix isolated mesitylene, as a function of matrix. Samples were photolyzed at 77K using the 249 nm line of a Lumonics excimer laser. The ratio was determined by monitoring the intensity of the fluorescence emission at the maxima of both the biradical (452 nm) and the monoradical (487 nm). The excitation wavelength was 296 nm throughout.[64]

Sample	Matrix	(49/48)
Mesitylene (3.0 x 10-3M)	3-methylpentane	0.42
Mesitylene (4.7 x 10-3M)	CH3OH	0.24
Mesitylene (4.7 x 10-3M)	CD3OD	0.32
Mesitylene (5.0 x 10-3M)	n-hexane	0.54
Mesitylene (5.0 x 10-3M)	n-hexane (d-14)	1.26
Mesitylene (5.0 x 10-3M)	methylcyclohexane	0.69

It was possible to design experiments to demonstrate that the monoradical and biradical were formed by secondary photolysis of the phosphorescent triplet state of mesitylene. Exposure of an optically clear organic glass containing mesitylene at 77K to UV light, produces the phosphorescent triplet state of mesitylene. The green phosphorescence of the triplet is easily detected by the unaided eye for several seconds following excitation. If a glass containing mesitylene was subjected to the pulses of an excimer laser (KrF, 249 nm) operating with a high repetition rate, the emission spectra of the matrix isolated mono and biradical could be subsequently detected by spectrofluorimetry. However, if the repetition rate of the excimer laser is reduced such that the sample is exposed to one pulse every twenty seconds, the mono and biradical are not formed in appreciable amounts, as assayed by spectrofluorimetry. Under the latter conditions the phosphorescence decay of the excited triplet state is complete before the sample is exposed to a second 249 nm photon. This experiment demonstrates that both the monoradical as well as the biradical were formed by the secondary photolysis of an intermediate having a lifetime of several seconds under these conditions. The phosphorescent triplet state of the hydrocarbon is the most likely candidate for this intermediate.

Laser flash photolysis (249 nm) of mesitylene in a hydrocarbon solution at ambient temperature produces the triplet excited state of mesitylene whose transient absorption spectrum can be obtained (Figure 16) in the usual manner. The triplet excited state of mesitylene absorbs strongly between 350-450 nm. When mesitylene in glassy 3-methylpentane at 77K is subjected to 50 excimer laser pulses spaced at relatively long intervals (~ 20 sec.) the yield of mono and biradical is very small as mentioned previously. This yield becomes at least 10 fold larger when the sample is simultaneously irradiated with an excimer laser (249 nm) operating at a repetition rate of 20 seconds and a nitrogen laser (337 nm) pulsed at a high repetition rate. Similar results are obtained when the nitrogen laser is replaced with a beam of 405 nm radiation produced from a mercury arc lamp. Photolysis of the samples with the nitrogen laser or 405 nm radiation alone does not produce measurable yields of either mono or biradicals. The ratio of mono to biradical formed from the two beam experiment is identical to the ratio of these species observed when mesitylene is irradiated with 249 nm laser light alone.

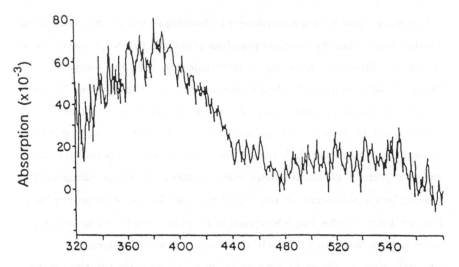

Figure 16. The transient absorption spectrum of triplet mesitylene produced by
flash photolysis. (Reprinted with permission from J. Am. Chem. Soc.
(1990) 112, 733)

Simple energetic arguments based on the two-beam experiment provide

additional support for the Berson mechanism. The triplet energy of mesitylene is

80.3 Kcal/mol.[67] Thus after absorption of a second photon of 337 nm the doubly

excited triplet state of mesitylene has 165.3 Kcal/mol of energy, which is sufficient

to fragment the highly excited triplet state to a monoradical plus a hydrogen atom.

However, the energy is insufficient to simultaneously cleave two benzylic CH

bonds of mesitylene. We guess that this would require 170 Kcal/mol, assuming

that the second CH bond dissociation energy of mesitylene is the same as that of the

first.[66] As the second CH bond dissociation energy of meta-xylene and mesitylene

are unknown this argument is somewhat tentative. However, when the

wavelength of the second beam is 405 nm, the sum of the triplet energy of triplet

mesitylene plus the energy provided by a 405 nm photon (70.7 Kcal/mol) seems

insufficient to break 2 CH bonds in a purely photochemical process. The data are

most consistent with a mechanism in which cleavage of the first CH bond requires

2 photons but cleavage of the second benzyl CH bond occurs in a non

photochemical reaction.

V The Low Temperature Photochemistry of Benzylic Halides and Dihalides

We have mentioned previously that the yields of biradicals produced from
mesitylene and meta-xylene are relatively low. In these matrices the monoradical
is the major emitting species. It has recently been discovered that much higher
yields of biradicals can be realized from the photolysis of bis chloromethyl
compounds at low temperature.[68] A relatively intense and clean emission
spectrum of meta-xylylene free of the emission of the meta methylbenzyl radical is
produced by brief photolysis (245 nm) of 1,3 dichloromethyl benzene (Figure 17).

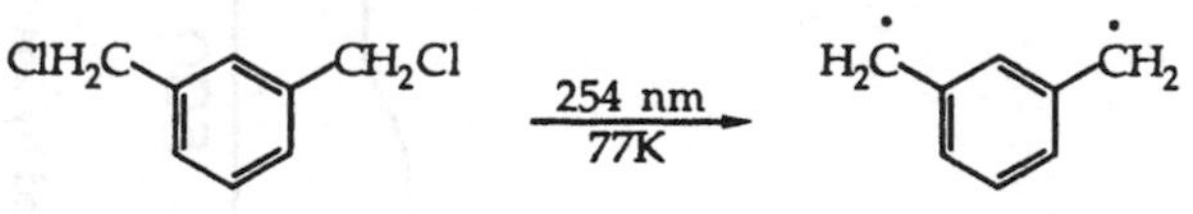

It is clear that the biradical is not formed from the precursor in a simple
monophotonic process. A photon of 254 nm does not have sufficient energy to
simultaneously cleave two carbon chlorine bonds.[66] A 254 nm photon does in fact
have enough energy to cleave a single carbon chlorine bond to produce a benzyl
radical, chlorine atom pair. However, a Berson type of mechanism seems very
unlikely in this system. This would require that the nascent chlorine atom
abstracts a chlorine from the meta-chloromethyl position to form a biradical plus
molecular chlorine. Thermodynamically, it is far more likely that a chlorine atom
will abstract a hydrogen atom from the matrix or from a benzyl CH bond to form
HCl and the chloro meta-xylylene biradical 53.

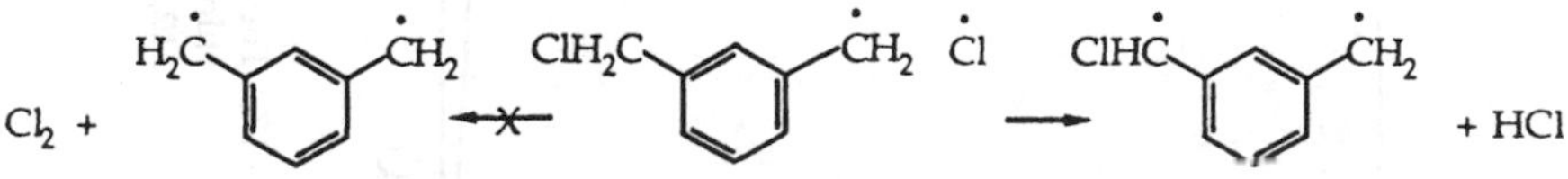

This latter process should also be kinetically favored because the abstraction of
hydrogen to form HCl can proceed rapidly at 77K by quantum mechanical

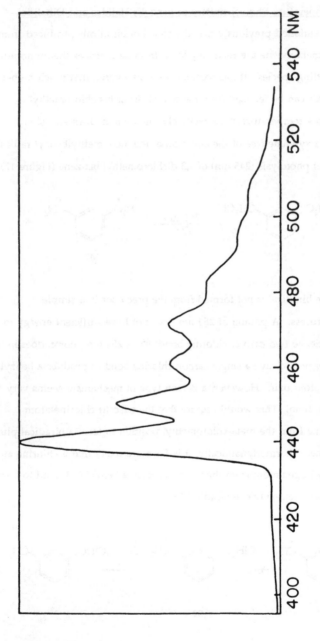

Figure 17. The emission spectrum of the meta-xylylene biradical produced by 254 nm photolysis of 1,3 bis chloromethyl benzene in ethanol at 77K. (Reprinted with permission from J. Am. Chem. Soc. (1990) 112, 733)

tunneling. A tunneling reaction is much less likely for the more massive chlorine atom abstraction process, which would form Cl_2. The flourescence spectrum of the chlorinated derivative of meta-xylylene 53 is not known. However, we expect that it would be significantly different from the emission spectrum of meta-xylylene itself as there is a considerable difference between the fluorescence spectrum of the meta-xylylene and the mesitylylene biradicals. The spectrum which is obtained following photolysis of the dichloride is in fact identical to that produced from the hydrocarbon in Migirdicyan's experiment. This argues strongly against a thermal reaction of the putative monoradical 52 to form the meta-xylylene biradical which can be subsequently detected by spectrofluorimetry.

Energetic requirements suggest that the formation of the biradical from the dichloride precursor is multiphotonic. This in turn implies that a species is produced by photolysis of the dichloride which suffers secondary photolysis to produce the biradical. Two obvious candidates for secondary photolysis are the excited triplet state of the dichloride and the monochloromethyl benzyl radical. At the present time it is not known for certain which, or in fact if either, of these intermediates might be formed upon photolyis of the dichloro precursor.

However, laser flash photolysis provides evidence for a transient intermediate whose absorption spectrum is consistent with a substituted benzyl mono radical 52. Putative radical 52 could be formed by the homolysis of a single carbon-chlorine bond. Neither the excited triplet state of the dichloride nor the putative monochloromethyl radical can be detected by their luminescence. Thus if either of these species are in fact intermediates in the process linking 51 to 53 then they must be non emissive intermediates.

Haider has studied the low temperature photochemistry of dichlorodiphenylmethane.[69] Dichloride 54 immobilized in an optically clear organic glass at 77K was exposed to 254 nm radiation. Analysis of the glass by spectrofluorimetry revealed the formation of both mono chlorinated radical 55 and diphenylcarbene.

Upon photolysis of the glass with nitrogen laser radiation (337 nm) the intensity of the monoradical fluorescence at 522 and 550 nm decreased but that of the carbene at

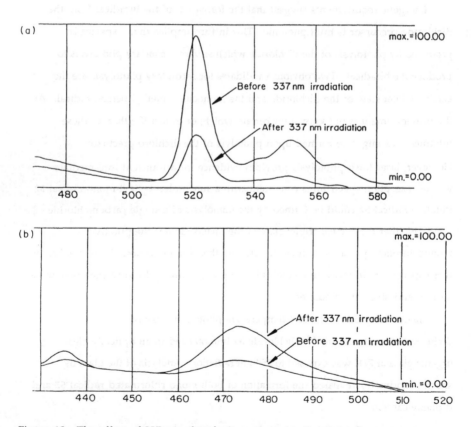

475 nm increased (Figure 18). In this system the data suggest the photochemical sequence shown below.

Figure 18. The effect of 337 nm photolysis on the intensity of the fluorescence intensity of (a) chlorinated radical 55 and (b) DPC. (Reprinted with permission from J. Phys. Org. Chem. (1989) 2, 623)

Biewer[70] has recently extended the dichloride precursor approach to naphthalenic systems. The 1,3; 1,6; and 2,7 dichloromethyl naphthalenes were synthesized and subsequently photolyzed with 254 nm light in ethanol glass at 77K. The emission spectra produced are shown in Figures 19, 20 and 21. The fluorescence bands observed can not be attributed to 1 or 2 naphthylmethyl type monoradicals, all of which emit at much longer wavelengths (595 and 605 nm respectively). The emission bands produced from photolysis of the dichlorides are in a reasonable region of the spectrum for naphthoquinodimethane biradicals. In fact a 1,8 NQM derivative (56) has been prepared from 57 in a low temperature

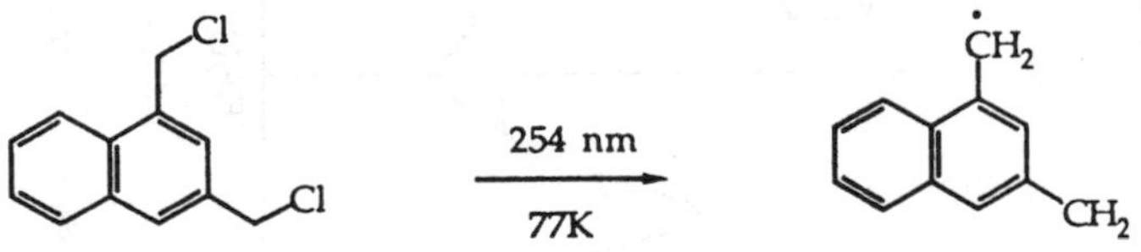

1,3 NQM

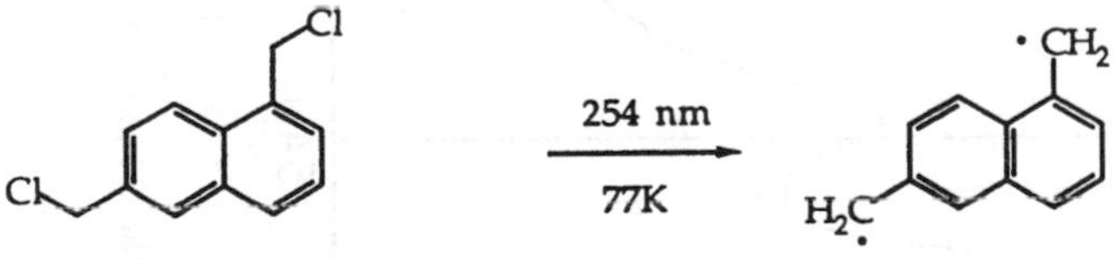

1,6 NQM

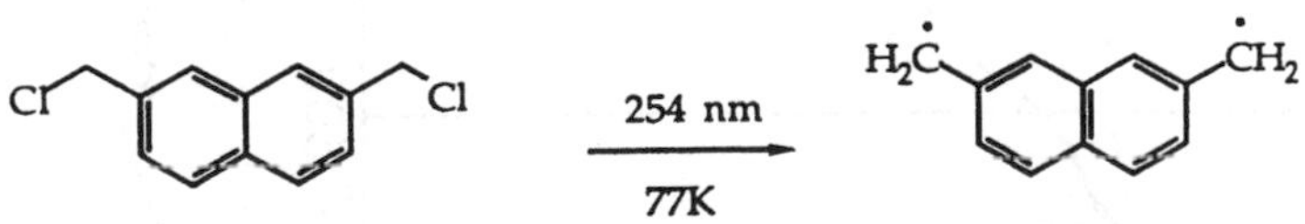

2,7 NQM

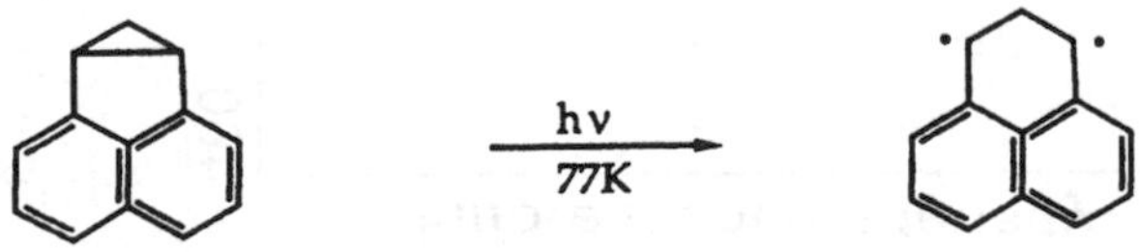

57   56

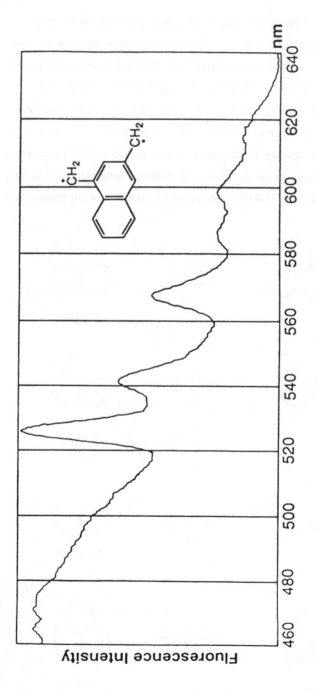

Figure 19. The emission spectrum of 1,3 napthoquinodimethane in ethanol at 77K (λ ex = 363 nm).

Table 12. Details of the fluorescence spectra of the naphthoquinodimethane biradicals at 77K[70]

λexc (nm)	NQM	Fluorescence Maxima(nm)	Spacing (cm^{-1})	Theory[d,72](nm)
363	1,3[a]	525[d], 542, 567.5	561, 1390	539
350	1,6[b]	545[d], 560, 589	492, 1375	542
348	2,7[a]	561.5[d], 575, 609	434, 1405	511
-	1,8[c]	512	-	509

a) 2MTHF   b) ethanol   c) An ethano bridged 1,8 NQM in EPA
d) zero-zero band

matrix, and has been found to emit at 509 nm at 77K.[71] Furthermore, Grisin and Wirz have calculated the zero-zero emission bands of the putative NQM biradicals produced in this work, and these are all in the 500 nm regime as well.[72]

The vibrational splittings observed in Figures 19, 20, and 21 are consistent with the formation of naphthalene derivatives.[73] The spacings are listed in Table 12, and they correspond to various stretching modes of the naphthalene molecule.

VI Final Remarks

Matrix isolation spectroscopy remains a powerful tool for the study of exotic molecules which are kinetically unstable in solution. Precisely because the matrix conditions impart a long lifetime to the species of interest, the matrix isolated radical, carbene, nitrene or excited state may itself undergo secondary photolysis. The photochemistry of these molecules presents challenges to the mechanistically inclined chemist, as well as opportunities to produce new species of interest for which traditional precursors for matrix isolation work are inconvenient.

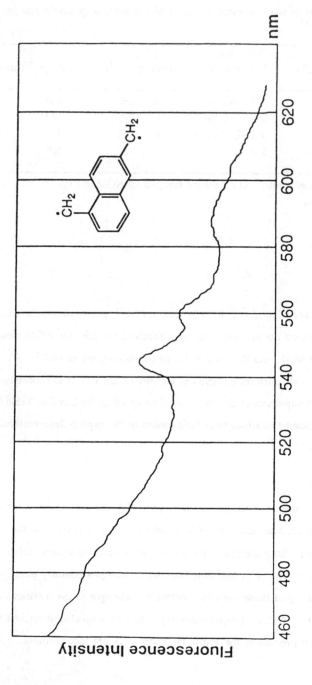

Figure 20. The emission spectrum of 1,6-naphthoquinodimethane in ethanol at 77K (λ ex = 350 nm).

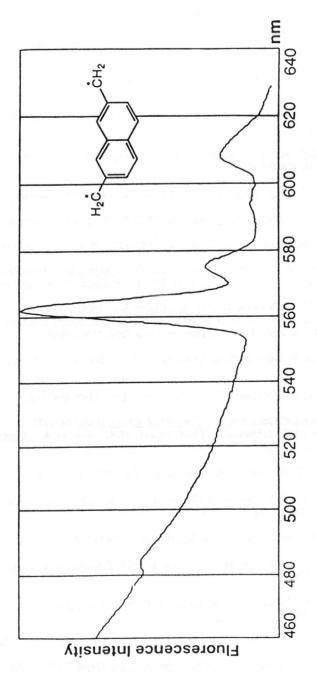

Figure 21. The emission spectrum of 2,7-naphthoquinodimethane in ethanol at 77K (λ ex = 348 nm).

References

1.   (a) Lwowski, W. "Nitrenes", Wiley, New York, N.Y. (1970).

2.   "The Chemistry of Diazonium and Diazo Groups" Patai S., ed., Wiley, New York, N.Y. (1978).

3.   "Chemistry of Diazirines, Volume I and II" Liu, M. T. H., ed. CRC Press, Boca Raton, Fla. (1987).

4.   Turro, N. J. "Modern Molecular Photochemistry," Benjamin Cummins, Menlo Park, CA (1978).

5.   Scaiano, J. C.; Accts. Chem. Res. (1982) 15, 252 and references therein.

6.   Wang, Y.; Eisenthal, K. B.; J. Chem. Ed. (1982) 59, 482 and references therein.

7.   Poliakoff, M.; Weitz, E.; Acc. Chem. Res. (1987) 20, 408 and references therein.

8.   (a) Trifunac, A. D.; Norris, J. R.; Lawler, R. G.; J. Chem. Phys. (1979) 71, 4380.
     (b) Trifunac, A. D.; Thurnauer, M. C.; Norris, J. R. Chem. Phys. Lett. (1978) 57, 471.
     (c) Closs, G. L., Redwine, O. D.; J. Am. Chem. Soc. (1985) 107, 4543.
     (d) Closs, G. L.; Forbes, M. D. E. in "Kinetics and Spectroscopy of Carbenes and Biradicals" Platz, M. S. ed., Plenum, New York, N.Y. (1990).

9.   Sheridan, R. S. Organic Photochemistry Vol 8, Dekker, New York, N.Y. (1987), 159.

10. Turro, N. J.; Cha., Y.; Gould, I. R. J. Am. Chem. Soc. (1987) 109, 2101.

11. (a) Scaiano, J. C.; Johnston, L. J.; Pure Appl. Chem. (1986) 58, 1273.
     (b) Scaiano, J. C.; Johnston, L. J.; McGimpsey, W. G.; Weir, D. Acc. Chem. Res. (1988) 21, 22.

12. Schrock, A. K.; Schuster, G. B.; J. Am. Chem. Soc. (1984) 106, 5228.

13. See Smith, P.A.S. in "Azides and Nitrenes" Scriven, E. F. V. ed., Academic Press, San Diego, CA (1984) p. 95.

14. Doering, W. von E.; Odum, R. A. Tetrahedron (1966) 22, 81.

15. Li, Y.-Z.; Kirby, J. P.; George, M. W.; Poliakoff, M.; Schuster, G. B.; J. Am. Chem. Soc. (1988) 110, 8092.

16. Meijer, E. W.; Nijhuis, S.; Vroonhoven, F. C. B. M. van J. Am. Chem. Soc. (1988) 110, 7209.

17. (a) Swenton, J. S.; I. Keler, T. J.; Williams, B. H.; J. Am. Chem. Soc. (1970) 92, 3103.
     (b) Leyshon, L. J.; Reiser, A.; J. Chem. Soc. Faraday Trans. II (1972) 1918

18. Smolinsky, G.; Wasserman, E.; Yager, W. A. J. Am. Chem. Soc. (1962) 84, 3220.

19. (a) Wasserman, E. Prog. Phys. Org. Chem. (1971) 8, 319.
    (b) Platz, M. S.; "Azides and Nitrenes" Scriven, E. F.V. ed., Academic Press, San Diego, CA (1984).

20. (a) Reiser, A.; Frazer, V.; Nature (London) (1965) 208, 682.
    (b) Reiser, A.; Wagner, H. M.; Marley, R.; Bowes, G.; Trans. Far. Soc. (1967) 63, 2403.

21. (a) Chapman, O. L.; LeRoux, J. P.; J. Am. Chem. Soc. (1978) 100, 282.
    (b) Chapman, O. L.; Sheridan, R. S.; LeRoux, J. P.; J. Am. Chem. Soc. (1978) 100, 6245.
    (c) Chapman, O. L.; Sheridan, R. S.; J. Am. Chem. Soc. (1979) 101, 3690.

22. (a) Kashiwagi, H.; Iwata, S.; Yamaoka, T.; Nagakura, S. Bull. Chem. Soc. Jpn. (1973) 46, 417.
    (b) Smirnov, V. A.; Brichkin, S. B.; Chem. Phys. Lett. (1982) 87, 548.
    (c) Waddell, W. H.; Feilchenfeld, N. B.; J. Am. Chem. Soc. (1983) 105, 5499.

23. Leyva, E.; Platz, M. S.; Persy, G.; Wirz, J.; J. Am. Chem. Soc. (1986) 108, 3783.

24. An alternative but in this case unlikely explanation for equivalent excitation and absorption spectra might be due to fast energy transfer from the absorbing to the emitting species.

25. (a) Dunkin, I. R.; Donnelly, T.; Lockhart, T. S.; Tetrahedron Lett. (1985) 359.
    (b) Dunkin, I. R.; Thomson, P. C. P.; Chem. Commun (1982) 1192.

26. (a) Young, M. J. T.; Platz, M. S.; Tetrahedron Lett. (1989) 30, 2199.
    (b) Banks, R. E.; Venayak, N. D.; Hamon, T. A.; Chem. Comm (1980) 900.

27. Schmidt, G. M. J. "Solid State Photochemistry" Verlag-Chemie; Weinheim, D. D. R. (1976).

28. Dunkin, I. R.; "Chemistry and Physics of Matrix Isolated Species" Andrews, L.; Moscowitz, M., ed. North Holland Press (1989) p.203.

29. (a) Wentrup, C.; Tetrahedron (1974) 30, 1301.
    (b) Crow, W.; Paddon-Row-M. N.; Aust. J. Chem. (1975) 28, 1755.

30. LeBlanc, B. F.; Sheridan, R. S.; J. Am. Chem. Soc. (1988) 110, 7250.

31. Leyva, E.; Young, M. J. T.; Platz, M. S.; J. Am. Chem. Soc. (1986) 108, 8307.

32. Leyva, E.; unpublished research at The Ohio State University.

33. Torres, M. J.; Zayas, J.; Platz, M. S. Tetrahedron Lett. (1986) 27, 791.

34. Kanakarajan, K.; Goodrich, R.; Young, M. J. T.; Soundararajan, S.; Platz, M. S.; J. Am. Chem. Soc. (1988) 110, 6536.

35. Leyva, E.; Munoz, D.; Platz, M. S.; J. Org. Chem. (1989) 54, 5938.

36. Soundararajan, N.; Platz, M. S.; J. Org. Chem. (1990) 55, 2034

37. McGlynn, S. P.; Azumi, T.; Kinoshita, M.; "Molecular Spectroscopy of the Triplet State" Prentice Hall, Englewood Cliffs, N.J. (1969) p. 261.

38. Schuster, G. B.; Adv. Phys. Org. Chem. (1986) 22, 311.

39. Dunkin, I. R.; Thomson, P. C.; Chem. Comm. (1980) 499.

40. Leyva, E.; Ph.D.; The Ohio State University (1986)

41. Leyva, E.; Platz, M. S.; Niu, B.; Wirz, J.; J. Phys. Chem. (1987) 91, 2293.

42. Schrock, A. K.; Schuster, G. B.; J. Am. Chem. Soc. (1984) 106, 5228.

43. Leyva, E.; Platz, M. S.; Tetrahedron Lett. (1987) 28, 11.

44. See Platz, M. S. in "Kinetics and Spectroscopy of Carbenes and Biradicals" Platz, M. S. ed., Plenum, New York, N.Y. (1990) p. 239, and references therein.

45. Brandon, R. W.; Closs, G. L.; Hutchison, C. A. Jr.; J. Chem. Phys. (1962) 37, 1878. and also see Closs, G. L.; "Carbenes Vol. II" Moss, R. A. and Jones, M. Jr., eds., Wiley, New York, N.Y. (1975) p. 159.

46. Murray, R. W.; Trozzolo, A. M.; Wasserman, E.; Yager, W. A.; J. Am. Chem. Soc. (1962) 84, 3213.

47. Closs, G. L.; Hutchison, C. A. Jr.; Kohler, B. E.; J. Chem. Phys. (1969) 51, 3327.

48. Doetschman, D. C.; Hutchison, C. A. Jr.; J. Chem. Phys. (1972) 56, 3964.

49. Trozzolo, A. M.; Gibbons, W. A.; J. Am. Chem. Soc. (1967) 89, 239.

50. (a) McMahon, R. J.; Chapman, O. L.; J. Am. Chem. Soc. (1987) 109, 683.
     (b) Chapman, O. L.; Johnson, J. W.; McMahon, R. J.; West, P. R.; J. Am. Chem. Soc. (1988) 110, 501.

51. West, P. R.; Chapman, O. L.; LeRoux, J. P.; J. Am. Chem. Soc. (1982) 104, 1779.

52. (a) Kesselmayer, M. A.; Sheridan, R. S.; J. Am. Chem. Soc. (1984) 106, 436.
     (b) Kesselmayer, M. A.; Sheridan, R. S.; J. Am. Chem. Soc. (1986) 108, 99.
     (c) Kesselmayer, M. A.; Sheridan, R. S.; J. Am. Chem. Soc. (1986) 108, 844.
     (d) Kesselmayer, M. A.; Sheridan, R. S.; J. Am. Chem. Soc. (1987) 109, 5029.
     (e) Kesselmayer, M. A.; Sheridan, R. S.; J. Am. Chem. Soc. (1988) 110, 7563.

53. Platz, M. S.; Acc. Chem. Res. (1988) 21, 236.

54. For a recent review see Platz, M. S. "Kinetics and Spectroscopy of Carbenes and Biradicals" Platz, M. S., ed. Plenum, New York, N.Y. (1990) p. 143.

55. Tomioka, H.; Izawa, Y.; J. Am. Chem. Soc. (1977) 99, 6128.

56. Eisenthal, K. B.; Turro, N. J.; Sitzmann, E. V.; Gould, I. R.; Hefferon, G.; Langan, J.; Cha, Y.; Tetrahedron (1985) 41, 1543.

57. (a) Grasse, P. B.; Brauer, B.-E.; Zupancic, J. J.; Kaufman, K. J.; Schuster, G. B.; J. Am. Chem. Soc. (1983) 105, 6833.
    (b) Griller, D.; Hadel, L. M.; Nazran, A. S.; Platz, M. S.; Wong, P. C.; Savino, T. G.; Scaiano, J. C.; J. Am. Chem. Soc. (1984) 106, 2227.

58. Leyva, E.; Barcus, R. L.; Platz, M. S.; J. Am. Chem. Soc. (1986) 108, 7786.

59. Griller, D.; Nazran, A. S.; Scaiano, J. C.; J. Am. Chem. Soc. (1984) 106, 198.

60. Barcus, R. L.; Wright, B. B.; Leyva, E.; Platz, M. S.; J. Phys. Chem. (1987) 91, 6677.

61. Zayas, J.; Platz, M. S.; J. Am. Chem. Soc. (1985) 107, 7065.

62. Schwartz, F. P.; Albrecht, A. C. Chem. Phys. Lett. (1971) 9, 163.

63. See Migirdicyan, E. and Baudet, J. J. Am. Chem. Soc. (1975) 97, 7400, and references therein.

64. Haider, K. W.; Migirdicyan, E.; Platz, M. S.; Soundararajan, N.; Després, A.; J. Am. Chem. Soc. (1990) 112, 733.

65. Personal communication from Professor J. A. Berson.

66. (a) Benson, S. W. "Thermochemical Kinetics" Wiley, New York, N.Y. (1968) p. 215.
    (b) Griller, D.; Kanabus-Kaminska, J. M.; "Handbook of Photochemistry" Scaiano, J. C., ed., CRC Press, Boca Raton, Florida (1989) p. 359.

67. Murov, S. L. "Handbook of Photochemistry" Marcel Dekker, New York, N.Y. (1973) p. 34.

68. Haider, K.; Platz, M. S.; Després, A.; Lejeune, V.; Migirdicyan, E.; Bally, T.; Haselbach, E.; J. Am. Chem. Soc. (1988) 110, 2318.

69. Haider, K.; Platz, M. S.; J. Phys. Org. Chem. (1989) 2, 623.

70. Biewer, M. C.; Platz, M. S.; Migirdicyan, E.; Després, A.; J. Am. Chem. Soc. submitted.

71. Gisin, M.; Rommel, E.; Wirz, J.; Burnett, M. N.; Pagni, R. M.; J. Am. Chem. Soc. (1979) 101, 2216.

72. Gisin, M.; Wirz, J.; Helv. Chim. Acta. (1983) 66, 5, 1556.

73. McClure, D. S.; J. Chem. Phys. (1956) 24, 1.

74. Wasserman, E.; Trozzolo, A. M.; "Carbenes Vol. II" Moss, R. A. and Jones, M., eds; Wiley, New York, N.Y. (1975)

Index